Wesley Smith

Proceedings of the Workshop on
HADRON STRUCTURE FUNCTIONS and PARTON DISTRIBUTIONS

Proceedings of the Workshop on

HADRON STRUCTURE FUNCTIONS and PARTON DISTRIBUTIONS

Fermi National Accelerator Laboratory
26 — 28 April 1990

Editors
D. F. Geesaman
J. Morfin
C. Sazama
W. K. Tung

Workshop on Hadron Structure Functions and Parton Distributions
Sponsored Jointly by:
Argonne National Laboratory, Fermi National Accelerator Laboratory, and
Superconducting Super Collider Laboratory

World Scientific
Singapore • New Jersey • London • Hong Kong

Published by
World Scientific Publishing Co. Pte. Ltd.
P O Box 128, Farrer Road, Singapore 9128
USA office: 687 Hartwell Street, Teaneck, NJ 07666
UK office: 73 Lynton Mead, Totteridge, London N20 8DH

Library of Congress Cataloging-in-Publication data is available.

Proceedings of the Workshop on
HADRON STRUCTURE FUNCTIONS AND PARTON DISTRIBUTIONS

Copyright © 1990 by World Scientific Publishing Co. Pte. Ltd.

All rights reserved. This book, or parts thereof, may not be reproduced in any form or by any means, electronic or mechanical, including photocopying, recording or any information storage and retrieval system now known or to be invented, without written permission from the Publisher.

ISBN 981-02-0334-9
     981-02-0524-4 (pbk)

Printed in Singapore by JBW Printers & Binders Pte. Ltd.

PREFACE

The Workshop on Hadron Structure Functions and Parton Distributions was held at Fermilab, April 26-28, 1990.

The purpose of the Workshop was to critically review the wide range of theoretical and experimental issues relating to short-distance hadron structure and parton distributions in the perturbative QCD framework, and to discuss open problems and promising future directions. This comes at a time when the QCD Parton Model formalism has become an essential part of high energy research ranging from precision tests of the Standard Model to the study of jet physics, and to the calculation of both signal and background for "new physics" at current and future accelerators. This requires detailed quantitative analysis far beyond the original simple parton model.

Although deep inelastic scattering Structure Functions have been standard agenda items at most high energy physics conferences in the last two decades, the determination of the universal Parton Distributions in the QCD framework in this new era is clearly no longer tied exclusively to the association between these two sets of quantities. Many new processes are becoming useful, and indeed indispensable, sources of detailed information on parton distributions in QCD-based global analyses. To incorporate these new processes in a consistent manner requires the application of perturbative QCD to the next-to-leading order and beyond, thereby opening new theoretical frontiers in several directions. The program of this Workshop underlined these new developments on both the experimental and the theoretical fronts.

As the first topical conference focused on this central theme in particle physics, the Workshop was designed to bring together experimentalists and theorists working on a wide range of related processes and problems. The active participation by 250 registered physicists from around the world, representing most major laboratories and universities, and their enthusiasm at the Workshop are testimony to the timeliness of this endeavor. The organizers would like to thank all speakers and participants for contributing to the success of the Workshop.

The Workshop was sponsored by Argonne National Laboratory, Fermi National Accelerator Laboratory, and the Superconducting Super Collider Laboratory. Special thanks should go to the Fermilab staff for the excellent administrative organization of the conference, especially to Treva Gourlay and Barbara Kristen for their administrative support, to Angela Gonzales for her artistic creativity in preparing our poster, to Cheryl Bentham, Barbara Burwell, Kristen Ford, Pamela Fox, Angelita Greviskes, Linda Olson-Roach, Carol Picciolo, and Suzanne Weber for their help both before and during the Workshop, and last but not least, a special note of thanks must be given to Olga Zdanovic for her data entry help during the six months spent in preparation for the Workshop.

The Editors
July, 1990

ORGANIZING COMMITTEE

William A. Bardeen, Fermilab (Co-Chairman)
Edmond L. Berger, Argonne
Friedrich Dydak, CERN
R. Keith Ellis, Fermilab
Donald F. Geesaman, Argonne
Joey Huston, Michigan State
John Huth, Fermilab

Harold Jackson, Argonne
Jorge Morfin, Fermilab
Milind Purohit, Princeton
Frank Sciulli, Columbia
Wu-Ki Tung, IIT (Co-Chairman)
Harry Weerts, Michigan State

TABLE OF CONTENTS

INTRODUCTION

Introduction to the Workshop 3
    F. Sciulli

Overview of Parton Distributions and the QCD Framework 18
    W. K. Tung

DEEP INELASTIC SCATTERING EXPERIMENTS

Recent Results from the New Muon Collaboration 41
    C. Peroni

Nucleon Structure Functions from ν_μ-Fe Scattering
at the Tevatron 50
    P. Z. Quintas (for the CCFR Collaboration)

Preliminary Results from E665 on Cross-Section Ratios at
Low x_{bj} Using H_2, D_2, and Xe Targets 58
    S. Aïd

Precise Extractions of the X and Q^2 Dependence of $R = \sigma_L/\sigma_T$,
F_{2p}, F_{2d}, and F_{2n}/F_{2p} from a Combined Analysis of SLAC
Deep Inelastic Electron Scattering Experiments 67
    A. Bodek

A QCD Analysis of High Statistics F_2 Data on H_2 and D_2 Targets
with Determination of Higher-Twists 76
    A. Milsztajn

Probing Nucleon Structure with ν-N Experiments 84
    S. R. Mishra

Present Status and Future of the Measurements of the Proton
and Deuteron Structure Function F_2 in Charged Lepton
Deep Inelastic Scattering 124
    M. Virchaux

Structure Functions, Parton Distributions, and QCD-Tests at HERA 139
    J. Bluemlein and G. A. Schuler

Comments on Precision Measurements of Nucleon
Structure Functions 165
    R. Wilson

Uncertainties in QCD Analysis of Structure Functions 169
    R. Lednicky

Derivation of Weak Neutral Current Structure Functions
from Distributions in a Hadronic Scaling Variable 180
    H. P. Borner

W/Z AND LEPTON-PAIR PRODUCTION AND PARTON DISTRIBUTIONS

Proton Structure from $p\bar{p}$ Colliders 187
    F. Halzen

W/Z and Lepton Pair Production at the Tevatron.................................. 205
    J. Hauser

Measurement of the Ratio $R = \sigma(W \to \mu\nu)/\sigma(Z \to \mu\mu)$
in UA1 at $\sqrt{s} = 0.63$ TeV 215
    M. W. van de Guchte

Recent Results from UA2 with Relevance to Hadron Structure
Function Determination... 224
    G. F. Egan (for the UA2 Collaboration)

What Can We Learn about Parton Distribution Functions
from the Drell-Yan Process?.. 234
    J. P. Rutherfoord

DIRECT PHOTON AND HEAVY QUARK PRODUCTION

The Gluon Structure Function from the Tagged Photon Lab 247
    M. V. Purohit

Photoproduction of Heavy Quarks, Preliminary Results from E687 254
    P. Lebrun (for the E687 Collaboration)

Next-To-Leading-Logarithm Calculations of Direct Photon Production 264
    J. F. Owens

Results on Direct-Photons from Fixed-Target Experiments 272
    C. Bromberg

Recent Results on Direct Photons from CDF 278
    R. M. Harris (for the CDF Collaboration)

Comparison of Direct γ Production in $\bar{p}p$ and pp Reactions at $\sqrt{s}$ = 24.3 GeV at $4 < p_T < 6$ GeV/c $(.3 < x_T < .5)$ 291
    L. Camilleri (for the UA6 Collaboration)

Direct Photon Cross Section Measured by UA2 297
    G. F. Egan (for the UA2 Collaboration)

GLOBAL FITS

A Global Analysis of Recent Experimental Results: How Well Determined Are the Parton Distribution Functions? 305
    J. G. Morfin

Parton Distributions from a Global Analysis of Data 338
    R. G. Roberts

Phenomenological Consequences of HMRS Partons 346
    A. D. Martin

SPIN STRUCTURE OF THE NUCLEON

Measurements of Nucleon Spin Structure Functions with Polarized Leptons 361
    K. Rith

Polarized Drell-Yan Experiments 378
    J. C. Collins

Prompt Photon Experiments Using Polarized Beams 384
    P. F. Slattery

Physics with Transversely Spinning Quarks 395
    X. Artru

THEORY

Parton Distribution Functions and Higher Order Corrections 405
    R. K. Ellis

Resummation of Large Terms in Hard Scattering 423
    G. Sterman

Multi-Parton Interactions and Inelastic Cross Section
in High Energy Hadronic Collisions 432
    D. Treleani

$1/Q^2$ Corrections to Deeply Inelastic Scattering
and Drell-Yan Cross Sections 442
    J. Qiu

SUMMARY

Experimental Horizons for Structure Function Measurements 453
    G. Wolf

Program 481

List of Participants 487

Author Index 499

Proceedings of the Workshop on
HADRON STRUCTURE FUNCTIONS and PARTON DISTRIBUTIONS

INTRODUCTION

INTRODUCTION TO THE WORKSHOP

FRANK SCIULLI
Columbia University and Nevis Laboratory
New York, NY 10027

Abstract: An attempt is made to explain why we are here. The speaker only partially succeeds.

1. Credo in Unum Theorie

   This workshop has the explicit purpose of examining the state of structure functions and parton distributions. You may well ask, "Why should we have a workshop on this subject?" My answers are

   a) Since protons and neutrons are the principle constituents of all known matter, we must know the "stuff" from which they are made! Indeed, we must understand how this stuff interacts as well!

   b) With knowledge of parton distributions, which describe the composition of nucleons, we can predict how often almost anything is made in the collisions of those nucleons. A rigorous calculation can be made for the cross-section of any final state, X, from the collision of nucleons A and B
$$\sigma (A + B \to X)$$
from the well-known formula

$$E \frac{d^3\sigma}{dp^3} = \sum \int dx_A \, dx_B \, \rho_{i/A}(x_A, Q^2) \, \rho_{j/B}(x_B, Q^2) \, E_{ij} \frac{d^3\sigma(i+j\to X)}{d^3p_{ij}}.$$

Miraculously, probabilities for any hard hadronic process will be known if the elementary cross-section on the right-side of this equation is known from first principles, and if the appropriate parton distributions have been measured or evolved from measurement. In the above equation, we need the probability, $\rho_{i/A}(x_i,Q^2)$, for finding a parton of type i, with momentum fraction, x_A, inside hadron A, as well as the probability for finding parton j inside B. (These probabilities are simply the structure functions divided by x.) From this equation comes all kinds of wonderful predictions about
   a) Drell Yan production (Saturday morning);
   b) Direct photon production (Thursday afternoon);
   c) Boson production (Saturday morning);
and many other processes about which we will hear.

Isn't this answer somewhat presumptuous? Why do we think we can do all this? We require some common ground, a collective agreement that all this is possible. For <u>this</u> workshop, such common faith would go something like

"I believe in the quark-parton model.
I believe in QCD;
I believe in calculations from perturbative QCD;
I believe in factorization;
...."

Such a Credo does not forbid disagreement. Anyone who has heard an argument about scale questions will immediately recall that disagreement is tolerated. There are even some fundamental differences in the perspective we collectively bring to these problems. The role of experiment is central to resolving disagreements; differences in experimental perspective might be generalized as follows:

Experimenters' creed: "I am skeptical about all experiments."

Theoreticians' creed: "I believe all experimental results; when they disagree, I will find some way to make them agree."

Phenomenologists' creed: " I believe in all experimental data implicitly; I am skeptical that experimentalists know how to use them!"

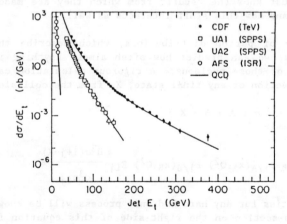

Figure 1: Inclusive E_t spectrum[1] from ISR, $Sp\bar{p}S$, and Tevatron data. The smooth curves are a leading order QCD calculation for $Q^2 = E_t/2$. Normalizations are absolute.

Such foolish joking disguises a very serious aspect: there exist fundamental bases on which most of us agree. The quark-parton model and QCD create a foundation from which we communicate, and factorization is an important extension of this base. The Credo above is well founded in experimental fact. A glance at figure 1 demonstrates that these bases are NO joke! The use of a single set of structure functions to fit properly experimental data from the reaction $p + p \to$ jets as a function of transverse energy over a range of eight decades in cross-section and a factor 30 in energy with no additional parameters is SERIOUS business. Something is working right.

On the other hand, the fundamental assumptions that go into this enterprise are very important. We should not blindly proceed to apply the Credo without understanding its origins and re-examining the experimental base. Hence, we begin this conference with a re-examination of our origins, to see from whence we came and how well our beliefs are supported in experimental fact.[2]

2. Deep Inelastic Scattering - the beginning

Twenty-five years ago, the beautiful idea[3] of quarks had already become accepted as a calculational tool for the strong interactions. It had grown out of the phenomenal successes of the SU(3) algebra as applied to spectroscopy of known hadrons. Building on Hofstadter's studies[4] of elastic electron scattering from nuclei, a great discovery[5] was made by Friedman, Kendall, Taylor and their collaborators at SLAC: the deep inelastic scattering of electrons from nucleons produced results that could be simply described when analyzed in terms of the Bjorken variable[6] $x = Q^2/2M\nu$. This phenomenon of scaling was beautifully and simply explained[7] by Feynman in the Parton Model. When it became clear that the hypothesized point-like constituents, partons, had properties characteristic of quarks, the model came to be called The Quark-Parton Model.

Figure 2 shows the scattering by a lepton of a point-like quark inside the nucleon with the exchange of either a virtual photon or charged/neutral weak boson. In the most elementary version of the model, the quarks are presumed free and the process is elastic. (This naive assumption was <u>always</u> known to be an approximation; after all, the quarks are bound in the nucleon.) The coupling strength and helicity dependence are determined by the well-established electroweak theory in conjunction with the known properties of quarks.

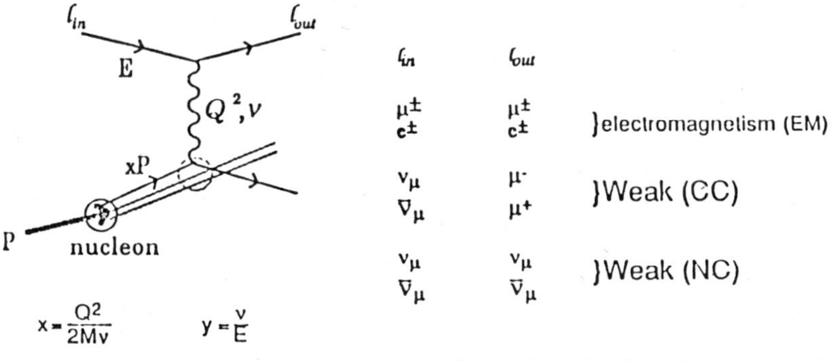

ℓ_{in}	ℓ_{out}	
$\mu^\pm$	$\mu^\pm$	} electromagnetism (EM)
$e^\pm$	$e^\pm$	
ν_μ	μ^-	} Weak (CC)
$\bar\nu_\mu$	μ^+	
ν_μ	ν_μ	} Weak (NC)
$\bar\nu_\mu$	$\bar\nu_\mu$	

$x = \dfrac{Q^2}{2M\nu}$ $y = \dfrac{\nu}{E}$

Figure 2

In this context, measurements from deep-inelastic scattering permit us to extract structure functions from measured cross-sections, differential with respect to the scaling variables: x and $y = \nu/E$. The structure functions are related in turn to distributions of quarks' momenta inside the nucleon (quark densities in momentum space).

Does this picture make qualitative sense? Clearly yes. To the best we can measure, energy dependence is consistent with the point-like behaviour expected of quarks. Tests of the quark properties, sumrules and other quantum number tests, agree as anticipated and are discussed in detail below.

Does the model work quantitatively in all respects? Clearly no. Scaling is not experimentally exact. However, perturbative QCD as the theory of strong interactions appears to describe the observed patterns of scale breaking. We clearly require QCD or something very like it at high Q^2. One question is how well perturbative QCD works at lower Q^2. We will hear two talks (Friday afternoon) regarding this question.

3. Tests of the Quark Parton Model and QCD

3.1 Neutral Current and Charged Current Comparison

The comparison of neutrino-nucleon data, the integrated neutral current versus charged current cross sections, provides a direct and exceedingly accurate measure of the electroweak parameters: $\sin^2\theta_W$ and ρ. The differential dependence on x comprises a test of consistency with the quark picture depicted in figure 1. This test has been carried out by two groups[8,9] and in both cases good agreement is obtained.

3.2 Neutrino - Muon/Electron Comparison

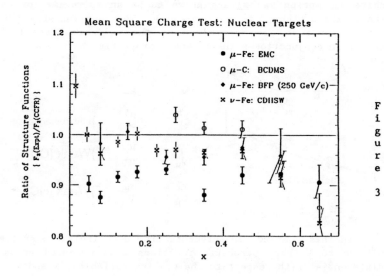

Figure 3

The ratio of charged lepton structure functions to neutrino charged current structure functions from the same nucleon provides a direct test of the mean square charge of the constituent quarks. With the predicted factor of 5/18 removed, the experimental comparison is shown[10] in figure 3. The data from three experiments are normalized to that of the CCFR narrow band data. The two neutrino experiments (CDHSW/CCFR) agree with each other to about one percent. The overall agreement for the three muon experiments with CCFR is .96 ± .04, with the EMC data lying considerably lower than the other two. If the EMC data is removed, the agreement is .98 ± .03. We conclude that the quark charges are verified at about the three percent level.

3.3 Callen - Gross Relation[11]

The smallness of the parameter $R = \sigma_l/\sigma_t$ follows from the spin 1/2 nature of quarks. While the value would be identically zero in the naive quark model, it should be finite and described by a well-defined function predicted from perturbative QCD. This will be discussed more fully in section 5. Suffice to say here that the value is small but finite, as expected for scattering from bound quarks.

3.4 Sum Rules

The sum rule predictions are very important; their present status is described here. In the following, N_u and N_d are the number of u and d quarks in the proton, respectively; e_u^2 and e_d^2 are their respective charges. Included are leading order QCD corrections, where appropriate.

3.41 Adler Sum Rule[12]

This prediction utilizes neutrino scattering structure functions from neutrons and protons:

$$S_A = \frac{1}{2} \int_0^1 \frac{dx}{x} \left[F_2^{\nu n} - F_2^{\nu p} \right] = (N_u - N_d)$$

The predicted value, one, is born out by experiment[13] at the twenty percent level:

$$S_A^{exp} = 1.01 \pm .20$$

3.42 Gottfried Sum Rule[14]

This utilizes muon or electron structure functions from neutrons and protons:

$$S_G = \int_0^1 \frac{dx}{x} \left[F_2^{\mu n} - F_2^{\mu p} \right] = (N_u - N_d)(e_u^2 - e_d^2)$$

The prediction of 1/3 is corroborated[15], with about fifty percent accuracy, at the one standard deviation level:

$$S_G^{exp} = 0.24 \pm .11$$

3.43 Gross - Llewellyn Smith Sum Rule[16]

Since the xF_3 structure function measures directly $\underline{x}$ times the density of valence quarks in the nucleon, the weighted integral predicts the total number of valence quarks, or three:

$$S_{GLS} = \int_0^1 \frac{xF_3^{\nu N}}{x} dx = (N_u + N_d)\left[1 - \frac{\alpha_s}{\pi}\right]$$

The factor in parenthesis is a leading order QCD correction. Assuming $\Lambda = 186 \pm 60$ MeV in the $\overline{MS}$ renormalization scheme, the prediction[2] is

$$S_{GLS}^{pred} = 2.74 \pm .06 .$$

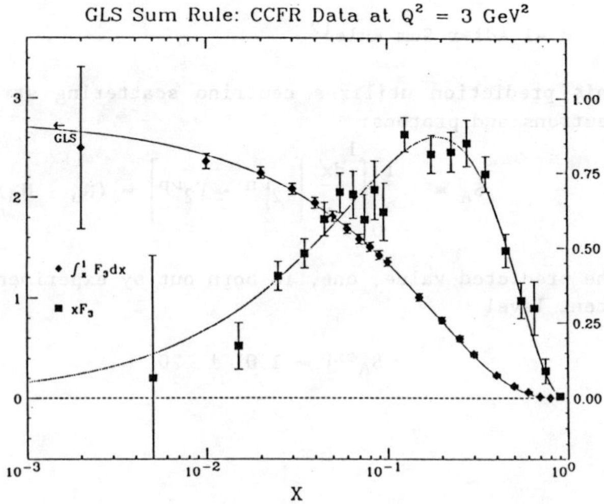

GLS Sum Rule: CCFR Data at $Q^2 = 3$ GeV^2

Figure 4

It is possible to measure this sum rule extremely well. Figure 4 shows the most accurate measurement[17] until today of the xF$_3$ structure function; the data also provide the best measurement of the sum rule. The average value from existing experiments[10] is

$$S_{GLS}{}^{exp} = 2.79 \pm .13 .$$

New data for xF$_3$ will be presented here by the CCFR group on Friday afternoon; the experimental error on the sum rule integral will be reduced even further.

3.44 Bjorken Sum Rule[18]

This sum rule provides a prediction for the "spin structure functions", defined in terms of the net asymmetry in the cross-section for incident muon spin parallel and antiparallel to the target nucleon spin. Evaluating the sum rule requires measurements on both polarized neutrons and polarized protons:

$$\int_0^1 [g_1{}^n(x) - g_1{}^p(x)]dx = \frac{1}{2}[(N_u{}^\uparrow - N_u{}^\downarrow) - (N_d{}^\uparrow - N_d{}^\downarrow)](e_u{}^2 - e_d{}^2)\left[1 - \frac{\alpha_s}{\pi}\right]$$

where, for example, $N_u{}^\uparrow$ is the number of up quarks with spin parallel to the incident lepton spin. The factor inside the first bracket on the right hand side of the equation is simply the expectation value of the quark spin, $<\sigma_z>$, which in turn is equal to the ratio of axial vector to vector coupling constants, G_A/G_V. The prediction[19] from the right-hand side is .191 ± .002. The only measurements to date come from polarized proton targets, the most recent[19] is shown in figure 5.

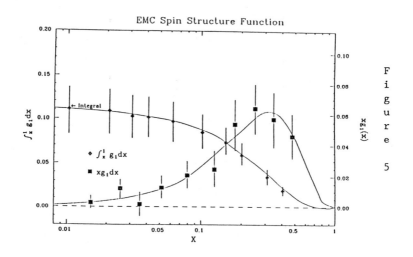

Figure 5

From the data in the figure, the integral is

$$\int_0^1 g_1^P(x)\, dx = 0.114 \pm .012 \pm .026$$

By <u>assuming</u> the Bjorken sum rule, the corresponding integral for neutrons is $\int g_1^n(x)dx = -.077 \pm .012 \pm .026$. The implication of this exercise is that, whatever assumptions one makes regarding strange quarks, it seems that the net spin of the nucleon has its origin primarily in something <u>other than valence quarks</u>. To those of us who grew up with the great successes of SU(3) and the quark model in the description of nucleon spectroscopy and magnetic moments, this came as somewhat of a surprise; however, there are explanations. An entire session on Friday morning is devoted to this important topic.

4. Quark Densities

The quark densities are the critical input quantities for predicting rates of the hadronic processes described in the first section. The most precisely measured of these are the "isoscalar densities", obtainable from heavy target data, which are the average of up and down densities:

quark $\quad q(x) = xu(x) + xd(x) + xs(x)$
antiquark $\quad \bar{q}(x) = x\bar{u}(x) + x\bar{d}(x) + x\bar{s}(x)$
valence $\quad q_V(x) = q(x) - \bar{q}(x)$.

These may be directly determined from neutrino scattering data. The valence distribution is shown in figure 4 and the antiquark distribution from CDHS wide band data and from CCFR narrow band data is shown in figure 6.

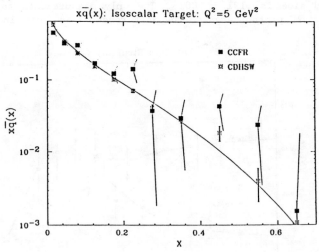

Figure 6

The separate quark densities (u, d, etc) require data obtained from light targets. In principle, these can be directly extracted from the y-dependent cross-sections of neutrino and antineutrino collisions on hydrogen. However, the precision of such experiments is not high. In practice, complementary information comes from electron/muon scattering data on hydrogen and deuterium combined with an assumption (from neutrino scattering) for the antiquark composition. In fact, there are some disagreements among the resulting quark densities.[10] Figure 7 shows the present data on the individual valence u and d distributions from neutrino (WA25,WA25,CDHSW) and muon (EMC) experiments. There are some obvious disagreements seen in the figure, particularly for the $d_{VALENCE}$ distribution. New analyses presented here by SLAC and BCDMS as well as the discussion session on Friday may shed light on this problem.

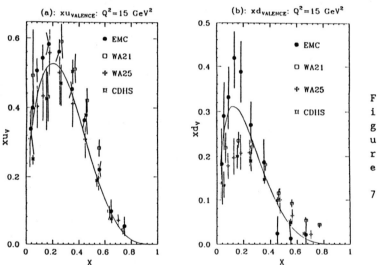

Figure 7

Neutrino scattering produces, about one percent of the time, two final state muons of opposite sign. These are known to be dominantly due to charm production. (The much smaller same sign dimuon rate is now well understood.[20]) The dimuon data permit extraction of the strange sea distribution in the nucleon. All such measurements are consistent with the parameter $\kappa = 2s/(\bar{u}+\bar{d})$, which measures the strange component relative the $\bar{u}$ and the $\bar{d}$ components, to be about 0.5. This will be discussed by Mishra in his review talk on Friday. The questions of interpretation implicit in such processes involving a heavy quark will be addressed by Tung in his talk this morning.

The very important question of the dependence of the quark distributions on the nuclear target[21] (the EMC effect), will be addressed on Friday with new results from the NMC group.

5. QCD Tests

In Quantum Chromodynamics, the important coupling between quarks and gluons is not constant, but a slowly varying function of the momentum transfer. In leading order, the dependence is

$$\alpha_s(Q^2) = \frac{12\pi}{(33 - 2N_f)\ln(Q^2/\Lambda^2)}$$

where N_f is the number of quark flavors and Λ is a constant. At next to leading order, a specific renormalization scheme must be adopted. The conventional choice these days is the MS-bar scheme, parametrized by the constant, $\Lambda_{\overline{MS}}$.

The principle difficulty in testing QCD quantitatively is to find tests which indeed test Quantum Chromodynamics rather than, say, the universality of structure functions. In deep inelastic scattering, there are two very specific tests which should, at high Q^2, be valid. These involve the R-parameter of the Callen-Gross relation and the evolution of non-singlet structure functions.

5.1 R = σ_l/σ_t

While non-perturbative terms could contribute to R, there <u>must</u> be a contribution calculable directly from perturbative QCD.[22] In leading order, it looks as follows:

$$R(x,Q^2) = \frac{\alpha_s(Q^2)}{2\pi} x^2 \int_x^1 \frac{dz}{z^3} \left[\frac{8}{3} F_2(z,Q^2) + 4f\left(1 - \frac{x}{z}\right) zG(z,Q^2) \right] \Big/ 2xF_1$$

Here the quark-gluon coupling, α_s, multiplies a known integral. The parameter, f, is the total number of flavors for neutrino scattering and is the sum of the squares of quark charges for muon/electron scattering. There exist some measurements of R at high Q^2 from neutrino and charged lepton scattering. The recent important measurements from SLAC[23] at low Q^2 have clarified the confusion that had existed in this low energy region.

These low Q^2 data also permit us to test the QCD prediction in a clean way. By assuming that the low Q^2 measurements of R are <u>all</u> due to non-perturbative terms, we obtain an upper limit on the contributions from such terms at high Q^2. This then permits a direct test of whether we are observing the contributions from the above equation.

The result of such an analysis[24] is that the available high Q^2 data are <u>consistent with, but no not require, QCD</u>. More precise high energy data are needed before we can make this quantitative test.

5.2 Non-singlet Structure Function Evolution

A second and very important direct test of QCD lies in the scale-breaking pattern of structure functions.[25] Data plotted versus log Q^2 at fixed x typically shows a logarithmic slope which is negative at large x and becomes more positive at smaller x. While this pattern is qualitatively what one expects from QCD, the quantitative prediction for the singlet structure function, F_2, depends on both the x-dependence of F_2 and that of $G(x)$, the gluon distribution. [This dependence of slope variation on $G(x)$ provides the principal method which is used at present for measuring G in this process.] The uncertain function, $G(x)$, creates problems for testing QCD with F_2 at small x. At larger x, the behaviour of F_2 is strongly coupled with the value of the quark-gluon coupling, α_s; again, the large x behaviour is a good way of measuring this value, or equivalently, Λ.

A most direct and simple test of perturbative QCD is to observe the Q^2-dependence of non-singlet structure functions at small x. Examples of such functions are xF_3 (from neutrino scattering) and $F_2^n - F_2^p$ (from muon/electron scattering). In the QCD evolution of these functions, the gluons enter not at all, and the small x behaviour is almost completely determined by the theory. In leading order, the predicted slope is given by

$$\frac{d\ F^{NS}(x,Q^2)}{d\ \ln(Q^2)} = \frac{\alpha_s(Q^2)}{2\pi} \int_x^1 F^{NS}(z,Q^2)\ P_{qq}(\frac{x}{z})\ dz$$

An important feature of this equation, as applied to xF_3, is that the right hand side goes through zero at a value determined primarily from the splitting function, P_{qq}, which is provided by perturbative QCD. Hence, independent of the quark-gluon coupling and only slightly dependent on the measured x-dependence of xF_3, we expect the logarithmic slope of xF_3 to go through zero at a well-defined and predicted value of x. This value is about $x = 0.12$.

Figure 8 shows measurements of this slope available up to the time of this meeting. The high statistics CDHSW data[26] simply do not show the predicted behaviour; some have argued that this failure of the test is due to systematic error rather than problems with QCD. The narrow band data[17] of CCFR have not had the statistical precision to make a good test. New xF_3 data at small x from the CCFR collaboration, with high statistics, will be presented on Friday afternoon bearing on this important question.

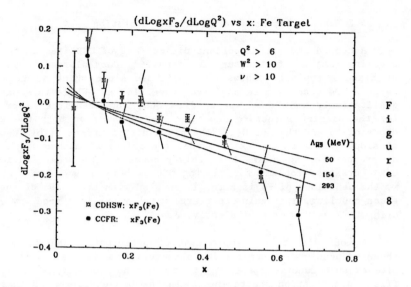

Figure 8

6. Gluon Distributions and Quark-Gluon Coupling

In the absence of high statistics pure non-singlet data, there has been an important contribution to testing QCD and measuring the quark-gluon coupling and gluon distributions by the BCDMS group[27], using muon-nucleon scattering data. In this analysis, they <u>assume</u> that the gluon distribution is small enough to ignore for x>.25, and that the sea distribution is similarly small. Under this assumption, the F_2 structure function at large x is a non-singlet, and it evolves as a non-singlet. While they cannot test the small x behaviour, they can use the large x data to measure the quark gluon coupling and extract the parameter, $\Lambda_{\overline{MS}}$. They obtain consistent results from hydrogen and carbon data; more recently, they have performed the analysis with deuterium. With data at smaller x from the light targets, they have proceeded to extract gluon distributions.

Other experiments have not yet corroborated these results. The anomalous behaviour of xF_3 from CDHS has already been mentioned. The Q^2-dependence of F_2 data from both CDHS and EMC do not show the behaviour expected from perturbative QCD[10]. It is possible that this is due to systematic errors, though the precise source of this error has not yet been quantitatively demonstrated.

There are several other candidate methods for measuring the quark-gluon coupling. In a recent review[28] by Altarelli, he has tabulated the values of $\Lambda_{\overline{MS}}$ from these various methods, which is reproduced here, together with the error, including his estimate of the theoretical error.

	α_s (34 GeV)	$\Lambda_{\overline{MS}}^{(4)}$ (MeV)	$\Lambda_{\overline{MS}}^{(5)}$ (MeV)
$R_{e^+e^-}$	0.14 ± 0.02	370^{+350}_{-220}	240^{+230}_{-140}
BCDMS	0.127 ± 0.006	220 ± 60	140 ± 40
Υ	0.123 ± 0.009	180 ± 80	120 ± 50
$e^+e^- \to$ jets	0.135 ± 0.015	330 ± 200	215 ± 130
γ structure function	0.120 ± 0.016	175 ± 125	115 ± 80

In general, methods outside of deep-inelastic are severely limited by theoretical and interpretational errors. Interestingly, there is general agreement within the assigned errors among the various techniques.

In the same review[28], the gluon distributions extracted for several data sets were shown and are reproduced here in figure 9. We have seen that, in many of these deep inelastic scattering data sets, there is a serious and fundamental problem with a QCD interpretation of the data. The BCDMS extracted curves and data points, using data consistent with QCD, are also shown. It is clear that we have much more to learn about the gluon distribution.

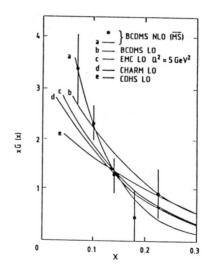

Figure 9

7. Benedictus

As a scientific enterprise, we should feel some pride! There is ample and strong reason to believe the fundamentals that provide us with the language by which we communicate and with the calculational tools so successful to our efforts. The quark distributions are being extracted, and these are being successfully applied to the hard hadronic processes of interest. Measurements of α_s are consistent among various methods.

Much remains, however. The fundamental assumptions must be experimentally verified to high precision at lower energies and must be checked at the highest attainable energies. The quark components should be delineated more cleanly with higher precision data. The gluon distribution should be known better, perhaps from several different processes. The spin properties of the nucleon are confusing; they must be measured and understood well. Certain regions of parameter space, particularly at small x and high Q^2, need to be explored.

The purpose of the workshop is to communicate, to see where we have been and where we are going. So then, let us proceed to communicate.

8. References

1. John Huth, "QCD and Jets", <u>Proceedings of the 1989 International Symposium on Lepton and Photon Interactions at High Energies</u>, Stanford University, ed by M. Riordan, p. 368.

2. For a more complete review, see S. R. Mishra and F. Sciulli, "Deep Inelastic Lepton-Nucleon Scattering", <u>Ann Rev Nucl Part Sci</u> 39:259 (1989).

3. M. Gell-Mann, <u>Phys Lett</u> 8:214 (1064); G. Zweig, CERN Reps. 8182/TH 401, 8419/TH 412 (1964), unpublished.

4. R. Hofstadter, <u>Electron Scattering & Nuclear and Nucleon Scattering</u>, New York: Benjamin (1963).

5. G. Miller et al, <u>Phys Rev</u> D5:528 (1972); A. Bodek et al, <u>Phys Rev Lett</u> 30:1087 (1973).

6. J. D. Bjorken and E. A. Paschos, <u>Phys Rev</u> 185:1975 (1969).

7. R. P. Feynman, <u>Phys Rev Lett</u> 23:1415 (1969); <u>Photon-Hadron Interactions</u>. Reading, Mass: Benjamin (1972).

8. T. Mattison, PhD Thesis, MIT, Dec 1986.
   F. Taylor, <u>Proc. Neutrino '88</u>, Boston (1988).

9. J. V. Allaby, et al, Phys Lett B213:554 (1988).

10. See reference 2 and references quoted therein.

11. C. G. Callan and D. J. Gross, Phys Rev Lett 22:156 (1969).

12. S. L. Adler, Phys Rev 143:1144 (1966).

13. D. Allasia et al, Phys Lett B135:231 (1984); Z Phys C28:321 (1985).

14. K. Gottfried, Phys Rev Lett 18:1154 (1967).

15. T. Sloan, Proc Int Europhysics Conf on HEP, Uppsala, Sweden (1987).

16. D. J. Gross and C. H. Llewellyn Smith, Nucl Phys B14:337 (1969).
    M. A. B. Beg, Phys Rev D11:1165 ((1975).

17. E. Oltman et al, presented at DPF88, Storrs, CN. PhD Thesis Columbia University (Mar 1989), to be published.

18. J. D. Bjorken, Phys Rev 148:1467 (1966); Phys Rev D1:1376 (1970).

19. J. Ashman et al, Phys Lett B206:364 (1988).

20. B. A. Schumm et al, Phys Rev Lett 60:1618 (1988).

21. J. J. Aubert, et al, Phys Lett B123:275 (1983).

22. G. Altarelli and G. Martinelli, Phys Lett 76B:89 (1978);
    M. Glueck and E. Reya, Nucl Phys B145:24 (1978).

23. S. Dasu et al, Phys Rev Lett 61:1061 (1988).

24. S. R. Mishra and F. Sciulli, Col Univ Nevis Preprint 1422, to be published in Phys Lett.

25. G. Altarelli and G. Parisi, Nucl Phys B26:298 (1977).

26. B. Vallage, PhD Thesis, Saclay CEA-IV-2513 (Jan 1987), to be published.

27. A. C. Benvenuti et al, Phys Lett B195:91 (1987); Phys Lett B195:97 (1987).

28. G. Altarelli, Ann Rev Nucl Part Sci 39:357 (1989).

OVERVIEW OF PARTON DISTRIBUTIONS AND THE QCD FRAMEWORK

Wu-Ki Tung

Illinois Institute of Technology, Chicago, Illinois 60616
and
Fermi National Accelerator Laboratory P.O. Box 500, Batavia, Illinois 60510

ABSTRACT

The perturbative QCD framework as the basis of the parton model is reviewed with emphasis on several issues pertinent to next-to-leading order (NLO) applications to a wide range of high energy processes. The current status of leading-order and NLO parton distributions is summarized and evaluated. Relevant issues and open questions for second-generation global analyses are discussed in order to provide an overview of topics to be covered by the Workshop.

1 Introduction

Perturbative Quantum Chromodynamics (QCD) provides the theoretical basis for the intuitively appealing Parton Model. It furnishes a comprehensive framework for describing general high energy processes in current and planned accelerators and colliders. The fundamental formula – the Factorization Theorem – relates a typical *physical cross-section* to a sum over relevant basic partonic *hard cross-sections* (which can be calculated perturbatively) convoluted with corresponding *parton distributions* (which can, in principle, be extracted from a set of standard experiments at moderately high energies). With proper attention to their definition and a consistent convention, these distribution functions are *universal* – *i.e.*, they are independent of the physical process to which they are applied.

In leading-order (LO), the QCD parton framework reproduces original parton model[1] results, with scale-dependent parton distributions. Since the early 70's, this simple model has enjoyed spectacular successes in unifying the phenomenology of all sorts of high energy processes to about the 10–15% level within currently available x range, modulo some overall "K-factors" for processes such as lepton-pair production. Furthermore, it has been used as an indispensable tool to make projections for future physics at much higher energies and small x [2]. In these applications, the well-known leading order parton parametrizations[3,2] played an essential role.

In recent years the use of QCD parton model has developed to a stage at which much improved knowledge of parton distribution functions are clearly required along three distinct fronts:

(i) In physical processes which provide precise tests of the Standard Model, the precision of experimental measurements has improved dramatically on the one hand, and the relevant hard cross-sections have been calculated to the next-to-leading order (NLO) and beyond on the other. In order to make real progress, it is crucial to know the parton distributions to a comparable degree of accuracy. This, in turn, prescribes the use of NLO evolution of the distributions and requires a much more detailed comparison with experimental data in the extraction of these distributions, especially for the sea-quarks which are not well determined because of their relative small size. Relevant processes in this category are: electron, muon & neutrino deep inelastic scattering (DIS); lepton-pair production (LPP or DY), and (W, Z, γ) production. It has become increasingly clear that the lack of detailed knowledge of the parton distributions often constitutes the largest source of uncertainty in precise tests of the Standard Model in these processes (e.g. the determination of the Weinberg angle).

(ii) In the study of jet physics and associated production of (W, Z, γ) with jets, which are important on their own right as well as sources of significant background for "new physics", reliable knowledge of the gluon distribution is crucial to the predictive power of the QCD parton model. But the gluon distribution is, so far, not very well determined.

(iii) Physical processes at future colliders (SSC, LHC) involve partons at very small x, well beyond the currently measurable range (around $0.03 < x < 0.75$). In order to make quantitative predictions, it is important: (a) to gain more theoretical insight on the small-x behavior of parton distributions and on the interface of small-x physics to "soft" (e.g. Reggeon) physics; and (b) to determine phenomenologically the range of possible small-x extrapolation of parton distributions consistent with currently available data.

This review obviously cannot cover the entire landscape of activities on Parton Distributions and the Perturbative QCD Framework. In the first part, I shall outline the NLO QCD parton formalism and focus on three examples to illustrate some of its non-trivial features which must be taken into account in any quantitative applications. Although these examples are fairly simple conceptually, not all of them are known to practitioners of QCD phenomenology, resulting in frequent misunderstanding and sometimes misuse of the formalism. In the second part, I shall survey the existing parton distribution parametrizations, summarize the relevant issues confronting "second generation" global analyses of parton distributions, and comment on the proper use of these distributions.

It should be emphasized from the very beginning that the study of parton distributions now encompasses a full range of lepton-hadron and hadron-hadron processes; and, as implied by the above list of motivations, it is inextricably intertwined with all areas of high energy physics from the precision tests of the Standard Model to the search for "new physics". The aim of this review is to help establish a certain common ground and a common language for subsequent discussions in this Workshop, among participants with a diverse background.

2 QCD Parton Model in Next-to-Leading Order

For the sake of simplicity, I shall use a generic lepton-hadron scattering process as the talking point. All the issues I discuss also apply to hadron-hadron scattering, albeit in a somewhat more involved form. The generic process is of the form: $\ell + H \longrightarrow C + X$, where C either represents an identified final-state particle with specific attributes (such as heavy mass or large transverse momentum) or is null (in the case of total inclusive scattering). The "master equation" of the QCD Parton Model is the *factorization* formula[4] which reads:

$$\sigma^i_{H \to C}(q,p) = \sum_a f^a_H(\xi, \mu) \otimes \hat{\sigma}^i_{a \to C}(q, k, \mu) \tag{1}$$

where, as illustrated in Fig. 1, H is the target hadron label; a is the parton label; i is the electroweak vector boson helicity label; (q, p, k) are the momenta of the vector boson, the hadron, and the parton respectively; μ is a renormalization scale; and ξ is the fractional momentum carried by the parton with respect to the hadron. The symbol $\otimes$ denotes a convolution (over the variable ξ) of the parton distribution function f^a_H and the hard vector-boson-parton cross-section $\hat{\sigma}^i_{a \to C}$.

The hard cross-section $\hat{\sigma}^i_{a \to C}$ can be calculated in perturbative QCD:

$$\hat{\sigma}^i(\xi, Q/\mu, \alpha_s(\mu)) = \hat{\sigma}^i_0\, \delta(1-\xi) + \alpha_s(\mu)\, \hat{\sigma}^i_1(\xi, Q/\mu) + O(\alpha_s^2) \tag{2}$$

where we have suppressed the initial parton label a. The LO $\hat{\sigma}^i_0$ is a constant proportional to the square of the electro-weak coupling of the parton. To calculate the NLO hard cross-section $\hat{\sigma}^i_1$, one encounters divergences which must be subtracted in order to yield finite answers. The subtraction term, in effect, corresponds to that part of the NLO contribution pertaining to almost on-the-mass-shell and collinear parton lines which is already included in the LO term by virtue of the use of QCD-evolved parton distributions. Since the subtraction, hence the hard cross-section $\hat{\sigma}$, is <u>renormalization scheme</u> and <u>renormalization scale</u> *dependent* while the physical cross-section on the left-hand side of Eq. (1) must be *independent* of these theoretical

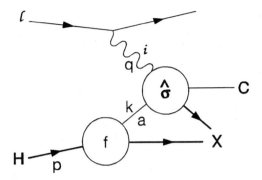

Figure 1: The QCD parton picture and the Factorization Theorem.

artifices, *the parton distribution functions f_H^a have to be scheme-dependent objects to match the definition of $\hat{\sigma}$.* We shall demonstrate in Sec. 2.2 that *the scheme-dependence of the gluon and sea-quark distributions can be very substantial* – contrary to conventional expectations. This can lead to important phenomenological consequences. (See also Sec. 2.3)

We begin by examining some basic issues concerning the QCD coupling function α_s in the presence of heavy quarks.

2.1 The QCD Coupling and Λ_{QCD}

The running coupling $\alpha_s(\mu)$ is the most basic of QCD quantities. In the well-known case of *all zero-mass quarks*, the standard formulas for α_s in LO and in NLO ($\overline{MS}$ scheme) are, respectively:

$$\alpha^{LO}(n_f, \mu/\Lambda) = \frac{1}{\beta_0 \log(\mu/\Lambda)} \tag{3}$$

$$\alpha^{NLO}(n_f, \mu/\Lambda) = \frac{1}{\beta_0 \log(\mu/\Lambda)} \left[1 - \frac{\beta_1}{\beta_0^2} \frac{\log\log(\mu/\Lambda)^2}{\log(\mu/\Lambda)^2}\right] \tag{4}$$

where n_f is the number of quark flavors and it enters the right-hand side through the constants,

$$\beta_0 = \frac{33 - 2n_f}{3} \qquad \beta_1 = \frac{153 - 19n_f}{3} \tag{5}$$

If all quarks are massless, the number n_f is fixed and the running coupling α_s is determined by a single parameter Λ – the "QCD lambda".

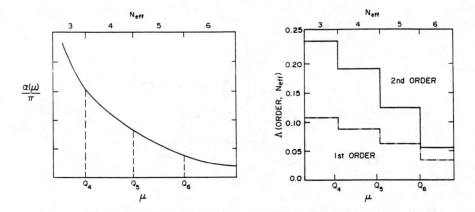

Figure 2: α_s and Λ_{QCD} as functions of μ and n_f^{eff} for a given coupling strength.

In the presence of massive quarks, the situation is quite different. According to the decoupling theorem, each heavy quark i with mass m_i is effectively decoupled from physical cross-sections at energy scales μ below a certain threshold Q_i which is of the order m_i. Thus, *the number of effective quark flavors n_f^{eff} is an increasing step function of the scale μ.* Under this circumstance, the specification of the running coupling α_s and the associated Λ_{QCD} is not as simple as before. Although this point is fairly well-known, there still exist considerable confusion and ambiguity about these parameters in both the current literature and in conference presentations. Hence, it is worthwhile to summarize the proper formulation of the problem explicitly.

The definitions of α_s and Λ_{QCD} in the presence of mass thresholds are not unique – they are renormalization-scheme dependent. A *natural* choice is based on the requirement that $\alpha_s(\mu)$ be a continuous function of μ, and that between thresholds, it reduces to the familiar $\overline{MS}$ α_s [5]. This requirement leads to the condition that $Q_i = m_i$ (instead of $2m_i$, or even $4m_i$ [2], as often chosen for phenomenological reasons). This choice has the additional desirable feature that the parton distribution functions so defined are guaranteed to be continuous across the thresholds. If Eq. (4) is to remain valid with α_s being continuous in μ, but n_f^{eff} a discontinuous function of μ, it is quite obvious that the effective value of Λ must also make discontinuous jumps with n_f^{eff} at heavy quark thresholds. The same remark applies if one uses the LO formula for α_s, Eq. (3). Fig. 2a shows a typical α_s vs. μ plot; and Fig. 2b shows the corresponding Λ_{QCD} as a function of μ (bottom scale) and n_f^{eff} (top scale).

Fig. 2a explicitly shows that the running coupling function of QCD $\alpha_s(\mu)$ can be unambiguously specified by giving its value at a (standard) scale, say μ_0. On the other hand, as shown in Fig. 2b, this same coupling function is associated with many *different* values of Λ_{QCD}, depending on the number of effective quark flavors and on whether the LO or NLO formula is used. Thus, if one prefers to define α_s by specifying a value of Λ_{QCD}, *it is imperative that one specifies the associated n_f^{eff} and the order (LO or NLO) explicitly.* In the recent literature, the second-order $\overline{\text{MS}}$ Λ_{QCD} with 4 flavors has increasingly become the standard choice.

2.2 Scheme-Dependence of Parton Distributions

We have mentioned in Sec. 2 that parton distribution functions $f_H^a(x,\mu)$ are renormalization scheme dependent beyond the leading order. In applications to various physical processes, the choice of scheme for the parton distributions must match that of the hard cross-section in the QCD parton model formula [4]. The same parton distribution in two different schemes differ by a well-defined expression which is *nominally* of one order higher in α_s. In this section, we shall point out some non-trivial consequences of the scheme-dependence of parton distributions when the nominal behavior is violated (for a reason). These cases are important in applications, but have not so far received much attention among users of the QCD parton formalism.

To be specific, we consider the often used "DIS" scheme [6] which is defined, in relation to the "universal" $\overline{\text{MS}}$ scheme (used by theorists to calculate hard matrix elements), through the following NLO formula for the W_2 structure function of virtual γ deep inelastic scattering (cf. Eq. (1)):

$$\begin{aligned} W_2^\gamma(x,Q) &= f_{\overline{\text{MS}}}^q \otimes \left[C_{2,q}^{(0)} + C_{2,q}^{(1)\overline{\text{MS}}} \right] + f_{\overline{\text{MS}}}^G \otimes C_{2,G}^{(1)\overline{\text{MS}}} + O(\alpha_s^2) \\ &\equiv f_{\text{DIS}}^q \otimes C_{2,q}^{(0)} + O(\alpha_s^2) \end{aligned} \qquad (6)$$

where $C^{(i)}, i = 0,1$ are the hard partonic structure function in LO and NLO ($\hat{\sigma}_{a\to C}^i$ of Eq. (1)), often called the Wilson coefficients in the current context [7]. This formula does not define the gluon distribution in the DIS scheme. It is conventional to require that the momentum sum rule be preserved and then fix the definition of the gluon distribution by generalizing the resulting condition to all moments [8].

The first point to be made about the DIS scheme is that its definition is designed to render simple the formula for W_2, *and only* W_2. Even in the same scheme, however, the other deep inelastic scattering structure functions – W_1, W_3 or $W_{\text{left}}, W_{\text{right}}$ – do contain non-trivial NLO contributions from both quarks and gluons. It is *not* true, as one might have surmised from the terminology, that all DIS structure functions become simple in the DIS scheme!

In the same vein, the above definition only applies to the *total inclusive* structure function W_2. In practice, one is often interested in semi-inclusive processes such as the production of jets at a given transverse momentum or the production of heavy flavors (charm, bottom, etc.). It would be quite *wrong* to assume *even in deep inelastic lepton-hadron scattering* that, by using the DIS scheme parton distributions, one can neglect NLO terms in the semi-inclusive processes. In fact, the scattering of the electroweak vector boson with gluons in the hadron (a NLO hard process) is *mainly responsible* for producing final state jets at non-vanishing transverse momentum and final state heavy quark flavors (by pair-production), especially at energies not too far above threshold. We will return to this point in the next subsection.

A moment's reflection should also reveal that, although it is theoretically allowed to absorb the entire NLO contribution to W_2 into the definition of the DIS scheme quark distributions (which assumes collinear, on-the-mass-shell partons), this convenience for the total inclusive process comes at the expense of apparent *over-subtraction* (since the NLO diagrams contain non-collinear, off-the-mass-shell quark configurations as well). Thus, in applications to semi-inclusive processes, the use of DIS scheme distributions requires some care, and may lead to counter-intuitive results.

We now show an example of the importance of specifying the scheme in which the parton distributions are defined. Let us consider the question of "hard" vs. "soft" gluons which is, so far, an unsettled issue. Without attempting to resolve this problem, we would like to show that its very formulation requires close attention to the defining scheme of the distribution. To wit, the difference between the $\overline{MS}$ and DIS definition of the gluon distribution is (cf. discussion following Eq. (6)):

$$f_{\overline{MS}}^{G}(x,Q) - f_{DIS}^{G}(x,Q) = f^{q_s} \otimes C_{2,q}^{(1)\overline{MS}} + f^{G} \otimes C_{2,G}^{(1)\overline{MS}} \qquad (7)$$

where q_s denotes the singlet quark distribution and, in keeping with the perturbative nature of this equation, the scheme label is dropped on the right-hand side. Assuming that one of the f^G's, say f_{DIS}^G is *soft* in one of the schemes – for example, it might behave like $(1-x)^\eta$ with $\eta \geq 6$ [9], then it approaches zero very fast as $x \longrightarrow 1$. However, since the first term on the right-hand side contains a convolution of the *valence quark* distributions with $C_{2,q}^{(1)\overline{MS}}$, it certainly is fairly "hard" – say, behaving like $(1-x)^\eta$ with $\eta \sim 4$ [13]. As a consequence, the same distribution in the other scheme ($f_{\overline{MS}}^G$ in this example) will necessarily be hard! Thus, *the "hardness" or "softness" of the gluon distribution is a very scheme-dependent concept*, which does not necessarily have an independent meaning. In fact, in the "large x" region (say, $x > 0.4$) where the gluon distribution in the DIS scheme is traditionally considered to be small compared to the valence quark ones, perturbative relations such as Eq. (7) becomes of questionable meaning since a NLO term on the right-hand side becomes larger than one of the LO terms on the left-hand side.

A similar situation exists for the sea-quark distributions *over the entire range of*

x. For this case we have:

$$f_{DIS}^{\bar{q}}(x,Q) - f_{\overline{MS}}^{G}(x,Q) = f^{\bar{q}} \otimes C_{2,q}^{(1)\overline{MS}} + f^{G} \otimes C_{2,G}^{(1)\overline{MS}} \tag{8}$$

It is well-known that, for moderate values of Q, the gluon distribution f^G is much larger numerically than the sea-quark distributions $f^{\bar{q}}$. For instance, in terms of the fractional momentum carried by the partons, the ratio is around 0.50 : 0.03 – a factor of greater than 10. Hence, the second term on the right-hand side of the equation can easily be of the same order of magnitude as the individual terms on the left-hand side in spite of the fact that the Wilson coefficient $C_{2,G}^{(1)\overline{MS}}$ formally carries one extra power of α_s. In other words, *the size of the sea-quark distributions can depend critically on the scheme in which they are defined*; and *it is not very meaningful to talk about a LO sea-quark distribution* since its definition is always coupled to the much bigger gluon distribution[10].

2.3 Order of Magnitude Estimates in QCD

The above observation on the relative order of magnitudes of sea-quarks and gluons is, of course, not surprising, since the sea-quarks are usually understood to arise from the splitting of the gluon. Actually, it is precisely because of this fact that the sea-quark distribution contains an implicit power of α_s with respect to the gluon distribution, at least for moderate values of Q. Thus, excluding the inactive heavy quarks below their thresholds, the parton distributions can be classified into two classes according to their numerical magnitude: (i) those of order 1 – the gluon and the "valence" quarks (u, d) (to be denoted by f^{large} below); and (ii) those effectively of order α_s – the active sea quarks (to be called f^{small}). In this subsection, we discuss important phenomenological consequences of this observation in certain classes of physically interesting processes.

Traditionally, in applying the perturbative QCD formalism to physical processes, the various terms which contribute to the right-hand side of Eq. (1) are classified as LO, NLO, etc., according to the perturbation expansion of the hard cross-section $\hat{\sigma}_{a \to C}^{i}$ only. In view of the large discrepancy in magnitude between the two classes of parton distribution functions f_H^i mentioned above, this traditional power counting can result in misleading conclusions: a "NLO" hard cross-section multiplied by an order-1 parton distribution function can be as important numerically as a "LO" one multiplied by an order-α_s parton distribution function. To make this point explicit, the perturbative QCD formula Eq. (1) can be reorganized, schematically, as follows,

$$\begin{aligned}\sigma_{phys} &= f^{\text{large}} \otimes \hat{\sigma}_{LO}^{l} + \left[f^{\text{small}} \otimes \hat{\sigma}_{LO}^{s} + f^{\text{large}} \otimes \hat{\sigma}_{NLO}^{l} \right] \\ &+ \text{numerically smaller terms} \end{aligned} \tag{9}$$

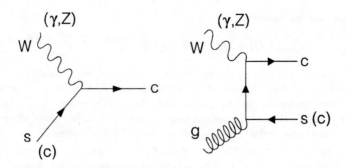

Figure 3: Hard-scattering mechanisms for heavy quark C production in deep inelastic charged-current (neutral-current) scattering.

This point becomes particularly important when the first term on the right-hand side is absent or suppressed because $\hat\sigma^1_{LO}$ vanishes or contains a suppression factor due to the electro-weak coupling. Then the traditional LO analysis which only keeps the "LO term", $f^{\text{small}} \otimes \hat\sigma^s_{LO}$, becomes totally inadequate since the "NLO term" in the same square bracket is of the same numerical order. A case in point is charm production in neutrino deep inelastic scattering. The conventional wisdom is that this process is dominated by the scattering of the weak W-boson on the strange quark in the hadron target, (Cf. Fig. 3a); and all existing data is analyzed according to this picture. The above discussion clearly suggests that the "NLO" contribution from the scattering of W on the gluons (Fig. 3b) can be just as important, hence must be included in a proper QCD formulation of the problem[10].

Interestingly, a diametrically opposite approach to this class of problem is also found in the literature, especially involving heavy flavor production in neutral-current processes. It invokes *only* the gluon contribution (the so-called "gluon-fusion" mechanism[11]) (cf. Fig. 3b) and ignores the LO quark diagram Fig. 3a. Although this may make sense in some restricted kinematic region just above the threshold of producing the heavy flavor pair, the LO quark scattering diagram must become increasingly important with increasing energy. The above discussion should make it clear that a consistent treatment must include both mechanisms if it is to be quantitatively reliable over the entire energy range.

We summarize the key point underlying the topics discussed in the last two subsections. Within the QCD parton model, the contributions from the sea quarks and from gluons are always inextricably intertwined. In spite of the conventional designations of LO and NLO respectively, they can be numerically comparable. The precise division between the two mechanisms is tied intimately to the choice of renormalization scheme used during the calculation. Although it is theoretically possible to minimize the contribution from one or the other mechanism to *one given quantity* (e.g. W_2) by a specific choice of scheme, both terms must be included in the analysis of all other physical quantities. In order to achieve consistent results, the choice of scheme must be specified explicitly in these applications.

3 Overview of Global Analyses of Parton Distributions

The global analysis of parton distributions refers to the quantitative comparison of experimental data from a wide range of physical processes with the QCD master equation, Eq. (1), for the purpose of extracting a set of universal parton distribution functions. These can then be used in other applications: to make "predictions" as well as to provide stringent tests of the self-consistency of the perturbative QCD framework itself or of the Standard Model in general. Since any compelling indications of inconsistency of the SM are signs of "new physics", and since even direct search for new physics must rely heavily on understanding of the background from conventional physics, the systematic analysis of parton distributions is intimately tied to all these ventures. To achieve this purpose, contemporary global analyses must incorporate all relevant modern high statistics experimental results and apply the NLO QCD formalism in a consistent manner as described in the first part of this review.

In the following, we: briefly review and assess the existing parton distribution parametrizations (Sec. 3.1); highlight the relevant phenomenological issues for modern quantitative global analysis (Sec. 3.2); summarize the current status of on-going programs of global analyses of parton distributions (Sec. 3.3); and remark on their proper use in phenomenological applications (Sec. 4).

3.1 Review of Parton Parametrizations

The list of widely used parton distributions is a long one. Prominent among these are the pioneering works of Feynman-Field and Buras-Gaemers; followed by the widely used distributions of Gluck-Hoffmann-Reya, Duke-Owens, and Eichten-Hinchliffe-Lane-Quigg [2,3]. These are all based on leading order QCD-evolved distributions extracted by comparison with data existing up to about 1983. In recent years, many second generation high statistics experiments have become available and more refined global analyses have been carried out to NLO by Martin-Roberts-Stirling [12], Diemoz-Ferroni-Longo-Martinelli [8], Aurenche-Baier-Fontannaz-Owens-Werlen [13], and

Morfin-Tung[14], reflecting the needs of the present time. Table I lists most of the currently used parton distributions and the experimental data on which the analyses were based.

	D-O[3]	EHLQ[2]	(H)MRS[12]	DFLM[8]	ABFOW[13]	M-T[14]
ν-DIS	CDHS	CDHS	CDHSW (CCFRR)	CHARM	—	CDHSW (CCFR)
μ-DIS	EMC	—	EMC,BCDMS	—	BCDMS	EMC,BCDMS
D-Y	E288,ISR	—	(E288),(E605)	—	—	E288,E605
Dir-γ	—	—	WA70	—	WA70	—

Table I: Parton Distribution sets and data used.
References to the experiments are: BCDMS[15], CDHS[16], CDHSW[17], CCFR[18], CCFRR[19], E288[20], E605[21], EMC[22], WA70[23]. Parentheses around entries indicate that the corresponding data were only used partially.

The experimental developments which have the most significant impact on recent analyses as compared to the previous LO ones are: (i) results of the high statistics CDHS[16] neutrino experiment, on which most earlier analyses were heavily dependent, has since been considerably revised and supplanted by the new CDHSW[17] results; (ii) the very accurate new data on muon scattering from the BCDMS[15] collaboration does not fully agree with earlier results, especially the previous "standard" EMC[22]. Both of these developments cause the predictions of the earlier parton distributions on deep inelastic scattering – the main source of information on these distributions – to disagree with the best current data up to 15-20%. Differences of this size correspond to many standard deviations in these high statistics experiments.

The disagreement between the BCDMS and EMC results has been a source of much uncertainty and discussion for the past two years. However, recent comprehensive studies[24], including the introduction of the new analysis of the SLAC-MIT experiments[25] as an independent check on the normalization, indicate possible ways to resolve the discrepancy. Several talks in this Workshop, including the first report on a new analysis of the EMC data[26], will shed further light on this issue.

To illustrate the general state of affairs, we show one plot on the comparison of current data with the results calculated from representative parton distribution sets in Fig. 4. This is typical of similar plots on the comparison of BCDMS and CDHSW data with the same parton distributions. Details can be found in some recent reviews[27][28]. Fig. 4 clearly shows the necessity for using up-to-date parton distributions which take into account of all relevant experiments in any QCD analysis where precision is required.

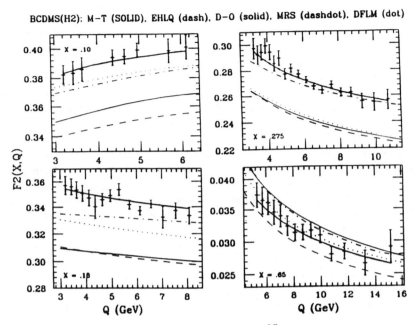

Figure 4: Comparison of a subset of BCDMS F_2 data[15] on hydrogen with calculated results from representative parton distribution sets.

3.2 Relevant Issues for Quantitative Global Analysis

A truly quantitative global QCD analysis of data involves many experimental and theoretical considerations which may affect the results, not all of which are generally known. We shall briefly summarize the important ones. A systematic discussion can be found in the report of the Structure Function and Parton Distribution group in the 1988 Snowmass Proceedings[27].

On the experimental side, in addition to the choice of physical processes and experiments to fit, all the following factors can affect the consistency and the correctness of the results: (i) the selection of data within a given experiment, according to kinematic cuts in Q^2, W or other variables. (How do the results depend on the values of these cuts?) (ii) Are systematic errors included in the fitting procedure? This is, in fact, a critical issue for the second generation analysis since: (a) errors on current high statistics experimental data are dominated by systematic errors; and (b) when data from several experiments are used in a chi-square or likelihood analysis, the fitting procedure is simply meaningless without including the systematic errors.

However, most existing global analysis *do not* include systematic errors.[1] (iii) Do the different experimental data sets apply the same "corrections" (e.g. "slow-rescaling", "isoscalar", etc.) to their data analyses; and, if not, how should one handle the differences?[27] Unfortunately, real differences can often be found in published data among different experiments, and they are usually overlooked. (iv) Finally, distinct physical quantities measured in the same experiments can have correlated errors. A proper fit to the data must take into account the correlations. No global analysis to date has attempted to incorporate this in a systematic way.

On the theoretical side, the parton distributions and QCD parameters one obtains from global fits can depend (in addition to the choice of LO or NLO formalism) on: (i) the functional form used for the initial distributions, especially if it happens to be too restrictive; (ii) the number of parameters which are allowed to vary when the fit to data is made. Unfortunately, there is no proven way to assure a correct choice on either of these considerations. A reasonable choice when fitting a given set of data may not remain so when additional experiments and/or physical processes are incorporated. Finally, results on these global analyses in the very small x region – a region of much interest in applications to "predict" the high energy behavior of standard and new physics – are heavily dependent on whether the small-x *extrapolation* is fixed by an assumed functional form or is characterized by parameters to be fit to existing data.

These issues, as summarized above, should not only be of concern to those working on global fits; the users of parton distributions must be aware of these considerations and their impact on the physics they are trying to extract from the physically measured quantities.

3.3 Current Status of Global Analyses and Open Issues

In this Workshop we shall hear progress reports on two of the currently active programs on global analysis of parton distributions [12,14] [29] [30]. It suffices to say that in spite of past efforts and the current detailed work, there are still many open questions which will require a combined effort of refined and expanded experiments, further theoretical clarifications, and continued phenomenological analysis to be resolved. The list includes:

(i) The gluon distribution: What is the proper or optimal definition? (Cf. Sec. 2.2 and [31]) How can it be determined in an unambiguous way? DIS without a reliable

---

[1] This is ironic, as most experiments devote more effort on understanding the systematic errors than on anything else.

measurement of the longitudinal structure function does not determine the gluon distribution well. Lepton-pair production data imposes better constraints (through combined effects of sea-quarks and gluons.) The "dedicated study" of the gluon distribution based on direct-photon production [13] show a great deal of promise, but is also subject to a number of experimental limitations (limited kinematic range, large errors, isolation criteria, etc.) and theoretical uncertainties (choice of scales, bremsstrahlung contributions, etc.). Can these be satisfactorily resolved? This will be discussed at this Workshop. Are there other comparable or better ways to determine the gluon distribution?

(ii) The sea-quark distributions: How should they be defined? (Cf. Sec. 2.2 and Sec. 2.3) What is the dependence of $f_{sea}^{q^i}$ on the flavor label i? Is it SU(3)-symmetric, SU(2)-symmetric, or asymmetric? Current data on charm production (dimuon final state) in DIS have been interpreted, in the LO picture, to indicate a non-SU(3)-symmetric sea [32]. However, the discussion of Sec. 2.3 suggests that this interpretation needs to be reassessed because of the inherent sear-quark – gluon mixing [10].

(iii) Small-x and large-x behavior of the parton distributions: In the familiar perturbative QCD formalism, the Q-dependence of parton distributions $f^i(x, Q)$ is governed by the evolution equation with calculable kernels; however, the x-dependence beyond currently measurable range is unknown except for some qualitative guidelines based on Regge-type of arguments at some unspecified scale. Much attention has been given to developing theoretical tools to extend our understanding of the parton model into the small-x region [33]. There are several distinct aspects to the "small-x problem": (i) the resummation of large $(\alpha_s \log(1/x))^n$ terms for fixed Q; (ii) the region of large $\log(1/x)$ *and* large $\log Q$; and (iii) the region of saturation of parton densities and the breakdown of the parton picture as we know it. Promising recent progress will be presented in this Workshop by Levin [34].

From the phenomenological point of view, one can ask: what reasonable constraints on the small-x extrapolation of the parton distributions can be obtained by global fits to current data? Shown in Fig. 5 are plots of the gluon distribution and the predicted structure function F_2 extending to very small x from two sets of NLO parton distributions which both fit current data well [14]. The difference between the extrapolated results is seen to be quite large. This difference can be resolved by direct measurement of F_2 (and hopefully $G(x,Q)$ as well) at HERA as well as by some suitable hadron-hadron collision process such as lepton-pair production (DY) at the colliders. Fig. 6 illustrates this point by showing the anticipated rapidity distribution of the lepton-pair at the Tevatron for $Q = 20 GeV$ using several choices of parton distributions, including the two mentioned above.

The behavior of parton distributions near $x \sim 1$ has also been of interest. The

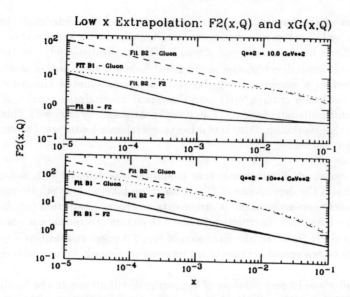

Figure 5: Small-x extrapolation of $G(x,Q)$ and $F_2(x,Q)$ based on two global fits to current data.[14]

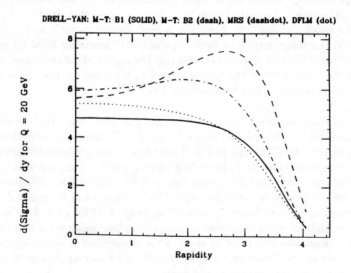

Figure 6: Rapidity distribution of Drell-Yan pairs predicted by some representative parton distributions.[14]

experimental consequences of this are concentrated in the relatively low energy region where they are also related to *higher twist* corrections to the conventional leading power-law QCD results [35].

In order to address the open questions described above, we need to go beyond the traditional reliance on DIS and lepton-pair production processes. While important progress will continue to be made in these areas, especially with the exciting expansion of experimental range offered by HERA, the coming of age of fixed-target direct photon experiments and quantitative measurements of an ever wider range of processes at the hadron colliders have opened up many more possibilities of determining the remaining uncertainties of parton distributions and testing the consistency of QCD. Foremost among these are W-, Z-production cross-sections (including rapidity and transverse momentum distributions), heavy flavor production, jet production (with or without associated vector bosons), as well as DY and direct-photon processes. The program of this Workshop in the coming days underlines this expanding horizon. As most of this work in has yet to be carried out, the quantitative analysis of parton distributions will have to be an on-going effort for some time to come.

4 Some Remarks on the Use of Parton Distributions

As the parton model advances from the original "naive" genre to the "QCD-improved" kind summarized here, its use requires certain reassessment. Here we touch upon some of these considerations, based on discussions of the previous sections.

The LO QCD parton model is still a very powerful and simple framework which describes a wide range of physical processes to within 10–20% accuracy. In these applications, it is sufficient to use LO parton distributions. In fact, it is preferable to do so, than to use the NLO distributions – as the latter are extracted by comparing complete NLO formulas with experiment. (A built-in error is always incurred by the mixed use of a LO hard cross-section with NLO parton distributions.) It is important, however, that the LO distributions used accurately reflect existing data where applicable. As discussed in Sec. 3.1, the most often used first generation parton distributions, unfortunately, strongly disagree with the recent high precision DIS and DY data.

For applications which require more accuracy, the LO formalism is known to be inadequate. When NLO hard cross-section formulas are used in these more refined cases, it makes no sense to use LO parton distributions, since the supposed extra accuracy gained in the use of the former is totally lost by the use of the latter (especially, in addition, if the latter does not fit known current data). Obvious and commonsensical as this remark may appear, the common practice of using well-known distributions under all circumstances in the "comparison of experiments with QCD" is only too apparent in the literature.

A common *myth* in contemporary phenomenology is to apply a variety of parton distributions to a given physical process and then cite the range of results obtained as the "theoretical error". There is no justification for such a practice, since: (i) the LO and NLO distributions are not designed to be used the same way; and (ii) some of these distributions are already known to disagree with current data, as mentioned above. In fact, as shown in Fig. 4, in many cases the deviations from the correct results are all to the *same* direction for most of these distributions rather than "bracketing" the right answer. Since existing data do not completely determine all the parton distributions, it is, of course, useful to explore the range of uncertainties due to this lack of knowledge both in performing precision tests and in making predictions for the future. One can do this properly by obtaining a range of parton distributions in the same global analysis with allowance made for the uncertain feature to be studied. This approach has been used in characterizing the small-x extrapolation of parton distributions in recent studies[12,14]. It can also be applied to other features such as the flavor dependence of the sea quarks[14].

In the use of NLO QCD formalism, it is important to make consistent choice of renormalization scheme in the hard cross-section formula and in the definition of the parton distributions used (cf. Sec. 2.2). Although the outcome of a given global analysis of parton distributions can, in principle, be presented in any scheme, most authors publish their results only in the scheme which was used in the original analysis. It is, therefore, incumbent on the user to make the necessary adjustment in order to bring about consistency. It is helpful to know that among the published NLO parton distributions, the MRS[12] and ABFOW[13] ones are in the $\overline{MS}$ scheme, whereas the DFLM[8] ones are in the DIS scheme. The new MT[14] distributions are given in both schemes.

All QCD "predictions" are subject to an uncertainty associated with the choice of renormalization and factorization scales. This uncertainty diminishes at very high energies or, in principle, when more and more higher order terms are included. At current energies and to NLO only, this scale-dependence can be substantial for certain processes. There is, so far, no clear consensus among theorists on whether there is some sensible method for making an intelligent (if not "correct") choice of scale for a given situation. Because of the practical importance of this issue, there will be continued discussion and debate about this topic in this Workshop and beyond.

Finally, the various current parton distribution sets differ considerably in: (i) the choice of input experimental data (cf. Table I); (ii) the treatment of experimental errors and corrections (cf. Sec. 3.2); and (iii) the selection of functional forms for the input distributions and the number of free parameters (cf. Sec. 3.2). Many implicit assumptions ride on the choice made in (iii) which users do not see explicitly. For critical applications, these factors must be examined carefully before definite conclusions are drawn.

5 Summary

Perturbative QCD stands as one of the main pillars of contemporary high energy physics. It plays a central role in pushing quantitative tests of the Standard Model to ever increasing accuracy (hopefully, even to its eventual limit); in studying the signals and backgrounds for new physics; and in providing hints for attempts to formulate non-perturbative solutions to strong interaction physics. Most applications of QCD are based on the Factorization Theorems which have a simple parton model interpretation. They rely on the use of the universal parton distributions to relate measurable quantities to the fundamental processes of the underlying theory.

This Workshop is concerned with a critical review of traditional approaches to the study of parton distribution functions (by DIS, DY, etc.) and, more importantly, a survey of the expanding scope of all the fields which contribute to and make essential use of the QCD parton model. This brief introductory talk underlines some of the important considerations which must be taken into account in applying this general framework to the expanded horizon; and highlights the current status and the open questions of the global analysis of parton distributions, in the hope of helping to establish a useful starting point for the specific sessions to follow in the next three days.

We have confined our survey mainly to NLO perturbative QCD. The thrust of most theoretical studies relating to this field is, of course, concerned with pushing the frontier beyond NLO QCD. Thus, in addition to reports on recent work on small-x and higher twist mentioned in Sec. 3.3, there will be talks on the resummation of large hard cross-sections [36,31], and on multi-parton processes [37]. Finally, there will be a session on spin-dependent parton distributions which pose some unique theoretical and experimental challenges which we have not had time to include in this talk.

Acknowledgement

Many of the issues described in this survey crystalized during discussions I have had with my collaborators M. Aivazis, J. Morfin, F. Olness and numerous other colleagues, too many to mention individually. Particular thanks are due, however, to John Collins for sharing his unusual insight in perturbative QCD which underlies much of this talk.

References

[1] R.P. Feynman, *Photon-Hadron Interactions*, Benjamin, (1972).

[2] E. Eichten *et al*, *Rev. Mod. Phys.*, **56**, 579 (1984) and Erratum **58** 1065 (1986).

[3] M. Glueck *et al*, *Z. Phys.* **C13**, 119 (1982); D. Duke and J. Owens, *Phys. Rev.* **D30**, 49 (1984).

[4] For a recent review and original references, see J. Collins, D. Soper and G. Sterman in *Perturbative Quantum Chromodynamics*, A. Mueller, Ed., World Scientific Pub., 1989.

[5] J.C. Collins, F. Wilczek, and A. Zee, *Phys. Rev.* **D18**, 242 (1978); J.C. Collins and Wu-Ki Tung, *Nucl. Phys.* **B278**, 934 (1986); See also W. Marciano, *Phys. Rev.* **D29**, 580 (1984).

[6] G. Altarelli, R.K.Ellis, & G.Martinelli, *Nucl. Phys.* **B143**, 521 (1978); and *ibid.* **B157**, 461 (1979).

[7] W. Furmanski & R.Petronzio, *Z. Phys.* **C11**, 293 (1982).

[8] M. Diemoz *et al*, *Z. Phys.* **C39**, 21 (1988).

[9] See, *e.g.* Eichten *et al*, Ref.2; Diemoz *et al*, Ref.8.

[10] M.A.G. Aivazis, F. Olness & Wu-Ki Tung, Fermilab-Pub-90/23 and IIT-90/10.

[11] J. Leveille, T. Weiler, *Nucl. Phys.* **B147**, 147 (1979); T. Weiler, *Phys. Rev. Lett.* **44**, 304 (1980); M. Arneodo, *et al*, *Z. Phys.* **C35**, 1 (1987).

[12] A.D. Martin, R.G. Roberts & W.J. Stirling, *Phys. Rev.* **D37**, 1161 (1988), *Mod. Phys. Lett.* **A4**, 1135 (1989); P.N. Harriman, A.D. Martin, R.G. Roberts & W.J. Stirling, Rutherford Lab. preprint RAL-90-007 (1990); RAL-90-018 (1990).

[13] P. Aurenche *et al*, *Phys. Rev.* **D39**, 3275 (1989).

[14] J.G. Morfin and Wu-Ki Tung, Preprint Fermilab-Pub-90/24, IIT-90-11.

[15] A.C. Benvenuti *et al*, *Phys. Lett.* **B223**, 485 (1989) and CERN-EP/89- 170,171, December, 1989.

[16] H. Abramowicz *et al*, *Z. Phys.* **C17**, 283 (1984); *Z. Phys.* **C25**, 29 (1984); *Z. Phys.* **C35**, 443 (1984).

[17] J.P.Berge *et al*, Preprint CERN-EP/89-103 (1989).

[18] Private communications from Sanjib Mishra.

[19] D.B. MacFarlane *et al*, *Z. Phys.* **C26**, 1 (1984).

[20] A.S.Ito *et al*, *Phys. Rev.* **D23**, 604 (1981).

[21] C.N. Brown *et al*, *Phys. Rev.* Lett. **63**, 2637 (1989).

[22] J.J. Aubert *et al*, *Nucl. Phys.* **B293**, 740 (1987).

[23] M. Bonesini *et al*, *Z. Phys.* **C38**, 371 (1988).

[24] See, e.g. J. Feltesse, *Proceedings of the XIV International Symposium on Lepton and Photon Interactions*, Stanford, August 1989.

[25] A. Bodek, in these Proceedings.

[26] S. Wimpenny, in these Proceedings.

[27] Wu-Ki Tung *et al*, in *Proceeding of the 1988 Summer Study on High Energy Physics in the 1990's*, S. Jensen, Ed., World Scientific, (1990).

[28] K. Charchula *et al*, DESY report DESY 90-019 (1990).

[29] See A. Martin and R. Roberts, these Proceedings.

[30] J. Morfin, these Proceedings.

[31] K. Ellis, these Proceedings.

[32] CDHS: H. Abramowicz, *et al*, *Phys. Rev.* Lett. **57**, 298 (1986); *Z. Phys.*, **C28**, 51 (1985); CHARM: J.V. Allaby, *et al*, *Z. Phys.* **C36**, 611 (1987); CCFR: K. Lang *et al*, *Z. Phys.* **C33**, 483 (1987); S.R. Mishra *et al*, in *Proceedings of 14th Rencontres de Moriond*, Mar. 1889.

[33] For a comprehensive review of pioneering works in this field, see: L. Gribov, E. Levin, and M. Ryskin, *Phys. Rep.* **100**, 1 (1983).

[34] E. Levin, these Proceedings; See also Proceedings of a forthcoming *Workshop on Parton Distributions at Small-x*, DESY, May, 1990.

[35] J.W. Qiu, these Proceedings.

[36] G. Sterman, these Proceedings.

[37] D. Treleani, these Proceedings.

DEEP INELASTIC SCATTERING
EXPERIMENTS

Recent Results from the New Muon Collaboration

C. Peroni
University of Torino and INFN
via P. Giuria, 1
I-10125 Torino
Italy

June 30, 1990

Abstract

Results are presented from a deep inelastic muon scattering experiment. The ratio of the structure functions F_2^n/F_2^p has been measured at incoming muon energies of 274 GeV and 89 GeV. The ratios $F_2^{He}/F_2^D(x)$, F_2^C/F_2^D and F_2^{Ca}/F_2^D have been measured at 197 GeV down to very low x ($\simeq 3 \times 10^{-3}$) and over a wide range in Q^2. The ratio $\sigma_{J/\psi}^D/\sigma_{J/\psi}^H$ has been measured at 274 GeV and found to be consistent with unity. Systematic errors are kept at the percent level by means of a complementary target arrangement. The ratio $F_2^n/F_2^p(x)$, together with the nucleon structure function obtained from a fit to the world data is used to calculate the Gottfried sum rule. The result is not consistent with the QPM prediction. The H, D J/ψ events have been used to extract the gluon momentum distribution $xG(x)$ for the free nucleon at small x.

Introduction

The results presented at this workshop cover part of the experimental program of the New Muon Collaboration (NMC, NA37 experiment at the CERN SPS).
The aim of the experiment is the investigation of a wide range of physics topics in deep inelastic muon scattering.
Notably, the program comprises [1]:

- a precise measurement of the neutron to proton structure function ratio;

- a detailed study of the EMC effect, including its A and Q^2 dependences;

- the measurement of the gluon distribution in different nuclei by means of J/ψ production;

- the R=σ_L/σ_T dependence on A;

- a low systematics determination of the nucleon absolute structure functions.

The experiment took data for four years (1986-1989) and the analysis is presently in progress. Preliminary results are available for the first three topics listed above.

1 The Experimental Apparatus

The experiment was carried out in the M2 muon beam line at the CERN SPS, using the NMC forward spectrometer to detect the scattered muons and the fast forward hadrons produced in deep inelastic scattering events.

The spectrometer was a modified and upgraded version of that used in the EMC experiment NA2'[2]. Beside replacing or refurbishing pieces of the apparatus that showed signs of aging, special efforts were made in order to extend the accessible x range (down to $x \sim 10^{-3}$) and to minimize systematic effects, with particular emphasis on those relevant to the determination of cross-section ratios and to the evaluation of absolute F_2's.

In addition to the standard trigger (T1), which accepts muons at scattering angles above 10 mrad, a small angle trigger (T2) which extends the acceptance down to 5 mrad was implemented. Both triggers cover the small x region ($x < 0.4$), the T2 events having smaller Q^2 and ν (and consequently smaller radiative corrections). Values of x above 0.4 are covered by T1 only.

Accurate measurements of cross-sections ratios on different target materials were allowed by a "complementary" target system. The two (or more) target materials were simultaneously exposed to the beam, one behind the other along the beam direction. This set-up was frequently exchanged with the "complementary" one where the two materials (A and B, say) were interchanged. Geometrical acceptance and efficiency corrections therefore canceled in the calculation of ratios; so did the integrated beam fluxes. The frequent exchange of the two target set-ups (every $\simeq$ 30 minutes) also allowed cancellation of possible time dependent changes in the apparatus acceptance and efficiency.

Following the above reasoning, the experimental cross-section ratio for materials A and B only depends on the measured number of events and on the ratio of the molar volumes κ:

$$\frac{\sigma_A}{\sigma_B} = \kappa \sqrt{\left(\frac{N_A}{N_B}\right)_{upstream} \left(\frac{N_A}{N_B}\right)_{downstream}} \ .$$

Corrections to the ratios were applied to subtract from the measured yields the contribution of higher order electromagnetic processes (radiative corrections), different for different target materials, the size of such corrections being larger at large ν, i.e. at low x. The uncertainty in the radiative corrections dominates the systematic error in the low x region.

Minor correction were also applied to the data to account for the smearing of the kinematic variables due to the finite apparatus resolution.

An increased accuracy in the measurement of the incoming muon momentum (at the $\simeq 0.3\%$ level) was obtained by means of an additional high resolution spectrometer placed downstream of the experiment.

Uncertainties in the beam and scattered muon momentum are the main contributions to the systematic error at high x.

2 The Neutron and the Proton Structure Functions Ratio

The x dependence of the ratio F_2^n/F_2^p was obtained from simultaneous measurement of deep inelastic muon scattering on hydrogen and deuterium. The data presented here were taken in separate runs at the (average) beam energies of 274 GeV and 89 GeV. The 274 GeV data sample represents about half of the total collected statistics at this beam energy

2 THE NEUTRON AND THE PROTON STRUCTURE FUNCTIONS RATIO

and cover the kinematic range $x = 0.004 - 0.8$ with $Q^2 = 1 - 5\ GeV^2$ for the lowest and $Q^2 = 10 - 190\ GeV^2$ for the highest x bin. The 89 GeV data cover the range $x = 0.004 - 0.7$ with $Q^2 = 0.5 - 1.1\ GeV^2$ for the lowest and $Q^2 = 7 - 26\ GeV^2$ for the highest x bin.

In the Quark Parton Model (QPM) F_2^n/F_2^p depends in the large x region on the ratio of valence quark distributions d_v/u_v. In the small x region, where sea quarks dominate, the ratio is affected by the residual valence contribution and by a possible flavour symmetry breaking in the sea. The measurement of the ratio F_2^n/F_2^p thus puts strong constraints on the flavour decomposition of the structure functions [3].

The parton distributions, especially in the small x region, are widely used to calculate hard scattering cross-sections in $p\bar{p}$, pp and ep collisions [4].

Neglecting nuclear binding effects, the cross section for scattering on deuterons is equal to the sum of the cross sections for scattering on free protons and neutrons. With the assumption

$$\frac{R^D}{R^H} = 1\ ,$$

one has

$$\frac{\sigma^D}{\sigma^H} = \frac{F_2^D}{F_2^H}\ .$$

Consequently, the ratio F_2^n/F_2^p is defined as:

$$\frac{F_2^n}{F_2^p} = \frac{F_2^D}{F_2^p} - 1\ .$$

No corrections for the Fermi motion of the proton and the neutron in the deuteron were made. These corrections are negligible at $x < 0.6$.

The results from the measurements with 274 GeV and 89 GeV muons are shown in fig. 1a,b. Previously, the ratio had been measured in high energy muon scattering at CERN by EMC and BCDMS [5,6] and it had also been measured at SLAC in electron scattering [7]. In fig. 1a the results are shown together with the BCDMS data. Similar Q^2 ranges are covered by the BCDMS experiment and by the present measurement with 274 GeV muons. The results are in good agreement in the region of overlap. NMC data extend to small x with small systematic errors. The BCDMS result is more precise in the region $x > 0.3$. The two results together cover a large x range with high accuracy.

The NMC result from 89 GeV muons covers approximately the same Q^2 region as the SLAC data (0.6 – 30 GeV^2). These two data-sets differ from the high energy results, suggesting a possible Q^2-dependence of F_2^n/F_2^p.

The NMC results extend below $x = 0.03$ covering the region where the sea partons dominate. The data points at low x are consistent with the QPM prediction $F_2^n/F_2^p = 1$ at $x = 0$.

The Gottfried sum rule [8] has been derived in the framework of the QPM under the assumptions of isospin symmetry for the proton and the neutron and flavour symmetry in the sea:

$$\int_0^1 \frac{F_2^p - F_2^n}{x} dx = \frac{1}{3}\ .$$

Perturbative QCD corrections to the sum rule are negligible [9].

2 THE NEUTRON AND THE PROTON STRUCTURE FUNCTIONS RATIO

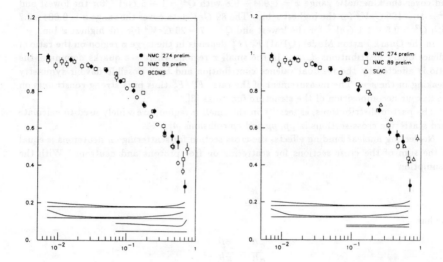

Figure 1: (a) F_2^n/F_2^p from NMC and BCDMS [6]; (b) F_2^n/F_2^p from NMC and SLAC [7]. The contours show the size of the systematic error

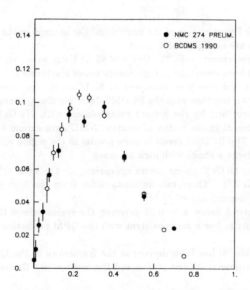

Figure 2: $F_2^p - F_2^n$ obtained from F_2^n/F_2^p at 274 GeV and F_2^D fit at $Q^2 = 15\ GeV^2$. The BCDMS results are also shown [6]

3 F_2^A/F_2^D FOR HE, C, CA

In order to test this prediction the difference $F_2^p - F_2^n$ was calculated from the measured ratio F_2^n/F_2^p and from F_2^D derived from a fit to most of the available muon and electron data [10]:

$$F_2^p - F_2^n = F_2^D \frac{1 - F_2^n/F_2^p}{1 + F_2^n/F_2^p}.$$

The difference is shown in fig. 2 together with the BCDMS result [6]. The two measurements are in good agreement, the NMC data extending to lower x. The value of the integral calculated in the measured x range is

$$\int_{0.004}^{0.8} \frac{F_2^p - F_2^n}{x} dx = 0.219 \pm 0.008 \,(\text{stat}) \pm 0.021 \,(\text{syst}).$$

The total systematic error on the integral was taken to be the quadratic sum of the the systematic error on the ratio F_2^n/F_2^p (0.016) and on F_2^D (0.013), the latter mainly due to normalization uncertainty.

The contributions to the integral on the low x and high x regions were estimated by extrapolating $F_2^p - F_2^n$ to $x = 0$ and $x = 1$. The contribution from the high x region is negligible. Extrapolating to $x = 0$, assuming a Regge-like behaviour

$$F_2^p - F_2^n \sim x^\alpha,$$

a contribution varying from 0.004 to 0.2 was calculated, depending on the value of α chosen: $\alpha = 0.7 - 0.3$. Even taking the extreme contribution of 0.2, the NMC result may indicate a violation of the sum rule at the 30% level.

Results from previous experiments [5], [6] have been claimed to be in agreement with the prediction of the Gottfried sum rule within their large errors.

3 F_2^A/F_2^D for He, C, Ca

Preliminary results on F_2^A/F_2^D for He, C and Ca are presented in fig. 3 as a function of x, averaged over Q^2. The results cover the kinematical range $x = 0.0035 - 0.7$ and $Q^2 = 0.7 - 90 \, GeV^2$. The inner error bars in the figure are the statistical errors only, the outer error bars include the systematic error added in quadrature. For F_2^{He}/F_2^D only statistical errors are shown.

The three structure function ratios show a characteristic x dependence. There is a depletion below unity at low x, which increases with decreasing x, more pronounced at larger A. The depletion is followed by an enhancement at intermediate x. Its onset moves towards higher x as the mass number increases. The enhancement reaches a maximum value of $\approx 2\%$ above unity for the three ratios. In the large x region ($x \gtrsim 0.3$) the present data are consistent with the well established depletion of the EMC effect [11].

Throughout the accessible Q^2 range no measurable Q^2 dependence of the EMC effect was found.

4 J/ψ Production on H, D

A sample of about 700 J/ψ's produced in deep inelastic scattering on hydrogen and deuterium was collected in the 274 GeV runs. The J/ψ was identified through its decay into

4 J/ψ PRODUCTION ON H, D

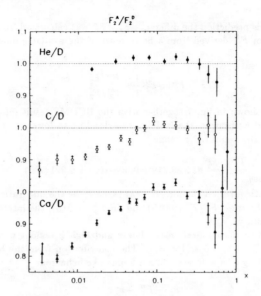

Figure 3: Structure function ratios F_2^A/F_2^D as function of x averaged over Q^2. The inner error bars are statistical only, the outer error bars include the systematic errors added in quadrature

two muons and its kinematics was fully reconstructed in terms of the lepton (Q^2, ν) and the hadron variables:

$$z = \frac{E_{J/\psi}}{\nu}$$

and p_T^2, the transverse momentum squared of the J/ψ with respect to the virtual photon direction. The μ^+, μ^- invariant mass distributions for the two targets are shown in fig. 4. Events were selected in the mass interval:

$$2.90 < M_{\mu^+\mu^-} < 3.30 \; GeV/c^2 \; .$$

The ratio of the J/ψ production cross-sections per nucleon from the two targets was derived with the method described in section 1:

$$\frac{\sigma_D}{\sigma_H} = 1.14 \pm 0.10$$

where the error is statistical only, the systematic one having been estimated to be negligible. The result is thus consistent with equal J/ψ production cross-sections for neutron and proton.

Deep inelastic leptoproduction of J/ψ offers the opportunity to investigate the gluon structure function of the nucleon. To this purpose the merged hydrogen and deuterium data were compared with the predictions of the colour singlet model [12] which describes the production of J/ψ via the fusion of a virtual photon with a gluon from the nucleon,

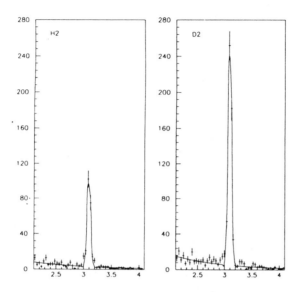

Figure 4: The μ^+, μ^- invariant mass distributions (in GeV/c^2) for the H and D targets

giving rise to a $c\bar{c}$ color singlet state with the J/ψ quantum numbers. The model was found to describe well the data within its range of validity ($p_T^2 > 0.1~GeV^2$ and $z < 0.9$), apart from an overall normalization factor, and was therefore used to extract [13] the gluon momentum distribution $xG(x)$, with the fraction of the nucleon momentum carried by the gluon, x, defined as

$$x = \frac{1}{s}(\frac{M_{J/\psi}^2}{z} + \frac{p_T^2}{(1-z)z}) \quad ,$$

where s is the center of mass energy squared in the photon-nucleon reference frame. The gluon momentum distribution $xG(x)$ is shown in fig. 5 together with the results of a similar analysis performed by EMC on J/ψ production on ammonia [13] and a fit to the gluon distribution extracted, again by EMC, from their open charm production data on iron, both in good agreement with the present result.

The NMC gluon distribution (on H and D) can be parametrized as

$$xG(x) = K\frac{n+1}{2}(1-x)^n \quad ,$$

with $K \simeq 2.5$ and $n = 5.08 \pm 0.86$.

5 Future Perspectives

The results on hydrogen and deuterium presented here are based on about 50% of the total statistics gathered by NMC at four beam energies: 90, 120, 200 and 280 GeV over a wide kinematic range. The deuterium data amount to several times the total statistics used by

REFERENCES

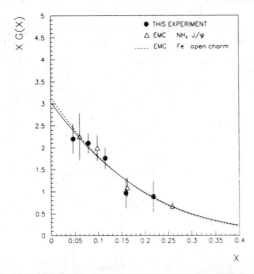

Figure 5: The gluon distribution $xG(x)$ obtained from J/ψ production. The continuous line is a fit to the data. Also shown are the EMC results on NH_3 [13] and a fit to the EMC open charm data [14]

BCDMS for their nucleon structure function analysis. The large NMC statistics, together with the tight control of systematic effects will allow a precise determination of F_2^p, F_2^n over a wide range in x, down to values of $\simeq 3 \times 10^{-3}$, thus making it possible to settle the EMC-BCDMS disagreement [15].

The heavy target results discussed here were obtained from a subsample of the data gathered by NMC on several materials, including a high luminosity sample of Sn/C data taken with a calorimetric target. The latter experiment was especially designed to study the A dependence of $R = \sigma_L/\sigma_T$ and to investigate the details of the Q^2 dependence of the EMC effect. The analysis of these data is in progress.

The comparison of J/ψ production cross-section per nucleon off different nuclei can be used to study the influence of the nuclear enviroment on the gluon distribution and therefore offers a complementary and independent way to investigate the EMC effect and to understand its origin. To this purpose the ratio of the J/ψ production cross-section per nucleon in Sn to that in C has been measured as a function of x using the calorimetric target, with a very small systematic error. The analysis is currently in progress.

References

[1] D.Allasia *et al.*, CERN/SPSC 85-18 SPSC/P210

[2] J.Ashman *et al.* (EMC), Phys. Lett. **B202** (1988) 603;
O.C.Allkofer *et al.* (EMC), Nucl. Inst. and Meth. **179** (1981) 445;
J.P.Albanese *et al.* (EMC), Nucl. Inst. and Meth. **212** (1983) 111;
C. Broggini, Ph.D. thesis, University of Neuchâtel (1989);

REFERENCES

D. Nowotny, Ph.D. thesis, University of Heidelberg (1989);

A. Simon, Ph.D. thesis, University of Heidelberg (1988);

C. Scholz, Ph.D. thesis, University of Heidelberg (1989)

[3] A.J.Buras and K.F.J.Gaemers, Nucl. Phys. **B132** (1978) 249;

M.Glueck, E.Hoffman and E. Reya, Z. Phys. **C13** (1982) 119;

D.W.Duke and J.F.Owens, Phys. Rev. **D30** (1984) 49;

E.Eichten *et al.*, Rev. Mod. Phys. **56** (1984) 579; for erratum see Rev. Mod. Phys. **58** (1986) 1065;

M.Glueck, E.Reya and A.Vogt, Dortmund University report DO-Th 89/20;

M.Diemoz *et al.*, Z. Phys. **C39** (1988) 21;

P.N.Harriman *et al.*, University of Durham report DTP/90/04, RAL/90/007 (1990)

[4] A.D. Martin, R.G.Robert and W.J.Stirling, Phys Lett. **B189** (1987) 220; Phys. Lett. **B206** (1988) 327; Phys. Lett. **B207** (1988) 205

[5] J. J. Aubert *et al.* (EMC), Nucl. Phys. **B293** (1987) 740

[6] A. C. Benvenuti *et al.* (BCDMS) , Phys. Lett. **B237** (1990) 599

[7] A. Bodek *et al.*, Phys. Rev. **D20** (1979) 1471

L.W. Whitlow, PhD thesis, Stanford University (1990), SLAC report 357 (1990)

[8] K. Gottfried, Phys. Rev. Lett. **18** (1967) 1174

[9] C. Lopez and F. J. Yndurain, Nucl. Phys. **B183** (1981) 157

D. A. Ross and C. T. Sachrajda, Nucl. Phys. **B149** (1979) 497

[10] Y. Mizuno, to be published

[11] R.G.Arnold *et al.* (SLAC E139), Phys. Rev. Lett. **52** (1984) 727

[12] R.Baier and R.Rueckl, Nucl. Phys. **B218** (1983) 289;

R.Baier and R.Rueckl, Nucl. Phys. **B201** (1982) 1

A.D. Martin, C.-K.Ng and W.J.Stirling, Phys. Lett. **B191** (1987) 200

[13] N.Dyce, PhD thesis, Lancaster University (1988)

[14] J. J. Aubert *et al.* (EMC), Nucl. Phys. **B213** (1983) 31

[15] see for instance M.Virchaux, these Proceedings

Nucleon Structure Functions from ν_μ-Fe Scattering at the Tevatron*

The CCFR Collaboration

P.Z.Quintas, K.T.Bachmann [1], R.E.Blair [2], C.Foudas [3], B.J.King,
W.C.Lefmann, W.C.Leung, S.R.Mishra, E.Oltman [4],
S.A.Rabinowitz, F.J.Sciulli, M.H.Shaevitz, W.G.Seligman, W.H. Smith[3]

Columbia University, New York, NY 10027

F.S.Merritt, M.J.Oreglia, B.A.Schumm[4]

University of Chicago, Chicago, Il 60637

R.H.Bernstein, F. Borcherding, H.E.Fisk, M.J.Lamm,
H.Schellman, W.Marsh, K.W.B.Merritt, D.D.Yovanovitch

Fermilab, Batavia, IL 60510

A.Bodek, H.S.Budd, P.de Barbaro, W.K.Sakumoto

University of Rochester, Rochester, NY 14627

* Presented by Paul Z. Quintas, Columbia University.

ABSTRACT

We present preliminary results for nucleon structure functions measured in high energy neutrino interactions. Included are new results for the Gross-Llewellyn Smith Sum Rule, $\int \frac{1}{x} x F_3 dx = 2.66 \pm .03\text{(stat)} \pm .08\text{(syst)}$, the ratio of cross-sections, $\sigma^{\overline{\nu}}/\sigma^{\nu} = .511 \pm .002\text{(stat)} \pm .005\text{(syst)}$, and an analysis of the Q^2 evolution of xF_3.

1 Introduction

High energy neutrino interactions provide an effective way to measure the structure of the nucleon. The differential cross-section for charged-current interactions is:

$$\frac{d\sigma^{\nu(\overline{\nu})}}{dxdy} = \frac{G^2 s}{2\pi} \left[\left(1 - y - \frac{Mxy}{2E}\right) F_2(x, Q^2) + \frac{y^2}{2} 2x F_1(x, Q^2) \pm y(1 - \frac{y}{2}) x F_3(x, Q^2) \right]. \quad (1)$$

By measuring the differential cross-sections for ν_μ-N and $\overline{\nu}_\mu$-N interactions, the structure functions F_2 and xF_3 can be extracted. These structure functions are related in the standard model to the momentum density of the constituent quarks, and are used to test both the quark-parton model and perturbative quantum chromodynamics (QCD).

---

[1] Present address: Widener Univ., Chester, Pa 19013.
[2] Present address: Argonne National Laboratory, Argonne, Il, 60439.
[3] Present address: Univ. of Wisconsin, Madison, Wi, 53706.
[4] Present address: LBL, Berkeley, Ca 94720.

We present new high statistics, high energy data from the CCFR collaboration. The data were taken in two runs in the Fermilab Tevatron Quadrupole-Triplet Beam (QTB) with neutrino energies up to 600 GeV. A sample of 3,700,000 triggers was reduced after fiducial and kinematic cuts ($E_\mu > 15$ GeV and $\theta_\mu < .150$) to 1,281,000 ν- and 270,000 $\bar{\nu}$-induced events. This is an order of magnitude increase in statistics and a factor of two higher mean energy compared with the CCFR Narrow Band Beam (NBB) samples [1][2].

2 Calibration

The measurement of the structure functions depends critically upon understanding the energy resolution and calibration of the detector. The CCFR detector[3] consists of target calorimeter carts which measure hadron shower energy and muon angle followed by a toroid spectrometer which measures muon momentum. The polarity of the toroid was alternated during the running periods to take data both focusing and defocusing the muons. The detector was calibrated using charged particle test beams (independent of the neutrino running).

A hadron beam was directed into the target carts at different energies and different positions. The beam was momentum analyzed to high precision and had a narrow width (a few per cent). This data was used to calibrate the apparatus to better than 1% and determine the detector resolution function. Figure 1A shows the hadron beam data at one energy setting (100 GeV) and the parameterization obtained from the data.

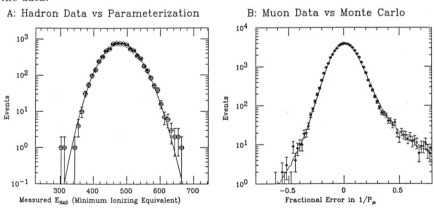

Figure 1: E_{had} and E_μ Resolutions

Muons from the test beam were used to calibrate the toroid spectrometer. Figure 1B shows the fractional difference in $1/P_\mu$ for data and Monte Carlo generated events. The data is from one energy setting (120 GeV) of the test beam. The Monte Carlo calculation propagates the muon with full dE/dx loss (restricted and catastrophic energy losses: ionization, pair-production, and bremsstrahlung). Again the data and Monte Carlo match in mean as well as shape. Both the single-scatter and the catastrophic-loss tails are correctly modeled as well as the multiple Coulomb scattering which dominates the shape.

Certain aspects of the scaling violations are more sensitive to the difference in hadron and muon energy scales than to overall energy miscalibrations. The relative calibration of E_{had} to E_μ can be checked by plotting $\frac{<E_{vis}>^{DATA}}{<E_{vis}>^{MC}}$ as a function of $y = E_{had}/E_{vis}$. If the hadron and muon energy scales are the same, the ratio will be unity for all y. If not, the deviation of the slope from zero will measure the difference in energy scales. Figure 2 shows the relative calibration for focusing and defocusing events. Both are well within 1% of zero on average.

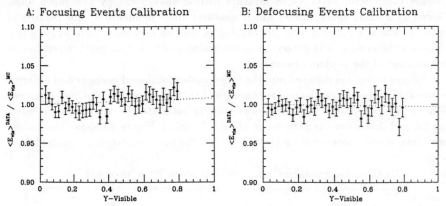

Figure 2: Relative Calibration for Focusing and Defocusing Events

3 Relative Flux Extraction

In the QTB there was no direct measurement of the secondary produced particles. Instead the neutrino flux had to be extracted from the charged-current data. From the events we can extract the relative flux, i.e. the ratio of fluxes at different energies and the ratio of neutrino to antineutrino flux. However, there is no direct measurement from this data of the absolute flux. Hence we normalize the data to the world average of the neutrino total cross-section in iron target experiments, $\sigma^{\nu N} = .68 \times 10^{-38} \, cm^2 \, E_\nu(GeV)$[4][5].

The principal method of relative flux extraction used here is the fixed ν-cut method[6]. In the limit $y \to 0$ (or $\nu = E_{had} \to 0$), the differential cross-section

$d\sigma/d\nu$ approaches a constant which is independent of energy and the same for neutrinos and antineutrinos. Thus after integration over all x and over ν from zero up to ν_0, the cross-section becomes independent of energy so that the relative flux is determined by counting events; i.e.

$$\Phi(E) \propto N(E, \nu < \nu_0) + \mathcal{O}(\frac{\nu_0}{E}) \qquad (2)$$

where $\mathcal{O}(\frac{\nu_0}{E})$ indicates corrections of order ν_0/E arising from helicity induced y-dependent terms. The parameter ν_0 is chosen small enough to minimize corrections yet large enough to provide sufficient statistics. For our ν_0 cut of 20 GeV, there are 410,000 ν and 140,000 $\bar{\nu}$ events used for determining the relative flux.

Two other methods check the relative flux extraction. The first is the y-intercept method[7] which relies on the same principle but uses nearly the entire kinematic range. Events are binned at different energies as a function of y_{vis}. At each energy the value of the fitted intercept dN/dy at $y = 0$ is proportional to neutrino energy times flux. This method agrees with the fixed ν-cut method to 1.5%. The third method is the overlapping x and Q^2 bin method [1]. The structure functions calculated at different energies but in the same x and Q^2 bin should be equal. Applying this constraint provides another measure of the relative flux. The values so obtained agree with the fixed ν-cut flux to a few per cent.

The ratio of antineutrino to neutrino total cross-sections can be directly obtained, since the ratio of their fluxes is measured. This ratio is relatively free of systematic errors because it is a double ratio. The new data presented here include the first measurement of this quantity above 200 GeV (figure 3). The average over the energy range of 30–400 GeV is $.511 \pm .002(\text{stat}) \pm .005(\text{syst})$.

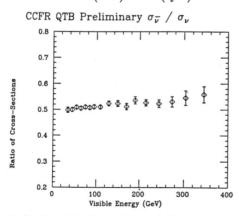

Figure 3: Ratio of Antineutrino and Neutrino Cross-sections

4 Structure Function Extraction

For the structure function analysis additional cuts of $E_{had} > 10$ GeV, $Q^2 > 1$ GeV^2, and $E_\nu > 50$ GeV were applied. After these cuts, there remained 990,000 ν- and 165,000 $\bar{\nu}$-induced events. The accepted events were separated into twelve x bins from .015 to .850 and sixteen Q^2 bins from 1 to 500 GeV^2. Eq. 1 takes the form

$$\frac{d\sigma^{\nu(\bar{\nu})}}{dxdy} = (a)F_2(x,Q^2) \pm (b)xF_3(x,Q^2) \tag{3}$$

after a number of assumptions discussed below. Integrating this cross-section times the flux over each x and Q^2 bin gives the number of events in that bin. We thus have sets of two equations in two unknowns to be solved for the structure functions at the bin centers, x_0 and Q_0^2.

$$\Delta N^\nu = \left(\int a\Phi(E)^\nu dE\right)(F_2(x_0, Q_0^2)) + \left(\int b\Phi(E)^\nu dE\right)(xF_3(x_0, Q_0^2)) \tag{4}$$

$$\Delta N^{\bar{\nu}} = \left(\int a\Phi(E)^{\bar{\nu}} dE\right)(F_2(x_0, Q_0^2)) - \left(\int b\Phi(E)^{\bar{\nu}} dE\right)(xF_3(x_0, Q_0^2)) \tag{5}$$

We assume $R \equiv \sigma^L/\sigma^T = R_{QCD}$ (the function predicted by pertubative QCD[8]). We apply corrections for the 6.85% excess of neutrons over protons in iron. We assume that the charm sea is zero and the strange sea is κ times the non-strange sea ($\kappa = .44$) [9]. We correct for threshold production of the heavy charm quark by the slow rescaling model[10][11] with $m_c = 1.31$ GeV[9]. Radiative effects follow De Rújula et al.[12], and the cross-section is corrected for the massive W-boson propagator.

5 Gross-Llewellyn Smith Sum Rule

The Gross-Llewellyn Smith Sum Rule[13] predicts that the integral of $\frac{1}{x}xF_3$ is the number of valence quarks in the nucleon. The value of three following from the naive parton model is modified by perturbative QCD and higher twist effects:

$$S_{GLS} \equiv \int \frac{1}{x} xF_3(x,Q^2) dx = 3\left[1 - \frac{\alpha_s(Q^2)}{\pi} + \mathcal{O}(\frac{1}{Q^2})\right] \tag{6}$$

The integral is evaluated at $Q^2 = 3$ GeV^2, which is the mean Q^2 of the lowest x bin, the bin which contributes most heavily to the integral. In each x bin a fit is made to $xF_3(x,Q^2)$ and interpolated or extrapolated to 3 GeV^2. These values of $xF_3(x, Q^2 = 3$ GeV$^2)$ are in turn fitted to $Ax^B(1-x)^C$ and this fit is integrated to determine the sum. The value obtained is $2.66 \pm .03(\text{stat}) \pm .08(\text{syst})$[14]. For $\Lambda = 250$ MeV the prediction for S_{GLS} is 2.63.

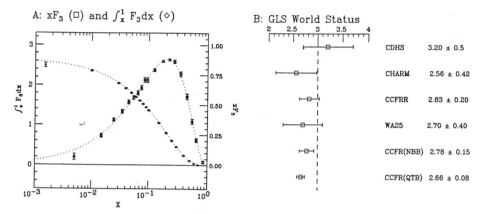

Figure 4: CCFR Preliminary GLS Measurement and World Status

Both the value of $xF_3(x, Q^2 = 3 \text{ GeV}^2)$ (squares, right-side scale) and of the integral from x to 1 (diamonds, left-side scale) are shown in figure 4A. The results from this experiment and from previous ones [15] are illustrated in figure 4B. This new result is about a factor of two more precise than the previous best measurement of this quantity.

The overall error is small because of our good measurement at small x values. The systematic errors are dominated by the uncertainty in the overall level of the neutrino cross-section (2.2%), our measurement of the relative neutrino-antineutrino flux (1.5%), and the muon energy calibration (1.0%). The hadron energy and muon angle resolution uncertainties do not contribute significantly to the systematic error.

6 Structure Function Evolution

In pertubative QCD, structure functions are expected to evolve according to the equations

$$\frac{dF^{NS}(x,Q^2)}{d\ln Q^2} = \frac{\alpha_S(Q^2)}{\pi} \int_x^1 P_{QQ}^{NS}(z) F^{NS}(\frac{x}{z}, Q^2) dz \qquad (7)$$

$$\frac{dF^{S}(x,Q^2)}{d\ln Q^2} = \frac{\alpha_S(Q^2)}{\pi} \int_x^1 (P_{QQ}^{S}(z) F^{S}(\frac{x}{z}, Q^2) + P_{GQ}(z) G(\frac{x}{z}, Q^2)) dz \qquad (8)$$

where the P_{IJ} are the "splitting functions" predicted by the theory.

The non-singlet structure function (xF_3) evolution depends only on itself and the known splitting function. The singlet equation is more complicated: the evolution of F_2 is coupled with that of the gluons, $G(x)$. We only discuss the xF_3 evolution here.

The prediction for the slope (equation 7) involves a product of two terms, one of which depends on Λ_{QCD} (α_s) and one which does not (the integral). The splitting function is independent of the value of the coupling constant to leading order so the behavior of the integral is known from the levels of the structure functions alone. The integral passes through zero at a point predicted by QCD, and hence the slope of xF_3 does so as well. Different values of Λ do not change this point but only vary the slope significantly when the integral is far from zero at high x. The low x behavior can be studied for consistency with QCD while the high x behavior provides the most sensitive measure of Λ. However the high x data points are also the most sensitive to the uncertainties in the energy scale and other systematic effects. Since our systematic studies are still in progress, we do not quote a value for Λ at this time.

The low x behavior is in itself quite interesting. There have been two previous measurements of the slope $\frac{dlnxF_3}{dlnQ^2}$. For various values of Λ there exists a family of curves, all of which pass through zero at the same point and fan out at high x (figure 5A). The CDHS collaboration has provided high statistics data which are statistically inconsistent with the curves for all values of Λ [16]. The CCFR NBB data are too limited in statistics to be conclusive [2].

Figure 5B shows our new data along with a sample curve. We observe that our low x behavior agrees well with the QCD prediction. This observation is independent of the possible calibration adjustments and independent of the value of Λ. This is the first confirmation of the QCD prediction for a non-singlet structure function.

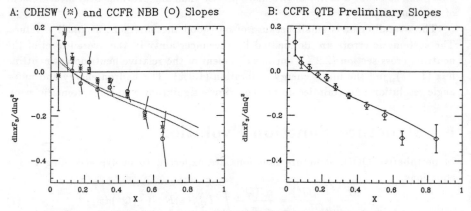

Figure 5: World Status and CCFR Preliminary xF_3 Evolutions

7 Conclusion

The CCFR collaboration has new high energy, high statistics structure functions. From this data we are able to extract the ratio of the neutrino and antineutrino cross-sections, $\sigma^{\bar{\nu}}/\sigma^{\nu} = .511 \pm .002 \pm .005$, the Gross-Llewellyn Smith Sum Rule,

$\int_0^1 \frac{1}{x} x F_3(x, Q^2 = 3 \text{ GeV}^2) = 2.66 \pm .03 \pm .08$. The data also provide the first observation of the non-singlet structure function evolution consistent with QCD.

We acknowledge the gracious help of the Fermi National Accelerator Laboratory staff and the dedicated efforts of many individuals at our home institutions. This research was supported by the National Science Foundation and the Department of Energy.

References

[1] D.B.MacFarlane et al. Z. Phys, **C26**:1, 1984.

[2] E.Oltman. *Nucleon Structure Functions from High Energy Neutrino and Antineutrino interactions in Iron*. PhD thesis, Columbia Univ., 1989.

[3] W.K.Sakumoto et al. *UR-1142; to be published in NIM*, 1990.

[4] R.E.Blair et al. Phys. Rev. Lett., **51**:343, 1983.

[5] P.Berge et al. Z. Phys., **C35**:443, 1987.

[6] P.Z.Quintas et al. In preparation.

[7] P.S.Auchincloss. *NEVIS-1394; to be published in Z. Phys.*, 1990.

[8] G.Altarelli and G.Martinelli. Phys. Lett., **76B**:89, 1978.

[9] C.Foudas et al. Phys. Rev. Lett., **64**:1207, 1990.

[10] R.M.Barnett. Phys. Rev. Lett., **36**:1163, 1976.

[11] H.Georgi and H.D.Politzer. Phys. Rev. D, **14**:1829, 1976.

[12] A.De Rújula et al. Nucl. Phys., **B154**:394, 1979.

[13] D.Gross and C.Llewellyn Smith. Nucl. Phys., B14:337, 1969.

[14] W.C.Leung et al. *Proceedings of the XXVth Rencontre de Moriond; NEVIS-1423*, 1990.

[15] S.R.Mishra and F.Sciulli. Annu. Rev. Nucl. Part. Sci., **39**:259, 1989.

[16] B.Vallage. *Détermination Précise des Fonctions de Structure du Nucléon dans les Interactions de Type Courant-Chargé de Neutrinos sur Cible de Fer*. PhD thesis, L'Université de Paris-Sud, 1987.

Preliminary Results from E665 on Cross-Section Ratios at Low x_{bj} Using H_2, D_2 and Xe Targets

Silhacen Aïd
Department of Physics and Astronomy, University of Maryland
College Park MD 20742, USA

for the E665 collaboration †

Abstract

Fermilab experiment 665 has taken deep-inelastic muon scattering data at a beam energy of 490 GeV/c, on H_2, D_2 and Xe targets. Two triggers have been used : a large scattering-angle trigger (LAT), sensitive to a minimum scattering angle of 3 mrad, and a small scattering-angle trigger which can accept a scattering angle down to 0.5 mrad. The neutron to proton ratio is reported for x_{bj} above 0.002, and it shows consistency with 1 as x_{bj} goes to 0. The Xe to D_2 cross-section ratio is reported for x_{bj} above 0.001 and it shows evidence of shadowing.

---

† Argonne National Laboratory, Argonne IL USA; University of California, San Diego, CA USA; Institute for Nuclear Physics, Crakow, Poland; Fermi National Accelerator Laboratory, Batavia, IL USA; Albert-Ludwigs-Universität Freiburg i. Br., W. Germany; Harvard University, Cambridge, MA USA; University of Illinois, Chicago, IL USA; University of Maryland, College Park, MD USA; Massachusetts Institute of Technology, Cambridge, MA USA; Max-Planck-Institute, Munich, W. Germany; University of Washington, Seattle, WA USA; University of Wuppertal, Wuppertal, W. Germany; Yale University, New Haven, CT USA.

1. The E665 experiment

The purpose of E665 is to study the scattering process $\mu + N \to \mu + X$, where N represents a nucleon and X a hadronic final state (Fig.1).

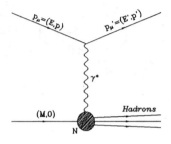

Fig.1

If $p_\mu = (E, \vec{p})$ and $p'_\mu = (E', \vec{p}')$ represent the 4-momenta of the incident and scattered muon respectively, and M the mass of the target nucleon, the relevant kinematic variables are

$$Q^2 = -(p_\mu - p'_\mu)^2$$
$$\nu = E - E'$$
$$y = \frac{\nu}{E}$$
$$x_{bj} = \frac{Q^2}{2M\nu}$$

and the cross-section for this process, in the one-photon exchange approximation is given by

$$\frac{d\sigma}{dQ^2 dx} = \frac{4\pi\alpha^2}{Q^4 x} \left(y^2 x F_1(x, Q^2) + (1 - y - Mxy/2E) F_2(x, Q^2) \right)$$

The spectrometer used for this purpose has been described elsewhere[1], but for the sake of completeness, we shall give a brief overview of its main features. A view of the apparatus is shown in Fig.2.

1.1 Beam

800 GeV/c protons from the Fermilab Tevatron interact with a beryllium target to produce pions and kaons which decay into muons, the average momentum of which is 490 GeV/c, with a spread of about 60 GeV/c. The beam tracks are

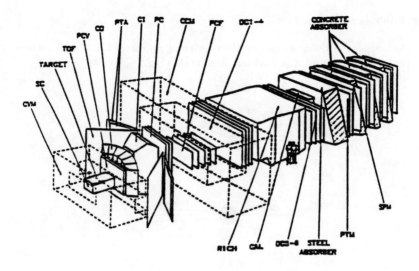

Fig. 2
View of the E665 spectrometer

reconstructed using a set of 24 wire-chamber planes, and the bending induced by an analyzing magnet is measured and the momentum determined, with a resolution of 0.5% at 500 GeV/c.

1.2 Forward spectrometer

The muon beam interacts with a target located inside a superconducting magnet (CVM). The charged hadrons and the scattered muon resulting from the interaction are detected by sets of proportional wire chambers and drift chambers. The momentum measurement is done by an other superconducting magnet (CCM) located downstream from the target, with a resolution of 2.5% at 500 GeV. Neutral particles are detected by an electromagnetic calorimeter.

1.3 Triggering system

Events are triggered by detection of the scattered muon in a set of four scintillator stations located behind a 3-metre thick hadron absorber. Two triggers have been used to get the data presented here. A large-angle trigger (LAT) requires a signal in at least 3 out of 4 scintillator walls and the absence of a signal in a fixed veto region centered around the beam. This trigger has reasonable acceptance for scattering angles above 3 mrad.

The small-angle trigger (SAT) is a floating-veto trigger. A fraction of the beam is projected to the absorber and the predicted impact point is compared with the actual muon position. If the latter falls outside a window centered around the beam projection the event is accepted. The minimum scattering angle to which this trigger has good acceptance is 0.5 mrad.

2. The neutron-to-proton ratio

2.1 Physics motivation

At low x_{bj}, the contribution to the deep-inelastic cross-section originates predominantly from the sea. The valence-quark contribution becomes negligible, and insofar as the sea distributions are the same in neutrons and protons, the ratio σ_n/σ_p, defined as

$$\sigma_n/\sigma_p = \frac{\sigma_D - \sigma_H}{\sigma_H}$$

(where σ_p is the proton cross-section, σ_n is the neutron cross-section, σ_D is the nuclear deuterium cross-section and σ_H the hydrogen cross-section, integrated over a given x_{bj} bin), should go to 1 as x_{bj} goes to 0. Previous measurements[2,3] suggest that this is indeed the case. The E665 contribution is at lower values of x_{bj} which have not been reached before.

2.2 Preliminary results

The measurement of the neutron-to-proton ratio has been performed at low x_{bj} using the large-angle trigger (LAT) and the small-angle trigger (SAT). The average Q^2 for the LAT is 15 GeV^2, with a minimum of 4 GeV^2, that of the SAT is 3.6 GeV^2, with a minimum Q^2 of 0.1 GeV^2. For the LAT data, a minimum x_{bj} cut of 0.005 was made, and a minimum x_{bj} of 0.002 was imposed on the SAT data. The results are shown in Fig.3. It can be seen that as x_{bj} approaches 0, the ratio is consistent with 1 within errors.

2.3 Estimate of systematic errors

Even though most systematic effects cancel out in the measurement of cross-section ratios, the fact that data for different targets were taken at different time periods introduces systematic effects due to variations in the running conditions. The sources of systematic uncertainties are :

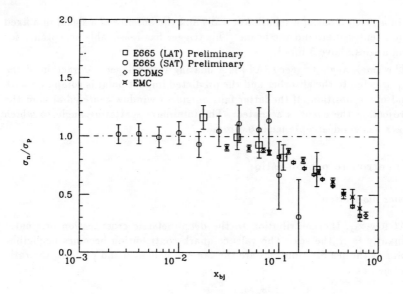

Fig. 3
Neutron-to-proton ratio as a function of x_{bj}
BCDMS points are from reference 3. EMC points from reference 2.

- *Beam normalization* : The correctness of the number of live beams was done by comparing "event" scalers, which count the number of beams during the live time of the experiment, against randomly preselected beams. When the prescale factor is taken into account, both numbers agree to better than 1/2%. In addition, the random beams allowed to factor out the beam-reconstruction efficiency. We currently estimate the normalization uncertainty to be around 1%.
- *Reconstruction efficiency* , due to variation in time of detector performance; this is currently under investigation, and we attribute to this correction a value not to exceed 10%.
- *Empty-target subtraction* : correction of the order of 1 to 2%.
- *Target density* : uncertainty estimated to be 1%.
- *Radiative correction* : the correction on the n/p ratio is less than 1%.
- *Fermi motion* : negligible at low x_{bj}.[3]

3. Xe to D_2 cross-section ratio

3.1 Physics motivation

Compared to the parton distributions of single nucleons, those of nuclei are modified by the nuclear environment. Previous measurements[4] have shown that at low x_{bj}, the deep-inelastic cross-section (per nucleon) on nuclear targets is smaller than that of single nucleons, a phenomenon known as shadowing. E665 has compared the xenon cross-section to the deuterium cross-section and observed a similar effect.

3.2 Preliminary results

At low x_{bj}, radiative processes are an important background to deep inelastic scattering in Xe. The traditional approach used by previous experiments was on the one hand to restrict the kinematic range of the data by essentially imposing an upper cut on the energy transfer ν, and on the other hand to apply corrections based on numerical calculations of the radiative contribution to deep inelastic scattering, using Mo and Tsai formalism or more accurate formulae such as those of Bardin et al., in which higher-order diagrams are taken into account. In addition to the aforementioned methods, E665 used an electromagnetic calorimeter to reject the electromagnetic background. Events are discarded if the fraction of energy deposited in the electromagnetic calorimeter to the energy transferred, ν, exceeds some threshold value. The ratios, corrected for radiative background using either a numerical integration or the calorimeter have been found to be compatible within 5%. The x_{bj} dependence of the cross-section ratio is shown in Fig.4. It decreases as x_{bj} goes to 0, showing evidence of shadowing. On the other hand (Fig.5), no strong Q^2 dependence is observed. Data points on Ca from NA28[4] are shown for comparison.

3.3 Estimate of systematic errors

The sources and estimates of systematic uncertainty are as follow :
- *Beam normalization* : 0.7%
- *Target density* : 0.4%
- *Trigger acceptance* : due to difference in target size. Less than 3%.
- *Empty-target subtraction* : 1%
- *Time-dependent efficiency* : 4%
- *Radiative corrections* : 1.4%

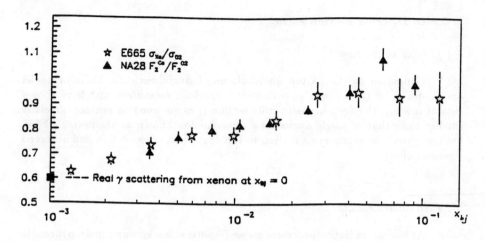

Fig. 4
Xe to D_2 cross-section ratio as a function of x_{bj}
for $Q^2 > 0.1$ GeV^2, $\nu > 40$ GeV and $y < 0.75$

The Xe photo-production point has been obtained by interpolating the 60 GeV data from ref.5 on C, Cu and Pb and then extrapolating the value thus found from 60 GeV to 150 GeV using the energy dependence of the Cu data.

4. Conclusion

The neutron-to-proton ratio has been measured by E665 in muon deep inelastic scattering at a beam energy of 490 GeV/c, for x_{bj} as low as 2×10^{-3} and has been found to be consistent with 1 as x_{bj} goes to 0. The ratio of Xe to D_2 cross-sections has been measured for x_{bj} down to 10^{-3} and shows evidence of shadowing, with no strong Q^2 dependence. Work is currently in progress to improve the understanding of systematic errors.

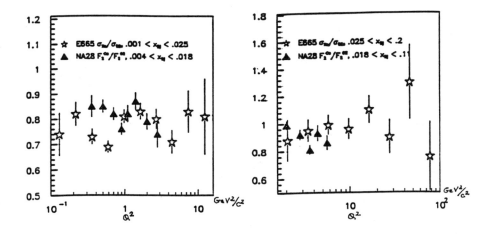

Fig. 5
Xe to D_2 cross-section ratio as a function of Q^2

5. Acknowledgements

The work of the University of California, San Diego was supported in part by the National Science Foundation, contract numbers PHY82-05900, PHY85-11584 and PHY88-10221; the University of Illinois at Chicago by NSF contract PHY88-11164; and the University of Washington by NSF contract numbers PHY83-13347 and PHY86-13003. The University of Washington was also supported by the U. S. Department of Energy. The work of Argonne National Laboratory was supported by the Department of Energy, Nuclear Physics Division, under Contract No. W-31-109-ENGGG-38. The Department of Energy, High Energy Physics Division, supported the work of Harvard University, the University of Maryland, the Massachussetts Institute of Technology under Contract No. DE-AC02-76ER03069 and Yale University. The Albert-Ludwigs-Universität and the University of Wuppertal were supported in part by the Bundesministerium für Forschung und Technologie.

6. References

1. M.R. Adams et al., E665 Collaboration, to be published in *Nucl. Inst. and Meth.*
2. A.C. Benvenuti et al., BCDMS coll., *Phys. Lett. B237(1990) 601*

3. J.J. Aubert et al., EMC coll., *Nucl. Phys.* B293(1987) 740
4. A. Arneodo et al.,*Nucl. Phys.* B333(1990) 1
5. D.O. Caldwell et al.,*Phys. Rev. Lett.* 42(1979) 553

PRECISE EXTRACTIONS OF THE X AND Q^2 DEPENDENCE OF $R = \sigma_L/\sigma_T$, F_{2p}, F_{2d}, AND F_{2n}/F_{2p} FROM A COMBINED ANALYSIS OF SLAC DEEP INELASTIC ELECTRON SCATTERING EXPERIMENTS

L.W. Whitlow[1], S. Rock[2], A. Bodek[3], E. M. Riodan[4], S. Dasu[3]

Presented by A. Bodek

[1] Stanford University, Stanford, California 94305
[2] The American University, Washington DC 20016
[3] University of Rochester, Rochester, New York 14627
[4] Stanford Linear Accelerator Center, Stanford, California 94309

Abstract

We report on a precise study of the x and Q^2 dependence of the proton, deuteron and neutron structure functions. In particular $R = \sigma_L/\sigma_T$, F_{2p}, F_{2d}, and the ratio F_{2d}/F_{2p} were extracted in a combined analysis of SLAC deep inelastic electron scattering experiments. Data from eight experiments were radiatively corrected with a new improved radiative correction formalism, and all experiments were normalized to the precise data of SLAC E140, thus yielding a single coherent data set. We find that $R_p = R_d$ as expected from QCD. The results for R are somewhat larger than the prediction of QCD with target mass effects, thus indicating the existence of additional dynamic higher twist effects. Using both SLAC and CERN data, we obtained a parametrization of R in the range $0.5 < Q^2 < 200$ GeV/c^2 and $0.1 < x < 0.86$. This fit in conjunction with the normalized hydrogen and deuterium cross sections was used to extract precise values of F_2. These data provide a constraint at low Q^2 for high Q^2 muon and neutrino experiments. A study of the Q^2 dependence of the ratio of deuteron and proton structure functions shows a slope which is consistent with QCD and explains the difference in this ratio between SLAC and CERN experiments.

Introduction

We report on the re-analysis of eight deep inelastic e-p and e-d electron scattering experiments at SLAC. The first of these experiments was done in 1970. The most recent experiment, SLAC E140 completed in 1985, was a high statistic experiment designed to extract $R = \sigma_L/\sigma_T$ from deep inelastic e-d and e-Fe cross sections. The key to the E140 minimization of systematic errors was the use of the new "Bardin/Tsai" procedures for calculating radiative corrections. In addition, the acceptance of the 8-GeV spectrometer used in E140 was remeasured using a wire float technique. The point to point relative systematic error of SLAC E140 were about 0.6% and the total overall uncertainty in the normalization of the cross section (from all sources) was 1.7%. The results of E140 can be summarized as follows: (1) R for iron and deuterium are the same as expected in QCD, and on average $R_{Fe} - R_D = 0.003 \pm 0.018$. (2) There is very little Q^2 dependence in the ratio of F_{2Fe}/F_{2d} for fixed values of x (EMC effect). (3) The x and Q^2 dependence of R only agreed with QCD at small x and large Q^2. The low Q^2 and high x behavior of R was better described by QCD with target mass corrections, but the data still required additional higher twist corrections.

The precise data of SLAC E140 was measured over a restricted kinematic range. In order to extend the kinematic range, we have reanalized previous SLAC experiments using the new radiative corrections procedures of SLAC E140. In addition, using overlap regions in the data, we cross normalized these experiments to the data of E140, which provided the precise absolute normalization. The goals of the reanalysis were: (1) Compare R for hydrogen and deuterium over a large kinematic range. (2) Extract R over a much larger kinematic range in the SLAC energy range and compare to theory. (3) Obtain an overall fit to R using the extended SLAC data, in conjunction with high Q^2 muon and neutrino data, to be used at all x and Q^2. (4) Use this fit to R and the reanalyzed precise SLAC cross sections to extract the structure functions over a large range and provide overlap with higher Q^2 experiment. (5) Provide an anchor at low Q^2 for the high Q^2 muon and neutrino experiments, and allow the separation of scaling violation from QCD and higher twist effects, and (6) Study the x and Q^2 dependence of the ratio of deuteron and proton structure functions in order to understand the experimental difference between the results at SLAC and the high Q^2 results from the CERN muon experiments.

Analysis and Results

We begin the global reanalysis by re-radiatively correcting all the cross sections as follows. The "internal" portion of these corrections is calculated using the exact prescription of Akhundov, Bardon and Shumeiko. The "external" portion of these corrections (due to straggling of the electrons in the target material) is calculated using the complete double integral formalism of Mo and Tsai. As a check of the external corrections, data taken with targets of 2%, 6% and 12% radiation lengths yield the same radiatively corrected cross section only if the complete formalism is used. As a check of the internal formalism, we found that the Bardin exact formalism agreed with a corrected version of the Mo/Tsai exact formalism. Differences of up to 5% were found between the the improved formalism versus with the previously used peaking approximation.

We normalize the data of previous seven experiments (SLAC E49a, E49b, E61, E87, E89a, E89b, E139) to that of E140 by fitting all cross sections measurements to a smooth model with floating normalization parameters for the earlier experiments. The resulting overall normalization uncertainty in the combined data set is 1.7% for deuterium and 2.1% for hydrogen. The normalized cross sections from all experiments are binned in intervals of x and Q^2 and a bin centering correction is applied. Then each value of R is extracted from cross sections measurements from up to to six experiments. The extracted values of $R_d - R_p$ display no Q^2 dependence. The average of $R_d - R_p$ over Q^2 versus x is shown in figure 1. It is consistent with zero as expected in QCD, with an overall average of $-0.001 \pm 0.009(\text{stat}) \pm 0.009(\text{syst})$.

Figure 2 shows the average of Rd and Rp versus Q^2 for fixed values of x. The data is mostly described by QCD with the addition of target mass corrections (dashed curve). The fit is above the curve indicating the need for higher twist corrections. The solid line shown in figure 1 is an empirical fit, which was used to extract values of F_{2p} and F_{2d} from all eight experiments. The hydrogen SLAC data are compared to the results of EMC in figure 3. Here the EMC data was multiplied by a factor of 1.07. A similar normalization factor is needed to get the EMC deuterium data to agree with SLAC. Figure 4 shows a comparison between the hydrogen SLAC data and BCDMS. There is overall agreement between BCDMS and SLAC, but there are some problems in the overlap region. There are correlations in the systematic errors of the BCDMS data (e.g. from B field) which are mainly important at small values of Q^2. A combined fit done by the BCDMS group indicates that a shift of about 1.5 standard deviation in the correlated systematic errors coupled with a renormalization of the BCDMS results by a factor of 0.98 can bring the SLAC and BCDMS results into reasonable agreement. Figure 5 shows old results for the ratio of neutron and proton structure functions F_{2n}/F_{2p}. There was a long standing difference between the results from SLAC and the higher Q^2 CERN experiments. Figure 6 shows new results for this ratio versus Q^2 for fixed values of x. Clear slopes is seen in the SLAC data, which match the high Q^2 CERN results. Figure 6 compares the values of these slopes in the SLAC Q^2 range to the predictions of QCD and QCD with target mass corrections. The slopes are consistent with the expectations from QCD.

Figure Captions

Fig. 1 Average of $R_d - R_p$ over Q^2 vs x.

Fig. 2 Average of R_p and R_d vs Q^2 for fixed values of x. Dotted line is QCD; dashed line is QCD + target mass corrections; solid line is empirical fit.

Fig. 3 Comparison of SLAC F_2 for hydrogen vs EMC data. The EMC results were multiplied by 1.07.

Fig. 4 Comparison of SLAC F_2 for hydrogen vs BCDMS data.

Fig. 5 Comparison of $F2n/F_{2p}$ (with Fermi corrections) between SLAC and CERN experiments.

Fig. 6 $F_{2d}/F_{2p} - 1 = $ "F_{2n}/F_{2p}" (with no Fermi corrections) vs $\ln Q^2$ for SLAC and BCDMS.

Fig. 7 Slopes of "F_{2n}/F_{2p}" vs $\ln Q^2$ for SLAC data only. Dashed line is prediction of QCD with target mass corrections.

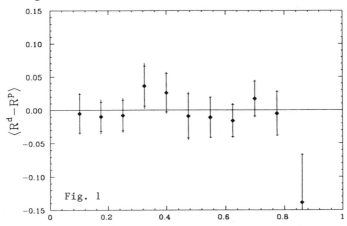

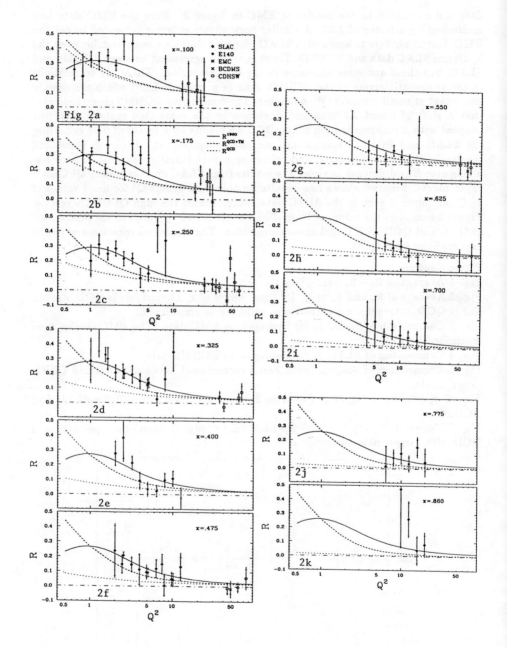

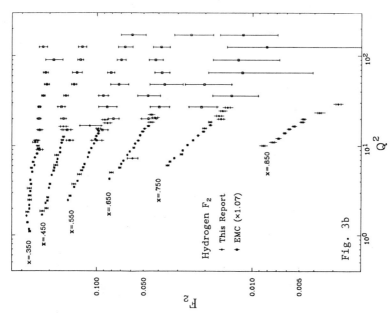

Fig. 3b Hydrogen F_2 This Report / EMC (×1.07)

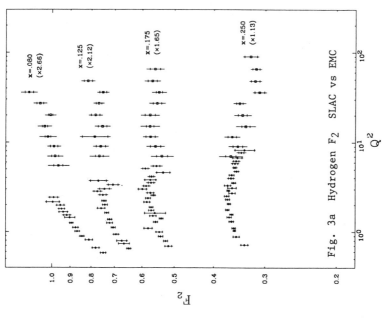

Fig. 3a Hydrogen F_2 SLAC vs EMC

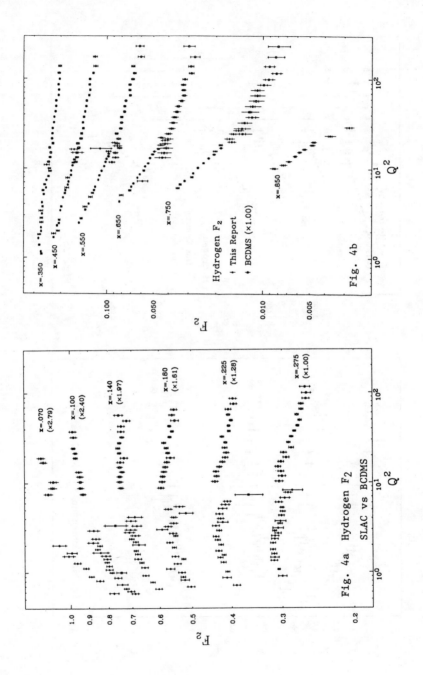

Fig. 4a Hydrogen F_2 SLAC vs BCDMS

Fig. 4b Hydrogen F_2

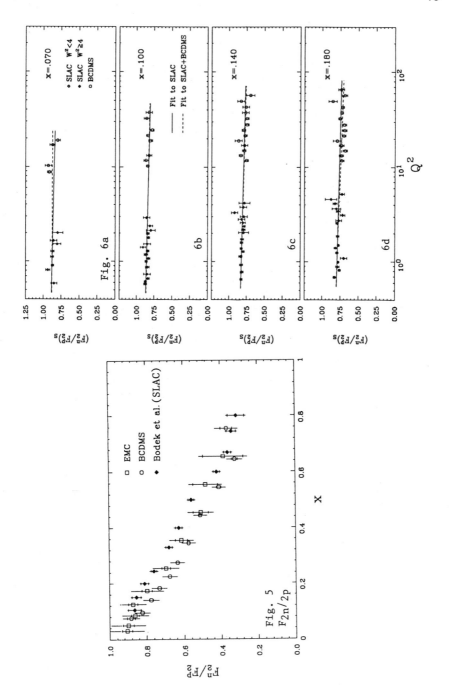

Fig. 6a, 6b, 6c, 6d

Fig. 5 F2n/2p

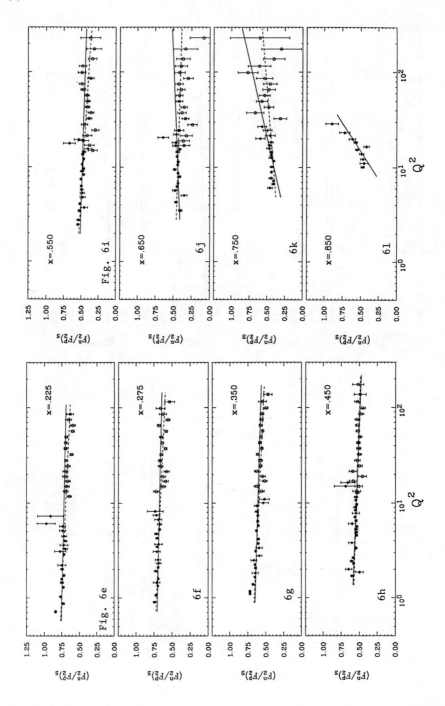

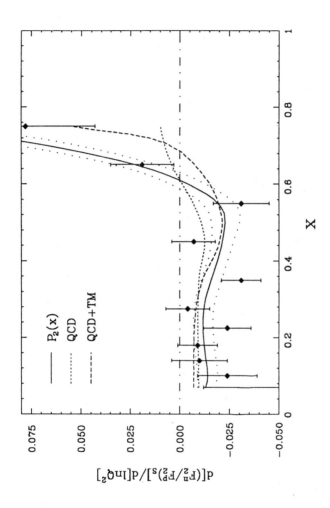

Figure 7 Slopes of "F_{2n}/F_{2p}" vx $\ln Q^2$ for SLAC data only. Dashed line is prediction of QCD with target mass corrections.

A QCD analysis of high statistics F_2 data on H_2 and D_2 targets

with determination of higher — twists

Alain MILSZTAJN
D.Ph.P.E., C.E.N. Saclay
91191 Gif-sur-Yvette Cedex (France)

ABSTRACT

We present the preliminary results of a QCD analysis of the high statistics BCDMS and SLAC F_2 data on D_2 and H_2. At high Q^2, the data are in good agreement with the predictions of perturbative QCD and lead to an improved measurement of Λ_{QCD}. At lower Q^2, a precise measurement of non-perturbative effects ("higher-twists") is obtained : they are very small below $x = 0.40$ and small, positive and increasing with x at higher x.

In the past months, the final results of the two highest statistics measurements of F_2 on deuterium and hydrogen targets have been presented. This set of results covers a wide kinematic range : 0.07 to 0.85 in x and 0.5 to 260 GeV^2 in Q^2. These data are thus well suited for a test of perturbative QCD as well as for a measurement of possible "higher-twist" (non-perturbative) effects in the Q^2-evolution of F_2. We present here the preliminary results of such a study (the evaluation of systematic errors is not yet completed).

1. Presentation of the data used in this analysis

The high Q^2 data (7 to 260 GeV^2) are those obtained by the BCDMS collaboration [1] with their muon scattering experiment using a toroidal iron spectrometer ; these data have already been used for QCD analyses of F_2 [2] at high Q^2, where non-perturbative effects are expected to be small, as well as for a determination of $F_2^p - F_2^n$ and a QCD analysis of this purely non-singlet structure function [3]. The low Q^2 data (0.5 to 30 GeV^2) come from a coherent global reanalysis of electron scattering data from a number of experiments at SLAC spanning the time period 1970 to 1985 [4]. The main improvements compared to previous publications are a better determination of $R(x,Q^2)$ and a correct treatment of radiative corrections. This allows to increase the useable kinematical range. In the present analysis, we have used all the published data, apart from the last x-bin (0.85), where only SLAC data exist.

77

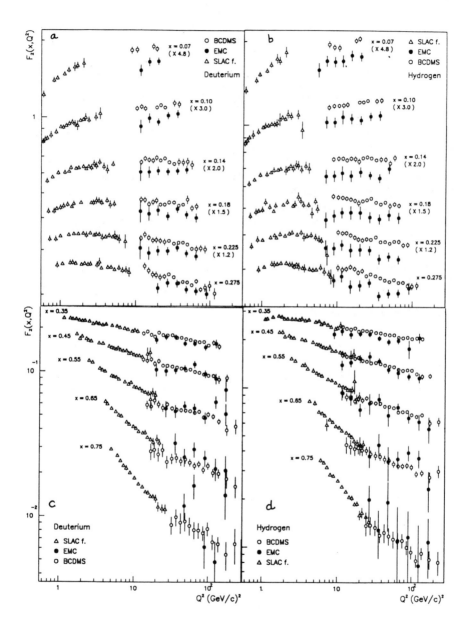

Figure 1: High statistics measurements of F_2 : a) and c) Deuterium target ; b) and d) Hydrogen target.

The data are shown in Figure 1a and b (for x < 0.30) and c and d (for x > 0.30), together with the earlier EMC data [5], interpolated where necessary to the x-bins used here. The errors shown on Figure 1 are "total" errors, *i.e.* statistical and systematic combined in quadrature. The total errors of the EMC data are typically three times larger than those of the BCDMS data. In addition to these point-to-point errors, there are global normalisation errors of 3%, 5% and 2% respectively for the BCDMS, EMC and SLAC data. We do not discuss here the comparison and compatibility of these data sets, which can be found in [6]. Of course, the quality of the QCD fits to BCDMS and SLAC data presented here is of relevance to this comparison. We want to emphasize a specific point from [6] that is important for the treatment of systematic errors in our fits : the kinematical region where the systematic errors are largest in the muon scattering data corresponds to high x (x > 0.50) and low Q^2. In this region, the systematic error originates predominantly from uncertainties on the calibrations of the measurement of the incident and scattered muon energy and on the resolution of the spectrometer. These three sources of error have a similar x and Q^2 dependence and can thus be combined quadratically into a one-standard-deviation 100%-correlated error which we call here the "main systematic error" of the BCDMS data (see [7] for full tables of errors and [6] for a more detailed discussion). Unfortunately, this dominant systematic error is largest precisely where the low Q^2 SLAC and high Q^2 BCDMS data overlap to some extent.

2. QCD fits to SLAC and BCDMS data

We have employed for these fits a computer program developed by members of the BCDMS collaboration [8] that has already been used to fit the predictions of perturbative QCD to the BCDMS data (see *e.g.* [2]). This program performs a fully numerical integration of the Altarelli-Parisi equations (both singlet and non-singlet in next-to-leading order) and achieves reasonable computing times by an intensive use of the vectorisation possibilities in the computation of the convolution integrals.

The free parameters in these fits correspond to

- a description of the x-dependence of the non-singlet and singlet part of F_2 ($F_2^{NS}(x,Q_0^2)$ and $F_2^{SI}(x,Q_0^2)$), and the gluon distribution $xG(x,Q_0^2)$; these parametrisations are taken at $Q_0^2 = 20\ GeV^2$

- the value of $\Lambda_{\overline{MS}}^{(4)}$ for four active quark flavours

- coefficients C_i (one per x-bin) describing the twist-four effects (HT) in the Q^2-evolution of F_2 , such that

$$F_2^{HT}(x_i,Q^2) = F_2^{LT}(x_i,Q^2)\ (1\ +\ C_i/Q^2)\ ,$$

where F_2^{HT} is the function that is fitted to the data and F_2^{LT} obeys the perturbative QCD Q^2-evolution according to the Altarelli-Parisi equations.

In addition to these phenomenological and theoretical parameters that enter in the fit, we include two "experimental" parameters describing the systematic errors

- one for the relative normalisation of the SLAC and BCDMS data sets

- one for the dominant systematic error of the BCDMS data discussed above. In that case, we have taken into account all correlation effects ; more explicitly, if $F_2(x_i, Q_j^2)$ and $\Delta F_2^{sys}(x_i, Q_j^2)$ are respectively the values of F_2 and of the (one standard deviation) dominant systematic error in each bin (x_i, Q_j^2), then the fitted quantity is $F_2(x_i, Q_j^2) + \lambda \Delta F_2^{sys}(x_i, Q_j^2)$, where λ is the free parameter describing the "amount of BCDMS dominant systematic error".

All the other sources of systematic error in the BCDMS and SLAC data are notably smaller (comparable to or smaller than the statistical error) and we have chosen to combine them quadratically with the statistical errors, and to use the resulting errors in the fits as if they were purely statistical. We thus ignore their possible correlations, but this is of minor importance given their sizes.

The QCD fits described above have been performed both with and without the inclusion of target mass corrections (TMC, from reference [9]). These corrections are computed numerically from the measured F_2's themselves and do not involve any additional free parameter. The results of the fits are summarized in Table 1.

Table *1*: Results of combined QCD (NLO) fits to BCDMS and SLAC data

	H_2, no TMC	H_2, TMC	D_2, no TMC	D_2, TMC
χ^2 / DOF	337 / 378	328 / 378	284 / 360	285 / 360
Λ ($\overline{MS}$, f = 4)	250 MeV	250 MeV	250 MeV	260 MeV
$xG(x, Q_0^2)$	0.48 $(1+7)(1-x)^7$	– idem –	– idem –	– idem –
BCDMS/SLAC rel. norm.	– 1.0 %	– 1.3 %	+ 0.2 %	+ 0.0 %
λ (BCDMS main syst.)	1.5	1.5	1.3	1.3

3. Discussion

We now comment on the general features of these (preliminary) fit results. The χ^2's are good – smaller than one per degree of freedom, partly because we have included some of the systematic errors in the total errors. They are slightly better with TMC included, but this is not very

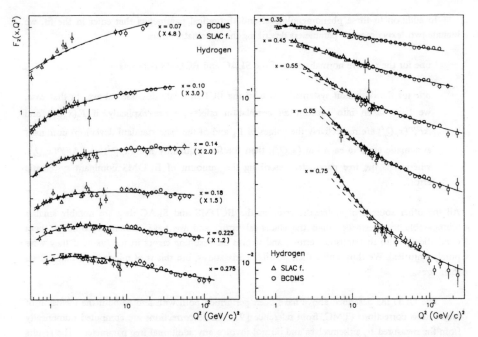

Figure 2: NLO QCD fit to SLAC and BCDMS H_2 data, with TMC. The solid line is the result of the fit ; the dashed line visualizes the Q^2-evolution without the higher-twist effects (leading-twist + TMC contribution only).

significant. As an example of the fit quality, we show in Figure 2 the fit to the H_2 data including TMC. It is clear that, apart from local minor problems (x = 0.07 and 0.275 for SLAC, x = 0.07 and 0.55 for BCDMS), the overall description of the data by the fit is good. For the D_2 data, the fits are even better. It is also clear, from the difference between the solid and dashed curve (see Figure 2 caption), that the influence of non-perturbative effects in the Q^2-evolution of F_2 are negligible above ~ 2 GeV^2 at low x (x < 0.30) and ~ 10 GeV^2 at higher x.

The value of $\Lambda^{(4)}_{\overline{MS}}$ is almost everywhere the same (it is here rounded to the nearest 10 MeV), and the total error on Λ is rather small (40 MeV) : it corresponds to an α_s measurement of 5% precision. This error is dominated by systematic errors, so that it is not reduced appreciably by combining the H_2 and D_2 fits. The central value of 250 MeV is in agreement with recent measurements of Λ at high Q^2 [2]. The gluon distribution estimated from the four fits are also very similar. The error on the exponent in the gluon distribution is typically ±2, and the fits are not sensitive to the gluon distribution above x = 0.30 . The relative normalisation of the two data sets is everywhere smaller than 1.5%, perfectly compatible with the absolute normalisation uncertainties of 3% and 2% on the BCDMS and SLAC data. The amount of BCDMS main systematic error (λ parameter) that corresponds to the the best χ^2 is of order 1.4 times the published errors.

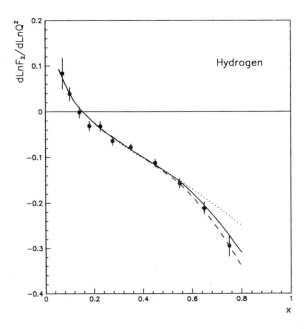

Figure 3: The logarithmic derivatives $dlnF_2/dlnQ^2$ at high Q^2 for the H_2 data. The dashed line is the QCD prediction from the fit with HT and TMC (solid line : TMC, no HT ; dotted line : no TMC, no HT) for Λ = 250 MeV.

We illustrate the good agreement between the measured Q^2-evolution of F_2 and the one predicted by perturbative QCD on Figure 3. On this Figure, the points represent the values of the logarithmic derivatives $dlnF_2/dlnQ^2$ for the hydrogen data at high Q^2 (larger than 8 to 20 GeV^2, depending on x), and the dashed line is the prediction obtained from the QCD fit (with $\Lambda_{\overline{MS}}^{(4)}$ = 250 MeV). The solid line corresponds to the same fit result with the higher-twist coefficients C_t arbitrarily fixed to zero : this line is also in good agreement with the Q^2-evolution of the data. The dotted line corresponds to the same fit with no higher-twists *and* no target mass corrections : the difference is visible for x > 0.55. Our conclusions on the deuterium data are similar.

In Figure 4, we show the values of the coefficients C_t, both for D_2 and H_2, with and without inclusion of TMC's. The x-dependences of these higher-twist terms are very similar in H_2 and D_2 data. They are small for x < 0.40, and even almost compatible with zero for fits with TMC's. As the C_t parameters are nearly mutually uncorrelated in the fits, this fact is of clear phys-

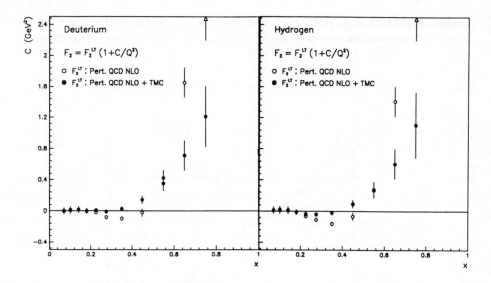

Figure 4: The higher-twist coefficients C_i as a function of x for D_2 and H_2 data. Full (open) circles are for fits with (without) TMC.

ical significance. For x > 0.40, the higher-twist terms increase with x, as expected; they are clearly smaller in the case of fits including TMC's. Because of the high statistical power of these data, this determination of higher-twist terms in deep inelastic scattering is presently the most precise. We consider *remarkable* the fact that the inclusion of TMC's in the fits reduces everywhere the magnitude of the higher-twist terms needed to describe the data, especially so at low x where this reduction is almost a cancellation.

The behaviour of these higher-twist terms has two interesting consequences : first, concerning the large x domain (x > 0.25), the values of $\Lambda_{\overline{MS}}$ resulting from QCD fits on high Q^2 data (Q^2 > 20 GeV^2) are not significantly affected by these higher-twist terms ; second, concerning the lower x domain, the higher-twist influence on the Q^2-evolution of F_2 is so small that even data at rather low Q^2 (down to 1 GeV^2) can be used in the estimation of the gluon distribution. This last point is of academic interest here, because of the Q^2-gap in the data at low x, but it will be important when data are obtained in this domain and at lower x.

4. Summary

We have presented combined QCD fits to the two highest statistics F_2 data on hydrogen and deuterium targets. These data are in good agreement and are complementary : the high Q^2 data of

BCDMS allows to test the perturbative QCD predictions and the low Q^2 data of SLAC leads to a precise determination of the magnitude of non-perturbative effects in the Q^2-evolution of F_2. The data are well described over the whole Q^2-range (0.5 to 260 GeV^2) by the perturbative QCD fits including target mass corrections and higher-twist terms ; these terms are very small or negligible at low x (x < 0.40) and they are small, positive and rise with x at higher x.

References

[1]   BCDMS collaboration, A.C. Benvenuti et al., Phys. Lett. 223B (1989) 485;
      BCDMS collaboration, A.C. Benvenuti et al., Phys. Lett. 237B (1990) 592.

[2]   BCDMS collaboration, A.C. Benvenuti et al., Phys. Lett. 223B (1989) 490.

[3]   BCDMS collaboration, A.C. Benvenuti et al., Phys. Lett. 237B (1990) 599.

[4]   L. Whitlow, Ph.D. Thesis, SLAC − REPORT − 357 (1990), and
      A. Bodek, contribution in these proceedings.

[5]   EMC collaboration, J.J. Aubert et al., Nucl. Phys. B259 (1985) 189;
      EMC collaboration, J.J. Aubert et al., Nucl. Phys. B293 (1987) 740;
      these data are presently being reanalysed (S.J. Wimpenny, private communication).

[6]   M. Virchaux, these proceedings, and A. Milsztajn et al., CERN-EP preprint in preparation (1990).

[7]   The full data and error tables corresponding to [1] can be found in preprints CERN-EP/89-06 and CERN-EP/89-170.

[8]   M. Virchaux, Thèse, Université Paris-7 (1988); A. Ouraou, Thèse, Université Paris-11 (1988).

[9]   H. Georgi and D. Politzer, Phys. Rev. D14 (1976) 1829.

Probing Nucleon Structure with ν-N Experiments

S.R. Mishra

Columbia University, New York, NY 10027

Abstract

A review of nucleon structure function measurements in high energy neutrino-nucleon scattering experiments is presented. The status of tests of Quark-Parton Model sum rules and QCD scaling violation predictions are examined. Future endeavors for precision tests of the Standard Model in a Second Generation Deep Inelastic Neutrino Experiment are outlined.

Probing nucleons with high energy neutrinos uniquely elucidates their substructure. In $\nu_\mu(\bar{\nu}_\mu)$-N scattering, quark and antiquark currents have distinct y dependence, and hence flavor selection is an inherent feature of such experiments. The content and shape of the strange sea of the nucleon can only be explored in neutrino-induced opposite sign dimuons. Furthermore, the parity violating amplitude of ν-N scattering yields the non-singlet structure function xF_3, whose evolution with respect to Q^2 can be unambiguously predicted by perturbative QCD, and thus provides one of the cleanest tests of this theory. We point out the enormous improvement upon these past measurements in a future neutrino experiment with an order of magnitude more statistics and a corresponding alleviation of systematic precision. A better understanding of the nucleon substructure implies and impels a better measurement of the electroweak parameter $sin^2\theta_W$ in ν-induced neutral current (NC) interactions: a channel with unique sensitivity to the electroweak radiative correction. Finally, a variety of rare processes and possible new phenomena could be explored with unprecedented sensitivity in a new experiment.

The outline of this article is as follows: First, we review the procedure of normalization, absolute and relative, in ν-experiments. Two new measurements were reported at this workshop by the CCFR collaboration under this subtopic. Second, the status of tests of the Quark Parton Model predictions in neutrino experiments is presented. The new measurement of the Gross-Llewellyn Smith sum rule by CCFR has attained a 3% precision. Third, we compare measurements of quark densities in various neutrino experiments. Here, recent measurements of the valence quark and antiquark distributions were reported by the CCFR collaboration. Fourth, scaling violations and QCD tests are reviewed, where the new CCFR xF_3 data show a Q^2 evolution *consistent* with the theoretical prediction: this was lacking in earlier data. Finally, we point out the improvements upon these measurements, and measurements in the electroweak sector, in a second generation-neutrino experiment.

A: Normalization

Counting the number of neutrinos (antineutrinos) incident upon the target yields the total cross section. A linear rise of the total cross section with neutrino energy implies that neutrinos interact with free, point-like constituents of the nucleon. Furthermore, by comparing the *absolute* magnitude of the total quark and antiquark densities ($\approx F_2(x, Q^2)$) measured in neutrino interactions with those in muon scattering, we discern the mean square charge of the interacting quarks. Finally, the measurement of the total momentum fraction carried by the quarks and antiquarks ($= 1/2$) provides a clear indication of the gluonic content of the nucleon: the missing momentum is carried by the gluons.

A.1: Absolute Flux Normalization

A.1.1: Direct Flux Measurement

Traditionally, incident neutrino flux magnitude and energy have been determined by monitoring the mesons (π/K) that give rise to neutrinos. Although several experiments indicated that the cross section's energy dependence is linear, there were discrepancies in the constant of proportionality (σ^ν/E_ν).[1] Most notable was the disagreement between the CDHS[2] and the CCFRR[3] collaborations. The CDHSW collaboration repeated the experiments with significant modifications in the flux measurement, and their new results[4] agreed with the higher CCFRR value. Measurements of σ^ν/E_ν as a function of E_ν by the CCFRR and the CDHSW collaborations are shown in Fig.1. The near independence of the systematics in the experiments permits one to take a weighted average of the cross sections:

$$\frac{\sigma^{\nu N}}{E_\nu} = [0.680 \pm 0.015(\text{stat.}+\text{syst.})] \, E_\nu 10^{-38} cm^2/GeV \quad --(a)$$
$$\frac{\sigma^{\bar{\nu} N}}{E_{\bar{\nu}}} = [0.339 \pm 0.009(\text{stat.}+\text{syst.})] \, E_{\bar{\nu}} 10^{-38} cm^2/GeV \quad --(b) \qquad (1)$$

The inherent difficulties of direct flux measurement, involving numerous measurements (such as determination of particle composition (π, K, p), energy, position, and angular divergence of the hadron beam), and using techniques with unexplained background (such as Cerenkov counter), induces one to conclude that the precision on these measurements is likely not to improve; unless an entirely new technique of direct flux monitoring is employed.[5]

There is, however, one indirect way of extracting the total neutrino flux which holds promise: the inverse muon decay.

A.1.2: Inverse Muon Decay: $\nu_\mu + e^- \to \mu^- + \nu_e$

The ν_μ-electron charge current (CC) scattering offers unambiguous tests of the Standard Model. These include measurements of the Lorentz structure of the weak current, the helicity of the interacting neutrino, the scalar coupling of the muon and the electron to their neutrinos, and the energy dependence of the cross section. The theoretical prediction for the cross section of this process is: $\sigma(\nu_\mu + e^- \to \mu^- + \nu_e) = G_F^2 s/\pi = 17.2 \times 10^{-42} E_\nu cm^2/GeV$, where $s = 2m_e E_\nu \gg m_\mu^2$. By counting the number of inverse muon decay (IMD) events, and using the theoretical cross section, the total neutrino flux can be determined.

Experimental studies[6,7,8] of IMD suffer from statistical limitations since the cross section of this process is a thousand times smaller than cross section of the inclusive

CC process. The most precise published measurement (by CCFR) of IMD was at a 7.5% level. The CCFR collaboration[9] has presented an improved measurement at this workshop (see Fig.2) from their tevatron quadrupole triplet beam (QTB) data; they report:

$$R_{IMD} = \frac{\sigma(\nu_\mu + e^- \to \mu^- + \nu_e)}{\sigma(\nu_\mu + N \to \mu^- + X)} = [0.1245 \pm 0.0057(stat.) \pm 0.0026(syst.)] \times 10^{-2} \quad (2)$$

The status of world measurements are summarized in Table 1. Though the present IMD error(5 %) is twice as large as the direct flux measurement error (2.2 %), the former is not encumbered by systematic limitations. In a new neutrino experiment,[10] with an order of magnitude more statistics, the IMD measurements could be substantially improved. It should be noted that the systematic uncertainty, due largely to $\overline{\nu}_\mu$-to-ν_μ flux normalization, is primarily limited by $\overline{\nu}_\mu$ statistics.

A.2: Relative Flux Normalization

Tests of scaling violation and perturbative QCD do not depend upon the *absolute* normalization; but they do depend upon how accurately the relative flux from energy-to-energy and from antineutrino-to-neutrino is determined; i.e. upon the *relative* normalization. It is, therefore, imperative to establish means of relatively normalizing neutrino interactions which are least prone to systematic uncertainties, and only nominally depend upon assumptions involving nucleon substructure.

Determination of relative flux normalization involves forming appropriate ratios of neutrino-induced events, from a specific kinematic region, separated into different energy bins and neutrino helicities. These ratios yield *relative flux between various energy bins* and *relative flux between ν_μ and $\overline{\nu}_\mu$ samples*. The flux is, thus, determined *up to one overall normalization constant.* This constant must be obtained from other measurements. At present, the direct flux measurement (being the most accurate) is used; in future (given commensurate statistics) we might be able to use the inverse muon decay process. The CCFR collaboration has explored three distinct ways of determining the relative normalization,[11] which, subsequently, are used in the extraction of structure functions. These methods, along with an additional one used by the CHARM II collaboration,[12] are briefly enumerated below:

A.2.1: Fixed ν_0 Method: The dynamics of neutrino-nucleon scattering imply that the number of events in a given energy bin with $E_{had} = \nu < \nu_0$ is proportional to neutrino (antineutrino) flux in that energy bin up to corrections $\mathcal{O}(\nu_0/E_\nu)$ and $\mathcal{O}(\nu_0/E_\nu)^2$; i.e.

$$\begin{aligned}\mathcal{N}(\nu < \nu_0) &= C\Phi(E_\nu)\nu_0 \left[\mathcal{F}_2 - \frac{\nu_0}{2E_\nu}(\mathcal{F}_2 \mp \mathcal{F}_3) + \frac{\nu_0^2}{6E_\nu^2}(\mathcal{F}_2 \mp \mathcal{F}_3)\right] \\ &= C\Phi(E_\nu)\nu_0 \left[\mathcal{A} + (\frac{\nu_0}{E_\nu})\mathcal{B} + (\frac{\nu_0}{E_\nu})^2\mathcal{C} + \mathcal{O}(\frac{\nu_0}{E_\nu})^3\right]\end{aligned}$$

In the above, $\mathcal{N}(\nu < \nu_0)$ is the number of events in a given energy bin (E_ν) with hadronic energy less than ν_0, and $\mathcal{F}_i = \int_0^{\nu_0}\int_0^1 F_i dx d\nu$ (where F_i is F_2 or xF_3), and the constants $\mathcal{A}$, $\mathcal{B}$, and $\mathcal{C}$ are empirically determined for each energy bin from the ν_μ- and $\bar{\nu}_\mu$-data themselves. The estimated systematic precision of this method is $\approx 1.0\%$.

A.2.2: The y-intercept Technique: The fundamental premise of this method is that near zero hadronic energy transfer (y=$E_{had}/E_\nu \to 0$), the differential inelastic cross section divided by energy is a constant, independent of the incident neutrino energy or flavor.

$$\left[\frac{1}{E}\frac{d\sigma^\nu}{dy}\right]_{y=0} = \left[\frac{1}{E}\frac{d\sigma^{\bar\nu}}{dy}\right]_{y=0} = \text{Constant} \tag{3}$$

Due to non-scaling processes occurring near zero hadron energy transfers, the y-intercepts are obtained (in each energy bin) by fitting the data in the *entire* kinematic region *without* the lowest y-bin, and extrapolating the fit to y=0. Figures 3(a,b) show $\frac{1}{E}\frac{d\sigma^{\nu,\bar\nu}}{dy}$ as a function of energy as measured by the CCFR collaboration in their narrow band beam data.[11] The systematic precision of this technique is estimated at 1.5% level.

A.2.3: <u>Overlapping x-Q^2 Bin Method:</u> Comparison of the number of events in a region of x and Q^2 sampled by all energies and neutrino helicities provides a consistency check on the relative flux measurement.[11,13] The validity of this method follows from the near independence of the structure functions on energy and neutrino helicity. This is a well known method employed in muon scattering experiments as well.

A.2.4: <u>Direct Relative Flux Determination Method:</u> In a sign selected beam, by counting the number of μ^+ (for $\bar\nu_\mu$ beam) and μ^- (for ν_μ beam) a measure of $\bar\nu_\mu$-to-ν_μ flux can be obtained. The CHARM II collaboration[12] has used this technique, among others, and attained a precision of 4% to 5%. In spite of its limited precision, the method does offer an independent consistency check on relative flux.

A.3: Measurement of $r = \sigma(\bar\nu N)/\sigma(\nu N)$

The ratio (r) of antineutrino to neutrino CC cross section is an important quantity in neutrino physics: it is used in determination of $sin^2\theta_W$ in $R^\nu = NC/CC$, in opposite sign dimuon analysis, in establishing the probe independence of quark densities (W-vs-Z) etc. Using the relative normalization techniques, the CCFR collaboration has reported a new preliminary measurement of r as function of neutrino energy at this workshop(Fig. 4).[14] The new measurement is in agreement with previous ones, and the world status of r is shown in Table 2.

B: Quark-Parton Prediction

B.1: Sum Rules

The invariant structure functions which parametrize the deep inelastic scattering cross section are uniquely related by the Quark Parton Model (QPM) to the densities of quarks constituting the nucleon. Quark Parton Model sum rules are, thus, consistency conditions that relate appropriate integral of measured quark densities to the total number and charges of the constituent quarks. We review tests of two important sum rules in neutrino experiments: Gross-Llewellyn Smith and Adler sum rules.

B.1.1: Gross-Llewellyn Smith Sum Rule

The Gross-Llewellyn Smith (GLS) sum rule is the most accurately tested of sum rules. It predicts that the number of valence quarks in a nucleon, up to finite Q^2 corrections, is 3.[15]

$$S_{GLS} = \int_0^1 \frac{xF_3^{\nu N}}{x}dx = 3\left\{1 - \frac{\alpha_s(Q^2)}{\pi} + \frac{\mathcal{G}}{Q^2} + \mathcal{O}(Q^{-4})\right\} \quad (4)$$

The QPM relates the parity violating structure function, xF_3, to the valence quark density of the nucleon; and the sum rule follows. The sum rule is modified by twist-2 and higher twist operators. The second term in the equation corresponds to the known perturbative QCD correction, while the third term corresponds to an estimate of higher twist (twist-4) contribution to the sum rule.[16] The sum rule therefore predicts, using $\Lambda_{QCD} = 200$ MeV and including only the twist-2 perturbative QCD correction, at $Q^2 = 3$ GeV^2, $S_{GLS} = 2.66$.

The (1/x) weighting causes the integral to be dominated by low-x events; 90 % of the integral comes from the region $x \leq 0.1$. The recent precision measurement of the sum rule by the CCFR collaboration[17,14] is pictured in Fig. 5. The sum rule is measured at $Q^2 = 3$ GeV^2 which is the mean Q^2 of their lowest x-bin (x=0.005), and the value is:

$$S_{GLS} = 2.66 \pm 0.029(\text{stat.}) \pm 0.075(\text{syst.}) \qquad (5)$$

The reported systematic error is enitrely dominated by the 2.2 % uncertainty in the total cross section. In a future experiment, where the absolute normalization will be determined from IMD events, the precision of the GLS sum rule should improve.

The 3 % accuracy of the GLS sum rule at $Q^2 = 3$ GeV^2 raises two theoretical concerns:

(a) Higher twist contribution: An perturbative estimate of the twist-4 contribution to S_{GLS} at $Q^2 = 3$ GeV^2, by Iijima and Jaffe.[16] yields less than 1% effect. A similar contribution is estimated by Shuryak and Vainshtein.

(b) Target mass effect: The target mass effect is operative at large values of x. Since S_{GLS} is dominated by low-x region, the effect of target mass is expected to be small.

An estimate yields less than 0.2 % correction to the sum rule.[18]

The status of the world measurement of S_{GLS} is illustrated in Fig. 6. It should be compared with the 20% accuracy of the Adler sum rule and the 50 % accuracy of the Gottfried sum rule.[19]

B.1.1: Adler Sum Rule

The Adler sum rule predicts the integrated difference between neutrino-neutron and neutrino-proton structure functions. Unlike the GLS sum rule this sum rule is exact, and is expected to be valid to all orders of perturbation theory. It states:[20]

$$S_A = \int_0^1 \frac{(F_2^{\nu n} - F_2^{\nu p})}{x} dx = 1 \qquad (6)$$

The WA25 (BEBC) collaboration[21] has used neutrino data on hydrogen and deuterium targets, and their measurement, averaged over $1 < Q^2 < 40$ GeV^2 and assuming Callan-Gross relation, yields:
$S_A = 1.01 \pm 0.08(stat.) \pm 0.18(syst.)$, consistent with the prediction.

B.2: Universality of Nucleon Structure Functions: $Z^0/W^{\pm}/\gamma$

An fundamental prediction of QPM is the probe independence of nucleon structure function. This universality is a consequence of the interpretation of the structure functions as quark densities. In ν-N charge current scattering, the structure function $F_2(x, Q^2)$ $[xF_3(x, Q^2)]$ is the sum [difference] of quark and antiquark densities. The corresponding expression for the netural current structure functions has an additional multiplicative factor which is a known combination of the left and right handed coupling of Z^0 boson. For μ(e) scattering off an isoscalar target, the measured structure function $F_2(x,Q^2)$ is the *same* as the $F_2^\nu(x,Q^2)$ except the quark densities are weighted by the square of their charges. The QPM universality of quark densities in ν-NC, ν-CC, and electromagnetic interactions predict that $q_Z(x) = q_W(x) = q_\gamma(x)$ and $\bar{q}_Z(x) = \bar{q}_W(x) = \bar{q}_\gamma(x)$. We proceed to examine these prediction among the data from various experiments.

B.2.1: Z^0 -vs- $W^\pm$ Structure Functions

A integral test of the QPM universality is the ratio, $\int q_Z(x)dx / \int q_W(x)dx$ be unity. Taylor[22] has observed that since the measured $sin^2\theta_W$ is the same in ν-q and ν-e channels, we can conduct the integral test using this fact and other known measurements; the above ratio is found to be unity within ±5% (the corresponding error for antiquark ratio is ±20%). The precision on this integral test will improve with the new measurement on $sin^2\theta_W$ in ν-e channel by the CHARM II collaboration.

A more detailed test of the probe independence comprizes a detailed comparison of the differential quark/antiquark distributions as measured in CC and NC channels. The measurement of the NC structure functions, however, is inherently difficult due to the intractibility of measuring the final state particles. First, measurements have to be conducted in a narrow band beam, where the event vertex provides a measure of the neutrino energy, therefore, are statistically limited. Second, the angle of the final state hadron shower has to be determined. The CHARM[23] and the FMM[24] collaborations have measured NC structure functions. These data demonstrate good agreement, within ±20%, with the CC structure functions (for details see Ref[s] 19, 22).

B.2.2: $W^\pm$ -vs- γ: Mean Square Charge Test

The ratio of F_2 structure functions for muon to neutrino scattering intimates the mean square charge of the interacting quarks (up to a small x dependent correction due to strange/charm asymmetry in the nucleon); i.e.

$$\frac{F_2^\mu}{F_2^\nu} = \frac{5}{18}\left[1 - \frac{3(s+\bar{s})}{5(q+\bar{q})}\right] \qquad (7)$$

The status of this comparison is illustrated in Fig.7 (for details see Ref. 19). There are discrepancies in F_2 measurements, notably among the muon experiments, at 5%-10% level. The EMC data are consistently lower by about 10%. Taking an average of all the data in Fig. 7, the ratio is 0.96 ± 0.04; excluding the EMC from the comparison yields 0.98 ± 0.03. The CCFR collaboration has new F_2 data (not presented so far) that might be able to clarify some of the normalization and shape discrepancies.

C: Parton Distribution Functions or Quark Densities

Precision measurements of parton distribution functions (PDF) or quark densities are important: they convey nucleon substructures, and offer compelling tests of QCD, they determine the parton luminosities in colliders which is one of the limiting factors in exploring new physics, they are instrumental in predicting cross sections at HERA where new measurements will be available shortly at two orders of magnitude higher Q^2, and any discrepancy might signal new physics. In neutrino scattering the in-built flavor selection provides a powerful means of extracting PDF's. Nevertheless, neutrino experiments on light targets (H or D) suffer from statistical precision. In the following, we briefly state the ν measurements on hydrogen, and dwell primarily upon measurements on isoscalar targets.

C.1: Quark Densities from ν-H Scattering

Neutrino measurements of quark densities from a hydrogen target are in agreement between the two experiments, CDHS[25] and WA21 (BEBC),[26] at about 15% level. Figure 8 shows the ratio of quark and antiquark components as measured by the two groups. (It should be noted that the CDHS data have been adjusted in overall normalization to reflect their most recent cross section measurement.[1])

C.2: Valence Quark Densities in the Proton

The present status of separate valence quark components, $xu_v(x)$ and $xd_v(x)$, is summarized in Fig. 9. As noted in Ref. 19, while there is general agreement on $xu_v(x)$ between the muon experiment (EMC) and neutrino experiments (WA21, WA25, and CDHS), there is a distinct discrepancy in the shape of $xd_v(x)$. The precise reason for the discrepancy is not known. It is hoped that the recent muon experiment data by the BCDMS and NMC collaborations[27] on hydrogen and deuterium might resolve this experimental conflict.

C.3: Valence Quark Densities in an Isoscalar Target

The valence quark density for an isoscalar target (i.e. average of neutron and proton), which is the non-singlet structure function, $xF_3(x,Q^2)$, is much more accurately determined in high statistics neutrino experiments. The CCFR collaboration[14] presented new measurements on $xF_3(x,Q^2)$ shown in Fig. 10 as a function of Q^2 in various x-bins. These are compared with the high statistics CDHSW[28] measurements of $xF_3(x,Q^2)$. The comparison conveys an overall agreement between the two data sets, *except* at low values of x where there is a clear indication of shape discrepancy. This has important ramifications for the test of scaling violation in

$xF_3(x,Q^2)$ as discussed below.

C.4: Antiquark Densities in Isoscalar Target

The antiquark densities as measured in light targets by three different groups, WA21, WA25, and CDHS, are in agreement as shown in Fig. 11 (for details see Ref. 19). The new measurement of $x\bar{q}(x,Q^2)$ by the CCFR collaboration[14] is shown in Fig. 12. The new data unequivocally show that $x\bar{q}(x) \neq 0$ up to $x \leq 0.40$.

C.5: Strange Quark Content of an Isoscalar Nucleon Sea

Neutrino-induced opposite sign dimuons, $\mu^-\mu^+$, offer the only direct measure of the content and the shape of the strange content $[s(x), \bar{s}(x)]$ of the nucleon sea. In addition, these events permit determination of the electroweak parameters V_{cd} (the Kobayashi-Maskawa matrix element: this is the only direct determination of this parameter), and m_c (the mass parameter of the charm quark: this is precisely the parameter which at present limits the precision of $sin^2\theta_W$ determination in ν-N scattering). The CDHS[29] and CCFR[30,31] collaborations agree in their determination of the fractional strangeness content of the nucleon sea ($\kappa = 2s/(\bar{u}+\bar{d})$); the average of the two measurements is:

$$\kappa = 0.52 \pm 0.07 \qquad (8)$$

A noteworthy feature of the CCFR data (see Ref. 31) is that the measured s(x) $[\bar{s}(x)]$ is somewhat softer than the non-strange sea (obtained from the single muon CC events). This is illustrated in Fig. 13. Two new developments are underway: (a) the CCFR physicists have quadrupled their sample of $\mu^-\mu^+$ events by including data from two separate runs (FNAL E744 and E770), and by imposing a softer muon momentum cut on the second muon ($E_\mu > 4 GeV$); (b) W.K.Tung[32] has performed a next-to-leading order calculation of ν-induced $\mu^-\mu^+$ production. We hope that these would help answer the question: is the strange sea <u>different</u> from non-strange sea? We point out the exciting prospect of investigating this question, and the question of the evolution of strange sea in a second generation experiment where 150,000 opposite sign dimuons might be expected.

D: Scaling Violations and Tests of QCD

The quantitative predictions of QCD are few; and the "best" means of measuring the strong coupling constant is a subject of some controversy. Nevertheless, within the framework of deep-inelastic scattering there are elegant and unambiguous QCD predictions that could be verified experimentally. In deep inelastic scattering there is <u>no</u> fragmentation uncertainty since one deals with inclusive final state hadrons; the scale, which is the four-momentum transfer Q^2, is well defined; the higher order corrections are small and there are no large radiative corrections; the scaling violations are well described by the Altarelli-Parisi[33] equations; and the measurements yield more than one structure function which is measured at different values of x and Q^2, and, thus, within the Altarelli-Parisi evolution scheme one deals with a system of tests. For these reasons, Altarelli states:[34]

"*In principle, deep inelastic leptoproduction is the most solid and powerful method for testing perturbative QCD and measuring α_s.*"

The two most elegant predictions of perturbative QCD are: Slopes of structure functions with respect to Q^2 as a function of x; and the absolute magnitude and dependence of $R(x, Q^2) = \sigma_L/\sigma_T$ on x and Q^2. Below we examine the status of these tests.

D.1: Measurements of $R(x,Q^2)$ -vs- QCD

The R parameter of deep inelastic scattering is defined as the ratio of the absorption cross section of the longitudinally to transversely polarized virtual boson, $R(x,Q^2) = \sigma_L/\sigma_T$, and is related to the structure functions F_2, and F_1 as:

$$R = \frac{\sigma_L}{\sigma_R} = \frac{F_2 - 2xF_1}{2xF_1} \equiv \frac{F_L}{2xF_1}. \tag{9}$$

where F_L is the longitudinal structure function, and the other symbols have their usual meaning.[19] Perturbative QCD predicts the magnitude of R and its dependence on x and Q^2 (due to gluon radiation and quark pair production) to be:[35]

$$R(x,Q^2) = \frac{\alpha_S(Q^2)}{2\pi} \frac{x^2}{2xF_1(x,Q^2)} \int_x^1 \frac{dz}{z^3} \left[\frac{8}{3} F_2(z,Q^2) + 4f(1-\frac{x}{z})zG(z,Q^2) \right], \tag{10}$$

where f is the number of flavors if the incident lepton were a neutrino, and the sum of the squares of quark charges if the incident lepton were a muon or an electron; $G(z,Q^2)$ is the gluon structure function. Numerous experiments have measured $R(x,Q^2)$ and claimed consistency with the theoretical prediction.

Using recent measurements at SLAC[36] and a simple model for higher twist effects, authors in Ref.37 argue that the present cumulative deep inelastic scattering data are <u>consistent</u> with but <u>do not demonstrate</u> R=R_{QCD}. Figure 14 illustrates that a higher-twist parametrization of $R(x,Q^2)$, based upon the SLAC data, can indeed explain measurements at higher values of Q^2.

Precise measurements of $R(x,Q^2)$ at sufficiently high Q^2 (e.g. $Q^2 > 20 GeV^2$) in next generation deep inelastic experiments [10,38] will provide a compelling test of the perturbative QCD.

D.2: Evolution of Non-singlet Structure Function

The Q^2 evolution of structure functions is described by the Altarelli-Parisi equations. It is the simplest for a non-singlet structure function [e.g. $xF_3(x,Q^2)$] since it is independent of the gluon structure function or $R(x,Q^2)$. The AP equation for xF_3 predicts:

$$\frac{dlnxF_3(x,Q^2)}{dlnQ^2} = \alpha_s(Q^2)\psi(x,Q^2) \qquad (11)$$

The term $\psi(x,Q^2)$ involves an integral of $xF_3(z,Q^2)$ for $z > x$; the integral is evaluated using a known splitting function $\mathcal{P}_{qq}$ (which has been calculated to next-to-leading order). Thus, the only unknown on the right hand side of the above equation is the strong coupling constant: the logarithmic slope of xF_3 is proportional to α_s at each x. In Fig. 15, a set of QCD curves are shown; a specific value of Λ pertains to a specific curve. Since the non-singlet evolution does not depend upon gluons, antiquarks, or $R(x,Q^2)$, the theoretical prediction for the slopes at small values of x is unambiguous. Furthermore, the theoretical prediction is relatively independent of the value of Λ for $x < 0.20$ (Fig. 15 illustrates a "fixed-point" for the QCD curves). Prior to this workshop, the experimental situation, however, was unclear or contradictory. Figure 15 also shows the slopes extracted from the CDHSW data[28] with $Q^2 > 6 GeV^2$. Data show a different evolution at low values of x when compared with the theory: in particular none of the curves agrees with the data. The CCFR(NBB)[39] data agree with the prediction, however, the statistical accuracy is not as high as the CDHSW's. Furthermore, the kinematic reach of the two data sets are such that imposing a harder Q^2 cut (e.g. $> 20 GeV^2$) makes the tests inconclusive.

The new CCFR (QTB) data,[14] along with an illustrative QCD curve, are shown in Fig. 16. The xF_3 evolution agrees well with the theoretical prediction. And the measured Q^2 region of the CCFR data is such that imposing a higher Q^2 cut (see Fig. 17) does not vitiate the statistical precision at higher values of x (which *are* sensitive to the value of Λ). The CCFR collaboration has not yet reported a value of Λ; they are currently finalizing their calibration. The data, however, hold promise for a good measurement of Λ.

D.3: Evolution of Singlet Structure Function

A review of singlet structure functions is discussed by M.Virchaux in these proceedings. We note (for details see Ref. 19) that there are some experimental conflicts in F_2-evolution: the BCDMS data show lovely agreement with the theory; nevertheless, the EMC and the CDHSW data on F_2-slopes are steeper than the prediction. We hope that the new high statistics F_2-data (at high Q^2) by the CCFR collaboration should be able to clarify the conflict on structure function evolution.

Finally, we point out once the QCD evolution is unequivocally verified in the non-singlet evolution, the singlet evolution permits the extraction of gluon structure function. In neutrino experiments, *the simultaneous evolution of F_2 and xF_3 permits a very powerful constraint on the gluon degrees of freedom.* This is illustrated in Fig. 18; the extracted gluon exponent, η_g [$g(x) \propto (1-x)^{-\eta_g}$], is more precise when F_2 and xF_3 are evolved simultaneously than when F_2 is evolved by itself. The reason is that in a simultaneous evolution $\bar{q}(x) = (F_2 - xF_3)/2$ is constrained, which in turn permits a more accurate determination of the gluon. The above illustration is with CCFR(NBB) data which is statistically limited. It should be interesting conducting this study with the CCFRQTB data.

E: Determination of the Electroweak Parameter: $sin^2\theta_W$

Improved measurements of structure functions, and a better understanding of the hadron substructure, induces one to reexamine the precision measurement of the electroweak parameter, $sin^2\theta_W$, in ν-N scattering. A deeper understanding of the electroweak sector crucially depends upon the precision of $sin^2\theta_W$.[40] The power of a precision determination of $sin^2\theta_W$ in $R^\nu = \sigma(NC)/\sigma(CC)$ channel is most clearly seen by examining the electroweak radiative correction. We can write the W and the Z masses as:[41]

$$M_W = \frac{37.281 GeV/c^2}{(1-\Delta r)^{1/2} sin^2\theta_W} \qquad (12)$$

where the relation, $M_Z = \frac{M_W}{cos\theta_W}$ defines $sin^2\theta_W$ in the Sirlin scheme, and the quantity Δr describes the radiative correction. Intimations of new physics will be signalled by Δr: the most important dependence is on the top quark mass (m_t), and it is here that the value of deep-inelastic neutrino measurement is most clearly seen. Precision measurements of W and Z masses can be used to extract Δr and $sin^2\theta_W$; and then compared to the value of $sin^2\theta_W$ from R^ν channel. If the radiative corrections, which depend upon m_t, are known the values should be consistent. Deep inelastic neutrino scattering is a unique benchmark: the functional form of the radiative correction makes the value of $sin^2\theta_W$ from R^ν only weakly dependent on m_t, whereas determinations of the electroweak parameter from boson masses, ν_μ-e scattering, or atomic parity violation, have a sizeable and similar dependence. If m_t is known, a precise measurement of $sin^2\theta_W$ would provide a stringent test of loop correction over a two order of magnitude range of Q^2; if m_t is unknown, it would place strong limits on m_t. Hence in either case the complementary measurement of $sin^2\theta_W$ in R^ν strengthens and extends the scope of the collider measurements.

The argument above sets the necessary precision for a new measurement. If the error on $sin^2\theta_W$ from R^ν is much larger than the collider measurements, the additional information has little power. The current deep-inelastic $sin^2\theta_W$ error of ± 0.007 is inadequate. A error on $sin^2\theta_W$ of about 1% or ± 0.002 would provide an error on Δr equal to or *better* than the colllider measurement.

Determination of $sin^2\theta_W$ in ν-N scattering with the stated precision is an important goal of a second generation neutrino experiment. The key to achieving this is to determine $sin^2\theta_W$ among statistically exclusive kinematic regions of the ν-N event sample such that systematically, too, these measurements are as independent as possible. This is achievable in a high statistics experiment with suitable instrumentation. We sketch below some of the salient means of determining $sin^2\theta_W$ from a large ν-N event sample:

E.1: Determination R^ν from events with $E_{HAD} > 150 GeV$

The dominant theoretical limiting factor to $sin^2\theta_W$ determination, the uncertainty in the mass-parameter of the charm quark (m_c), could be eliminated by imposing a hard cut on the hadronic energy of the events (e.g. $E_{HAD} > 150 GeV$). At such high hadronic energy transfers R^ν is virtually insensitive to m_c. Other concerns, such as charge current background, ν_e contamination, smearing etc., have to be understood.

E.2: Determination R^ν from events with Intermediate E_{HAD}

Extraction of R^ν from intermediate region in E_{HAD} (e.g. $50 < E_{HAD} < 150 GeV$) would be sensitive to m_c. However, using the 150,000 $\mu^-\mu^+$ events (available in a new experiment), we should substantially reduce the current error on m_c.

E.3: Determination ρ^2 from $R^{\bar\nu}$

Extensions to the Standard Model involve a second parameter ρ^2, which should be determined *independently* from $sin^2\theta_W$. The parameter ρ^2 is best determined from $R^{\bar{\nu}} = \sigma(\bar{\nu}:NC)/\sigma(\bar{\nu}:CC)$, since $R^{\bar{\nu}}$ has a strong dependence on ρ^2, but is virtually independent of $sin^2\theta_W$. This could be accomplished in a sign-selected antineutrino beam.

E.4: Determination $sin^2\theta_W$ from Paschos-Wolfenstein Relation

A bonus of running in a sign-selected beam is the use of Paschos-Wolfenstein (PW) relation for measuring $sin^2\theta_W$ and ρ^2. The PW relation deals with the ratio of difference of ν_μ- and $\bar{\nu}_\mu$-induced NC and CC events. Thus, many backgrounds and systematic uncertainties diminish or cancel. The dominant concern is the $\bar{\nu}_\mu$-to-ν_μ normalization, which should be reliably measured in a high statistics experiment (see sec.A)

E.5: Determination $sin^2\theta_W$ from $\sigma(\nu:NC)$

The use of IMD events to obtain the absolute flux also permits determination of the total neutrino neutral current cross-section. So far, NC events have only been normalized to CC events; indeed dominant systematic concerns in $sin^2\theta_W$ determination all arise from the CC denominator. The precision of $sin^2\theta_W$ from the total NC cross section would depend upon the accuracy with which the quark/antiquark densities are measured. The latter *is* one of the prime goals of a new experiment.

The need to measure the electroweak parameters in ν-N experiments with high precision can hardly be overemphasized. These measurements will be a fitting and competitive counterpart to those at LEP, SLC and the Tevatron Collider, and could provide a signal of physics beyond the Standard Model.

F: Rare and New Physics

The statistical power of the new experiment offers a new regime for the study of rare processes (with rates $\leq 10^{-5}$ of the regular CC) and for searching new phenomena. Some of these are:[42] i) Inverse Muon Decay, ii) Measurement of V_{cd}, iii) Recoilless Dimuons, iv) Neutral Heavy Leptons, v) Neutrino Oscillations, vi) Search for Right Handed Currents, vii) Like Sign Dimuons, and viii) Trimuons.

Commencing with the discovery of scaling and scaling violation, to the discovery of neutral current, to the discovery of "open" charm, to the precision measurement of perturbative QCD and of $sin^2\theta_W$, deep-inelastic experiments have been a fertile ground for extending the horizons of our understanding. There are a few existing experimental conflicts, a few theoretical conundrums; perhaps some of the data in hand could shed light on these. Nevertheless, for a *substantial* improvement over our present understanding, a next generation of deep inelastic experiments are extremely important both in the breadth of topics pertaining to the Standard Model and regarding precision measurements of the parameters.

It is a pleasure to thank my gracious hosts at Fermilab who made the "Workshop on Hadron Structure Functions and Parton Distributions" a memorable experience. I extend my heartfelt thanks to Prof. W.K.Tung and Dr. J.Morfin for numerous discussions, and for their infectious enthusiasm for "PDF's". Innumerable contributions by my CCFR colleagues is gratefully acknowledged. Mr. W.C.Leung and Mr. P.Z.Quintas helped me a lot in preparing this talk. I thank Drs. M.Virchaux, A.Milsztajn, and A.Staude for discussions, and Mr. Eric Hyatt for critically reading the manuscript. This research was supported by funds from the National Science Foundation.

References

[1] F.Sciulli, "Review of Lepton Hadron Interactions", Proc. of 1985 International Symposium on Lepton and Photon Interactions, Kyoto, Aug.1985.

[2] H.Abramowicz et al., Z. Phys., **C17**, 282(1983).

[3] R.Blair et al., Phys. Rev. Lett., **51**, 343(1983).

[4] P.Berge et al., Z. Phys., **C35**, 443(1987).

[5] P.Auchincloss, S.R.Mishra and F.Sciulli, "Neutrino Tagging", Nevis Rep., 1985. More recently the technique of neutrino tagging has been considered by R.H.Bernstein (Fermilab) in connection with neutrino oscillation and $sin^2\theta_W$ measurements.

[6] S.R.Mishra et al., Phys. Rev. Lett., **63**, 132(1989).

[7] J.Dorenbosch et al., Z. Phys., **C41**, 567(1989).

[8] N.Armenise et al., Phys. Lett., **84**, 137(1979).

[9] S.R.Mishra et al., "Inverse Muon Decay at the Fermilab Tevatron", Nevis Preprint # 1424, Apr(1990); Submitted to Phys.Lett.B.

[10] S.R.Mishra, "A 2nd Generation Neutrino Deep Inelastic Scattering Experiment at the FNAL Tevatron"; presented at the workshop, "Fermilab in the 1990's", at Breckenridge, Colorado. A summary is to appear in the conference proceedings, ed. R.Brock and H.Montgomery et al.

[11] P.S.Auchincloss et al., "Measurement of the Inclusive Charge Current Cross Section for Neutrino and Antineutrino Scattering on Isoscalar Nucleons", Nevis Preprint # 1394, Mar(1990); to appear in Z.Phys.C. For details of the Fixed ν_0 Technique, see P.Z.Quintas et al., in preparation.

[12] D.Geiregat et al., Phys. Lett., **B232**, 539(1989).

[13] P.S.Auchincloss et al., Proc. 12th Int. Conf. on Neutrino Physics and Astrophysics, Sendai, ed. T.Kitagaki and H.Yuta, pp.351(1986). Also see, D.MacFarlane et al., Z. Phys., **C26**, 1(1984).

[14] P.Z. Quintas et al., see these proceedings.

[15] D.J.Gross and C.Llewellyn-Smith, Nucl. Phys., **B14**, 337(1969);
M.A.B.Beg, Phys. Rev., **D11**, 1165(1975).

[16] B.A.Iijima, M.I.T. Preprint CTP993 (1983); R.Jaffe private communication.

[17] W.C.Leung et al., "A Precision Measurement of the GLS Sum Rule in ν-N Scattering at the Fermilab Tevatron", Nevis Preprint #1423; presented at XXV Rencontres de Moriond, Les Arc, Mar.1990; to be submitted to Phys. Rev. Lett.

[18] The S_{GLS}, with the target mass correction, takes the form,

$$S_{GLS} = \int_0^1 \frac{xF_3}{x} \frac{1 + 2(1 + 4M^2x^2/Q^2)^{.5}}{[1 + (1 + 4M^2x^2/Q^2)^{.5}]^2} dx \qquad (13)$$

[19] S.R. Mishra and F. Sciulli, "Deep Inelastic Lepton Nucleon Scattering", Ann. Rev. of Nucl. and Part. Sci., **39**, 259(1989).

[20] S.L.Adler, Phys. Rev., **143**, 1144(1966).

[21] D.Allasia et al., Phys. Lett., **B135**, 231(1984); Z. Phys., **C28**, 321(1985).

[22] F.Taylor, "Review of Nucleon Structure Functions", Proc. of Neutrino'88, Boston, ed. J.Schenps et al., pp.184, (1988).

[23] J.V.Allaby et al., Phys. Lett., **B213**, 554(1988).

[24] T.S.Mattison et al., "Nucleon Neutral Current Structure Functions", Submitted to Phys.Rev.D.

[25] H.Abramowicz et al., Z. Phys., **C25**, 29(1984).

[26] G.T.Jones et al., WA21(BEBC) Preprint 87, Rec. Jul(1987).

[27] M.Virchaux, see these proceedings.

[28] B.Vallage, Ph.D. Thesis, Saclay CEA-IV-2513, Jan., 1987.

[29] H.Abramowicz et al., Z. Phys., **C17**, 19(1982).

[30] K.Lang et al., Z. Phys., **C33**, 483(1987).

[31] C.Foudas et al., Phys. Rev. Lett., **64**, 1207(1990).

[32] W.K.Tung, private communication.

[33] G.Altarelli and G.Parisi, Nucl. Phys., **B26**, 298(1977).

[34] G.Altarelli, "Tests of QCD", Ann. Rev. Nucl. Part. Sci., **39**, 357(1989).

[35] G.Altarelli and G.Martinelli, Phys.Lett., **B76**, 89(1978); M.Gluck and E.Reya, Nucl. Phys., 145, 24(1978).

[36] S. Dasu et al., Phys. Rev. Lett., **60**, 2591(1988); Phys. Rev. Lett., **61**, 1061(1988). Also see, L.W.Whitlow, SLAC-Rep-357, March(1990).

[37] S.R.Mishra and F.Sciulli, "Do Present Deep Inelastic Scattering Data Demonstrate $R = R_{QCD}$?", Nevis Preprint # 1422; To appear in Phys. Lett. B, 1990.

[38] C.Guyot et al., "A New Fixed Target Experiment for a Precise Test of QCD", Saclay Print-88-0741, Sep.,1988.

[39] E.Oltman et al., "Nucleon Structure Functions from High Energy Neutrino

Interactions", Nevis Preprint #1425; Submitted to Z.Phys.C.

[40] U. Amaldi et al., Phys.Rev., **D36**, 1385(1987); D.Haidt, DESY 89-073, June(1989) and References therein. Also see J.Ellis and G.L.Fogli, CERN-TH.5511/89, Aug(1989). For experimental results, see H.Abramowicz et al., Phys.Rev.Lett., **57**, 298(1986) and A.Blondel, et al., Z.Phys.C, **45**, 361 (1990) from CDHSW , and from CCFR: P.Reutens et al., accepted for publication in Z.Phys.C. A discussion of the implications of measurements of R_ν on the determination of $\sin^2\theta_W$ in e^+e^- experiments is found in A. Blondel, CERN-EP/89-84, July(1989). A analysis of the current errors on $\sin^2\theta_W$, concentrating on slow-rescaling, can be found in R.Brock, "Deep-Inelastic Neutrino Measurements of $\sin^2\theta_W$", Proceedings of New Directions in Neutrino Physics, p. 57, ed. R.Bernstein, Fermilab. A number of papers on radiative corrections can be found in *Radiative Corrections in $SU_{2L} \otimes U_1$*, edited by B.W.Lynn and J.F.Wheater, World Scientific Press, Singapore, 1984. The Paschos-Wolfenstein relation is discussed in L. Wolfenstein and D. Wyler, Phys.Rev., **D15**, 670(1977).

[41] W.J.Marciano and A.Sirlin, Phys.Rev. **D11**, 2695 (1980).

[42] S.R.Mishra, "Rare Processes in ν-N Scattering", Proceedings of Neutrino'88, pp.259, Boston(1988), ed's J.Schneps et al., World Scientific, Singapore.

Table 1: **Status of IMD Measurement** The attained precision on IMD measurement by various collaborations is enumerated.

Collaboration	Precision on $\sigma(IMD)$
Gargamelle ('79)	22.0 %
CHARM ('89)	11.7 %
CCFR ('89)	7.8 %
CCFR ('90)	5.5 %

Table 2: **Measurement of** $r = \sigma(\bar{\nu}:CC)/\sigma(\nu:CC)$: The attained precision on r measurement by various collaborations is enumerated.

Collaboration	r
CCFRR ('83)	0.508 ± 0.035
CDHSW ('87)	0.494 ± 0.009
CCFR ('90)	0.511 ± 0.005

Figure Captions

Figure 1: The status of $\sigma(\nu_\mu : CC)/E_\nu$ -vs- E_ν, and $\sigma(\bar{\nu}_\mu : CC)/E_{\bar{\nu}}$ -vs- $E_{\bar{\nu}}$ by neutrino collaborations.

Figure 2: Distribution of $Q^2 = E_\mu^2 \theta_\mu^2$ for events with $E_{HAD} \leq 1.5$ GeV. The ν_μ events are shown with solid circles. The $\bar{\nu}_\mu$ events, scaled up by the relative ν_μ to $\bar{\nu}_\mu$ flux, are shown by solid line.

Figure 3: The CCFR(NBB) data showing (a) y-intercepts as a function of energy for ν_μ-induced events; and (b) for $\bar{\nu}_\mu$-induced events. The dotted line is a Monte Carlo prediction.

Figure 4: The new CCFR(QTB) data showing the measured ratio, $r = \sigma(\bar{\nu} : CC)/\sigma(\nu : CC)$ as a function of energy.

Figure 5: The GLS sum rule test by the new CCFR(QTB) data; $xF_3(x)$ are shown in "squares", $\int_0^x xF_3(x)dx$ are shown "diamonds".

Figure 6: The world status of GLS sum rule measurement.

Figure 7: Ratios of F_2 measurements among two sets of neutrino experiments [CDHSW and CCFR)NBB)], and three sets of muon experiments [BCDMS, BFP, and EMC]. (For details see Ref. 19.)

Figure 8: Ratios of the WA21 and CDHS data of quark (crosses) and antiquarks (solid circles).

Figure 9: Valence quark densities in a proton: (a) $xu_{VALENCE}$, (b) $xd_{\text{\textbackslash} ALENCE}$ as a function of x. The solid curve is the Tung et al parametrization.

Figure 10: The non-singlet structure function, $xF_3(x, Q^2)$, as a function of Q^2 in various x-bins. The data are from CCFR(QTB) (circles), and CDHSW (squares) groups.

Figure 11: The antiquark component of the protons. The curve is the Tung et al parametrization.

Figure 12: The antiquark, $\bar{q}(x,Q^2)$, as a function of Q^2 in various x-bins. The data are from the CCFR(QTB) collaboration.

Figure 13: Comparison of strange sea and non-strange sea: (a) $s(x)$, and (b) $\bar{s}(x)$. The non-strange sea is from the CCFR(NBB) data extracted at the mean Q^2 of the dimuon data in each x-bin.

Figure 14: Measurements of R at high Q^2: (a) Hydrogen target with BCDMS, and EMC data, and (b) nuclear target with BCDMS, EMC, and CDHSW data. The solid curve represents a "higher-twist" parametrization by the authors of Ref.19 (extracted using the SLAC data) and calculated using the mean Q^2 of the BCDMS data in (a); and the CDHSW data in (b). The dotted curve show a corresponding prediction of QCD with the inclusion of higher twist effect following the SLAC data.

Figure 15: The predicted logarithmic derivative of xF_3 with respect to Q^2 as a function of x are shown as curves for each specific Λ. The measured slopes are due to the CDHSW and the CCFR(NBB) collaborations.

Figure 16: The new CCFR(QTB) measurement of the logarithmic slopes of xF_3 with respect to Q^2 as a function of x. The curve is a sample QCD fit through the data.

Figure 17: The CCFR(QTB) measurement of xF_3 slopes with $Q^2 > 20 GeV^2$.

Figure 18: The extracted gluon exponent η_g with (a) F_2 evolution, and (b) with simultaneous F_2 and xF_3 evolution.

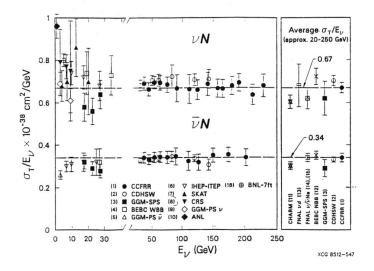

Fig. 1

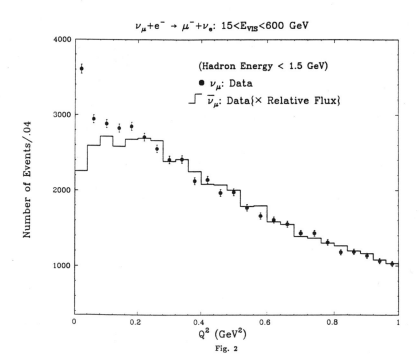

Fig. 2

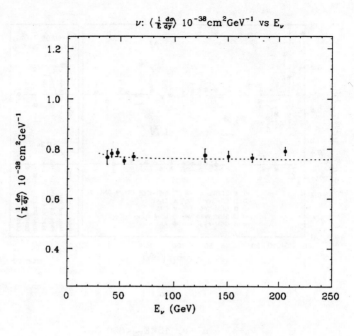

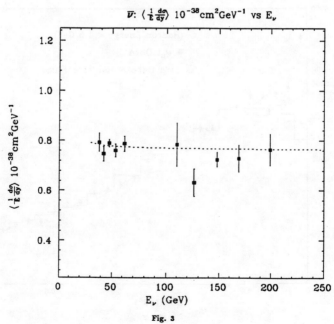

Fig. 3

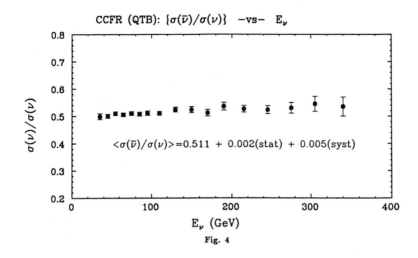

Fig. 4

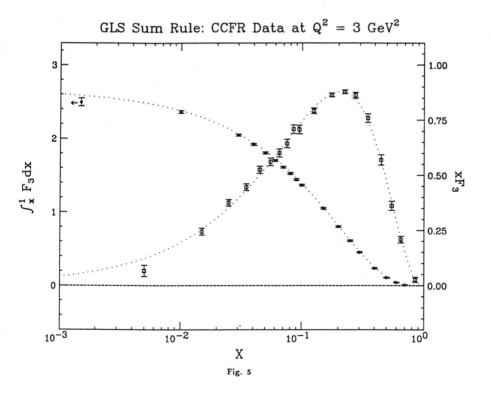

Fig. 5

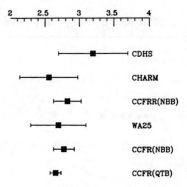

Status of GLS Measurement

Fig. 6

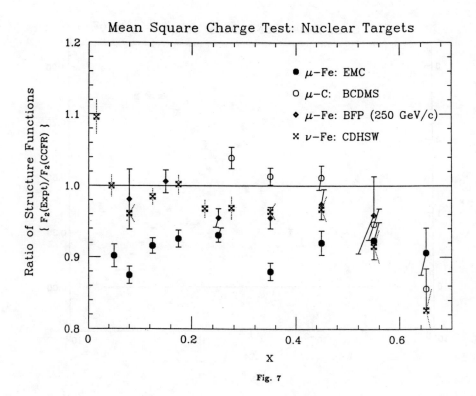

Fig. 7

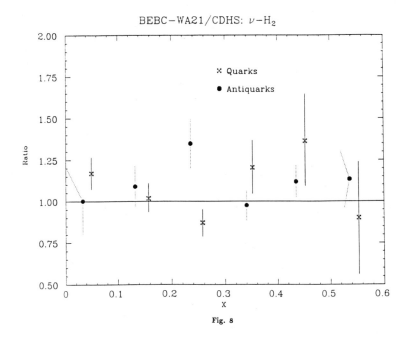

Fig. 8

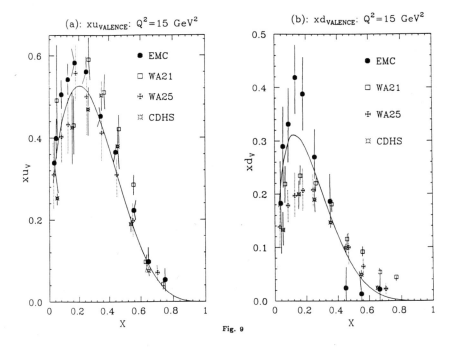

Fig. 9

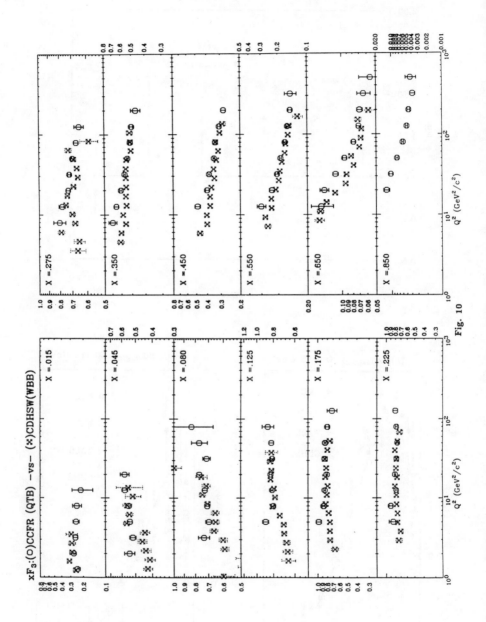

Fig. 10

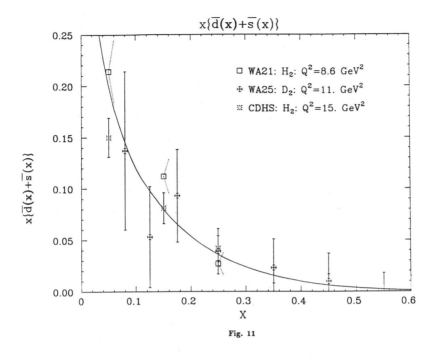

Fig. 11

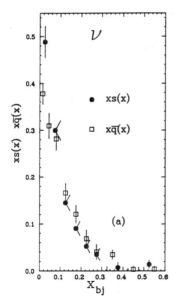
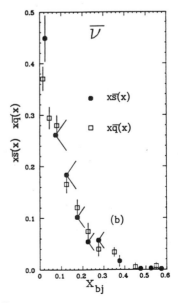

Fig. 13

120

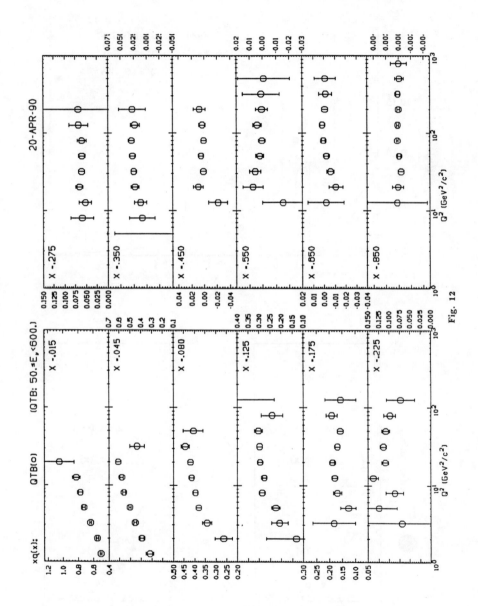

Fig. 12

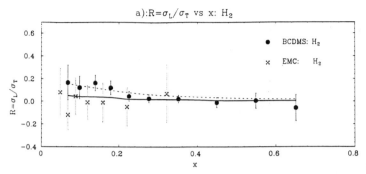

Fig. 14

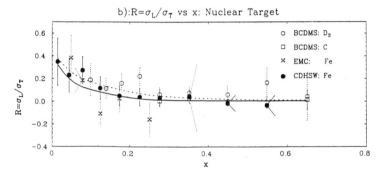

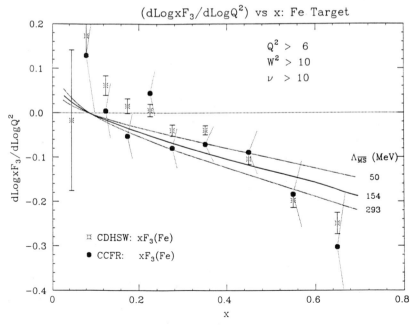

Fig. 15

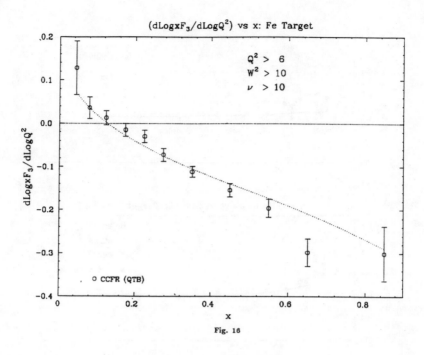

Fig. 16

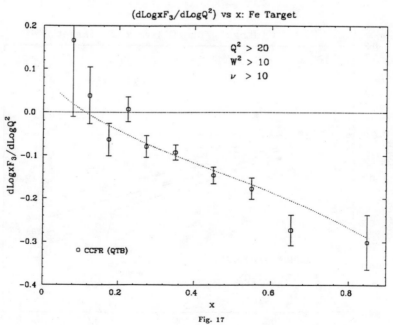

Fig. 17

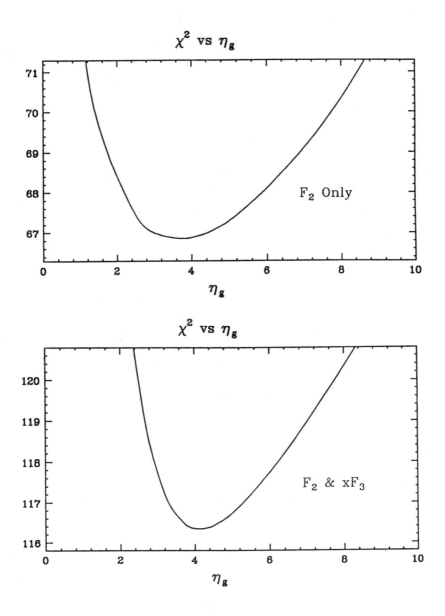

Fig. 18

Present status and future of the measurements of the proton and deuteron structure function F_2 in charged lepton deep inelastic scattering

M. VIRCHAUX
D.Ph.P.E., C.E.N. Saclay
91191 Gif-sur-Yvette Cedex, FRANCE

In this short presentation, I will concentrate on two specific topics of charged lepton deep inelastic scattering that seem to me to be of importance at the present stage. More general reviews can be found in [1].

The first topic is the relative clarification of the present experimental situation that can be derived from the recent reanalysis of many old data on deep inelastic electron scattering taken at SLAC. The two available high statistics high Q^2 measurements of the structure function F_2 on hydrogen and deuterium targets come from the CERN experiments run by the EMC and BCDMS collaborations. It is by now well known that these two results are in poor agreement. The SLAC data cover a region of lower Q^2 but their recent careful reanalysis has increased the kinematical range and improved the accuracy of the F_2 measurements that they allow. So it is certainly interesting to study how they compare to the CERN results.

The second topic concerns the future of charged lepton deep inelastic scattering. The main experimental effort which is being pursued presently in this field is the realisation of the big electron-proton collider HERA at Hamburg. It will provide unique opportunities to probe the proton structure at very high values of Q^2 (up to 40000 GeV^2) but will not be the ultimate "structure functions measuring machine". There is room for a new generation of fixed target experiments, complementary to HERA, that can provide data of unmatched precision in a wide kinematical range : a unique opportunity for detailed QCD studies.

1. Present status of the proton and deuteron structure functions measurements

Three different high statistics final results are now available. Two of these results come from the deep inelastic scattering experiments which shared the CERN muon beam, EMC (final publications in 1985 for hydrogen [2] and 1987 for deuterium [3]) and BCDMS (publications in 1989 for hydrogen [4] and 1990 for deuterium [5]). The third result comes from a recent global reanalysis of earlier data on electron scattering taken at SLAC between 1970 and 1985 [6]. As the SLAC electron beam was of much lower energy (4.5 – 21 GeV) than the CERN muon beam (100 – 280 GeV), the kinematic range of the SLAC measurement is at lower Q^2 and is almost disjoint from the ones of the two CERN experiments. The latter are, of course, directly comparable.

In the one photon exchange approximation, the deep inelastic cross section for charged leptons depends linearly on the structure function $F_2(x,Q^2)$ and, although non-linearly and very weakly, on the structure function $R(x,Q^2)$ which is equal to the ratio of absorption cross section for virtual photons of longitudinal and transverse polarization σ_L/σ_T. Experimentally, F_2 and R can be disentangled by measuring deep inelastic cross sections at the same (x,Q^2) point but at different values of the incident beam energy. In practice, as these cross sections depend only weakly on R, all measurements of R suffer from large experimental errors. Within these errors, there is no disagreement between the experimental results on R. The structure function $F_2(x,Q^2)$ is much more precisely measured than R and we will restrict our comparison to this quantity.

1.1 Presentation of the published data

EMC

The EMC results on F_2 were evaluated assuming $R \equiv 0$ as predicted by the naive parton model. For the measurement with a hydrogen target, F_2 is given separately for four different beam energies (120, 200, 240 and 280 GeV). The deuterium data were obtained at 280 GeV beam energy only. For each of these data sets, the different systematic errors are given at each (x,Q^2) point but only in absolute values. There is thus no complete information about the correlations of these errors. For hydrogen, a final F_2 obtained by merging the four data sets corresponding to the different beam energies is also given. The EMC merging procedure is rather complex in order to account for the possibility that F_2 values from different data sets differ at some (x,Q^2) points beyond the estimated errors. It allows to increase the final errors by modifying the weights of the points from different data sets according to the quality of their local consistency. The increase of errors in the merging procedure, and the unsufficient information about correlations for the individual data sets, explain why there is no information on the point-to-point correlations of systematic errors in the final F_2. The overall normalisation uncertainty of all EMC data is estimated at 5%. In the comparison presented here, we omit the EMC data below $x = 0.06$ (i.e. the first two x bins) since the SLAC and BCDMS data do not extend to such low values of x.

BCDMS

The BCDMS results have been evaluated using a perturbative QCD prediction R_{QCD} computed from the measured x dependence of F_2 and from an assumed gluon distribution [5]. Hydrogen data were obtained for 100, 120, 200 and 280 GeV beam energy and deuterium data for 120, 200 and 280 GeV. For both targets, the data from different beam energies are in statistical agreement and can be merged in a straightforward way, using weights computed from statistical errors only. For all beam energies, the effect of each source of sizeable systematic error on F_2 is given in form of a multiplicative factor for each data point. Systematic uncertainties have been estimated such that they correspond to a one standard deviation error. This presentation allows to account for all important correlation effects.

Qualitatively, the three principal sources of systematic errors (beam energy calibration, spectrometer calibration and spectrometer resolution) distort the measurements in a very similar way, the largest effect being at high x and low Q^2. This allows to simplify the treatment of the systematic errors by combining the point-to-point effects of these three main error sources quadratically into a one standard deviation, 100%-correlated error which we will refer to as the "main systematic error" of the BCDMS data. The correlation effect is illustrated in Figure 3 where the F_2 measurements at high x on hydrogen and deuterium are shown with total errors, *i.e.* statistical and systematic errors combined in quadrature, together with the same data points distorted by the main systematic error. This figure, which also displays the high x SLAC data, shows that the BCDMS point to point total errors are at large x, *i.e.* in the region where they are largest, totally dominated by 100%-correlated sytematic errors. It also illustrates how a one standard deviation effect can distort the data in this kinematic region in a very significant way. The lower x data are not represented in this figure because for $x < 0.40$ the BCDMS main systematic error is negligible. The overall normalisation uncertainty of all BCDMS data is estimated at 3%.

SLAC

In the SLAC reanalysis, data were merged from different exposures, recorded with different spectrometers and at various beam energies. The phenomenological parametrisation of R which was used to compute $F_2(x,Q^2)$ was obtained from a fit to the same data. Results are available for $0.0625 \leq x \leq 0.85$, at the central x values of both the EMC and the BCDMS bins. As for the EMC data, systematic errors are given as absolute values only and no information is available about their correlation. The systematic errors are, however, much smaller than for the muon data. The overall normalisation uncertainty is estimated at 2%.

1.2 Corrections for a consistent treatment of R

The three experiments discussed here all made different assumptions on R in their analysis. For a direct comparison of the different data sets, R has to be applied in a coherent fashion. For the high Q^2 muon data, we chose the QCD prediction for R which is used in the BCDMS analysis. This prediction is favoured by the BCDMS results over $R \equiv 0$ and is also compatible with the EMC results. For the SLAC data which are at low Q^2, the validity of the perturbative QCD prediction is questionable and we therefore retain their parametrization of the measured R. In the Q^2 region from about 8 to 25 GeV^2 which connects the SLAC data to the EMC and BCDMS data, this parametrisation turns out to be very close to the QCD prediction used by BCDMS, at least for $x < 0.50$. In this region, the F_2 measurements by SLAC and BCDMS are thus directly comparable. At larger x, the data are directly comparable in any case since the BCDMS and EMC data join the SLAC data with small values of y and are thus insensitive to R.

In order to treat all data in a consistent way, we therefore only need to correct the EMC F_2 data such that they correspond to the same values of $R = R_{QCD}$ as assumed in the BCDMS analysis. It is straightforward to compute this correction for data taken at a fixed beam energy like the EMC deuterium data. For the hydrogen results, corresponding to a merging of measurements taken at four different beam energies, one has to take into account the specificity of the EMC merging procedure (see *e.g.* [7]). The effect of this R_{QCD} correction to the EMC data is small everywhere and nonnegligible only for $x < 0.20$ at high Q^2.

1.3 The rebinning method

The x binning of the EMC results differs from the one of the BCDMS results at $x < 0.30$. At larger x, the binnings coincide. The SLAC results have been computed in both binnings. In order to provide a direct visual comparison of the results, we have corrected the data to make them correspond to the same binning. The data of one experiment (*i.e.* central values, statistical and systematic errors) were interpolated to the x bins of the other experiment, separately for each value of Q^2, with a third order polynomial. Because of the slow x dependence of the data in the rebinning region, any reasonable interpolation procedure gives almost identical results when compared to the size of the errors. We have rebinned both the BCDMS data in the EMC bins and *vice versa*. None of the conclusions made in this paper depends on the choice of binning.

1.4 Significance of the EMC/BCDMS discrepancy

After the corrections discussed above, a consistent comparison of the different data sets becomes possible (see Figure 1 of [8] in these proceedings). When comparing the three results, the most striking fact is the apparent discrepancy, far beyond the point to point total errors, between the EMC and BCDMS data at small x ($x < 0.4$). For both the hydrogen and the deuterium measurements, the BCDMS data in this region are about 10% higher than the EMC data. This is marginally compatible with the quoted normalization errors but, when considering the full x range of the data, also the x dependence of the F_2's appears to be different.

This discrepancy does not depend strongly on Q^2 and can thus be represented by averageing the ratios of the two F_2 measurements in the Q^2 region of overlap, separately for each x bin. In a quantitative assessment of this discrepancy, it is essential to have meaningful estimates of the errors on these Q^2-averaged ratios. As the point-to-point total errors are most often dominated by systematic errors, it is crucial to take into account their correlations which can be very strong. So, in order to compute the errors on the Q^2-averaged ratios of the EMC and BCDMS F_2 measurements we proceed as follows. We first compute the average ratio and its statistical error, calculating the weights from statistical errors only. We then compute with the same weights, for each systematic error source of each experiment in turn, an average of ratios where F_2 is distorted by the effect of the respective source of systematics; we call these averages distorted averages. The total systematic error on the average is obtained as the quadratic sum of all differences between the distorted averages and the undistorted one. We finally combine quadratically the statistical and total systematic errors to obtain the total error on the average. This method, in principle, requires all correlations between systematic errors to be known. As we have mentioned above, this is not the case for the EMC data. We therefore use the EMC errors under two extreme assumptions, first taking them to be totally uncorrelated along Q^2 in each bin of x, and then assuming that they are 100%-correlated. In this way we obtain upper and lower bounds on the true total systematic error of the ratio F_2^{BCDMS}/F_2^{EMC}.

The Q^2-averaged ratios of the BCDMS and EMC data are shown in Figure 1a and b for hydrogen and deuterium. The outer (inner) error bars correspond to the total errors on these ratios when one assumes that the systematic errors of the EMC measurements are 100%-correlated (totally uncorrelated) along Q^2. The total errors corresponding to the actual correlations in the systematics of the EMC result are bound to lie between these two extremes; we estimate the truth to be closer to the assumption of fully correlated errors. These total errors may also be significantly correlated along x.

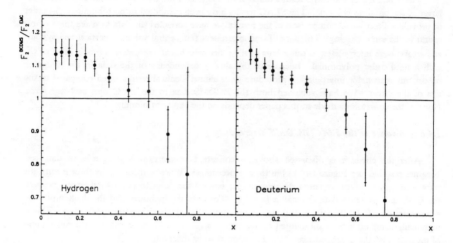

Figure 1: Q^2-averaged ratios of the BCDMS and EMC F_2 measurements.

From Figure 1, the ratio F_2^{BCDMS}/F_2^{EMC} is clearly decreasing with x. To reconcile the discrepancy between the two experiments with a simple normalization effect, the EMC data would need to be changed by at least twice the published systematic errors under very specific assumptions about their correlations, *i.e.* increasing all data points at small x and decreasing all data points at large x. Making less specific assumptions on the correlation, the systematic errors would need to be increased by larger factors. Alternatively, the BCDMS systematic errors would have to be increased by a factor of 8 or more. We therefore conclude that the BCDMS-EMC discrepancy is not compatible with a simple normalisation error.

In view of this discrepancy between the two CERN results, it is interesting to study how they compare to the reanalysed SLAC data.

1.5 Comparison of the SLAC data to the EMC and BCDMS results

It is obvious from Figure 1 of reference [8] that a direct comparison between the SLAC and CERN results is difficult because of the different Q^2 ranges covered by the three experiments. A quantitative comparison is possible only by fitting phenomenological parametrisations to the combined data. For this, we have employed the following technique :

i) To describe the Q^2 dependence of the data, we use a QCD-inspired parametrisation

$$\ln F_2 = A + B \ln[\ln(Q^2/\Lambda^2)] + \ln(1 + C/Q^2)$$

where the term $\ln(1 + C/Q^2)$ accounts for possible non-perturbative effects ("higher twists"). To avoid any bias from imposing on F_2 an algebraic form of the x dependence, we fit the above expression separately in each bin of x with free parameters A, B, and C. We use $\Lambda = 250$ MeV in all bins of x; the results of the comparisons reported here are insensitive to the choice of Λ (between 100 and 400 MeV).

ii) In order to obtain meaningful values of χ^2, the point-to-point correlations of the systematic errors have to be treated correctly. A simple way of doing this is to add in the fit as many free parameters λ_i as there are sources of independent systematic errors i. Each λ_i describes the linear distortion of the measured F_2 generated by each source of systematic error in units of one standard deviation. Statistical errors only are used for the fit but the data are allowed to vary as a function of all λ_i's: $F_2 \rightarrow F_2 + \sum_i \lambda_i \Delta F_{2i}$ and a term $\sum_i \lambda_i^2$ is added to the χ^2. In practice, we use an approximation to this procedure, treating only the correlation effects of the major systematic errors — where they are fully available (BCDMS) — and using instead of statistical errors "pseudostatistical" ones where statistical and smaller systematic errors are added in quadrature. For EMC, we use their systematic errors under the two extreme assumptions of full correlation or no correlation. For SLAC, as their systematic errors are of the same order of magnitude as the statistical ones and are relatively small compared to the BCDMS and EMC total errors, we use them as uncorrelated (we checked that this assumption does not affect any of our conclusions).

1.5.1 Relative normalisation of SLAC versus BCDMS and EMC data

When using this procedure to compare the EMC hydrogen data to the SLAC ones — chosen arbitrarily as a reference — it turns out that the best values of χ^2 are always (i.e. whatever the treatment of the EMC systematic errors) obtained for a global normalisation change of $+8\%$ ($\pm1\%$) of the EMC data. When comparing BCDMS to SLAC hydrogen data, the best agreement is obtained for a normalisation change of -2% ($\pm1\%$) of the BCDMS data. The corrections found for the deuterium data are similar, $+8\%$ for EMC and -1% for BCDMS. We recall that the estimated normalisation errors for both targets are 2% for SLAC, 5% for EMC, and 3% for BCDMS. The results of all three experiments after these renormalisations are shown in Figure 2a-b for the measurements on hydrogen and Figure 2c-d for those on deuterium.

1.5.2 Comparison of data after renormalisation

From Figure 2 it is clear that the low x ($x < 0.30$) results of EMC and BCDMS are no longer in disagreement. Although the agreement is not perfect — the hydrogen data of EMC still tend to be lower than the BCDMS ones — both data sets exhibit the same Q^2 dependence in each x bin and, assuming some correlation in the total errors, the Q^2-averaged F_2's of both experiments are compatible. Quantitatively, this is confirmed in Figure 1 where the first six points along x ($x < 0.30$) of the Q^2-averaged ratio F_2^{BCDMS}/F_2^{EMC} are roughly compatible with the BCDMS/EMC normalisation differences found above when taking SLAC as reference ($+10\%$ for hydrogen, $+9\%$ for deuterium).

We now compare the results from the CERN experiments to the SLAC measurements in this same low x region (Figure 2a and c). For $x < 0.15$, the SLAC and CERN data are nicely aligned

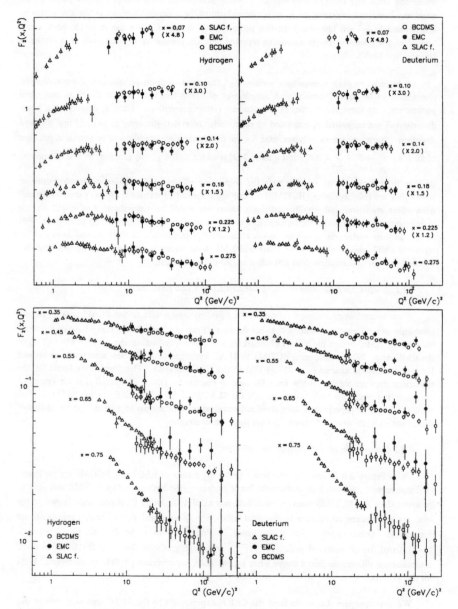

Figure 2: SLAC, EMC and BCDMS F_2 measurements after renormalisations (see text).

and can be well described by common fits. For $0.15 < x < 0.30$, a high Q^2 extrapolation of F_2 from the SLAC points would tend to pass below the CERN points. The QCD inspired fits of the Q^2 evolution indicate that this slight problem, which is significant only for the hydrogen data, simply arises from a tendency for the three or four high Q^2 SLAC points to drop rather abruptly. These points correspond to high $y = v/E$, a kinematical region where F_2 measurements are most sensitive to R and to radiative corrections. We conclude that, afer the renormalisations discussed above and apart from some minor problems, the results from the three experiments are in good agreement in the region $x < 0.30$.

At larger x, there exist small overlaps in the Q^2 ranges covered by SLAC and CERN measurements. In these regions of overlap, the EMC points are clearly in agreement with the SLAC ones whereas, for $x > 0.50$, the BCDMS points lie below the SLAC data for hydrogen as well as deuterium. For the muon data, this is precisely the region of small Q^2 and high x where the systematic errors are largest and are dominated by 100%-correlated effects. Taking these correlations into account, the observed difference between SLAC and BCDMS results in the regions of overlap actually corresponds to ~1.3 standard deviation of BCDMS main systematic error (see section 1.1). This is illustrated in Figure 3. These two data sets are thus in agreement.

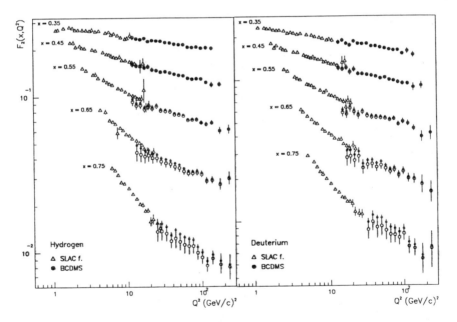

Figure 3: SLAC and BCDMS high x F_2 measurements with indication of systematic effects. The open circles represent the raw BCDMS data. The close circles represent the same data distorted by one-standard-deviation main systematic error (see text).

We want to stress that 1.3 standard deviation of BCDMS main systematic error can in no way account for the EMC-BCDMS discrepancy.

1.5.3 Can the new SLAC data resolve the EMC versus BCDMS discrepancy?

It thus turns out that at small x all three experiments are compatible (after renormalisation) and that, at large x and in the Q^2 region of overlap, the SLAC data agree with both CERN results. In view of the EMC-BCDMS discrepancy, this may seem surprising. It is however easily explained by the fact that this discrepancy actually originates from only a few EMC and BCDMS points that do *not* lie in the region of overlap between SLAC and CERN data. Indeed, the decrease of the ratio F_2^{BCDMS}/F_2^{EMC} for $x > 0.30$ in Figure 1 can be traced back to a few isolated data points where BCDMS and EMC results lie more than one standard deviation of total error apart. There are 13 such pairs of points in the hydrogen data and 14 in the deuterium data ; when these data points are removed, the EMC-BCDMS discrepancy essentially disappears. This is shown in Figure 4a and b where the displayed Q^2-averaged ratio F_2^{BCDMS}/F_2^{EMC} has been computed excluding these pairs of points.

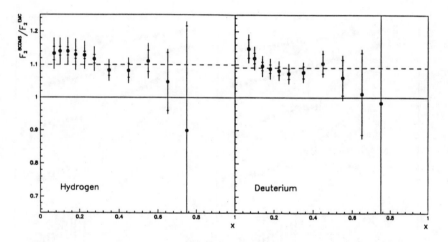

Figure 4: Q^2-averaged ratios of the BCDMS and EMC F_2's excluding disagreeing points.

These ratios are now compatible with the simple normalisation shifts discussed in 1.5.1 and represented on Figure 4 by dashed lines. In Figure 5a and b, we show the EMC-BCDMS disagreeing points (after renormalisation) together with the SLAC data for $x > 0.30$ (we recall that for lower x the three results are compatible). It is clear without the help of any fitting procedure, merely assuming a reasonable smoothness in the Q^2 evolution of F_2, that the SLAC data extrapolate better to the BCDMS than to the EMC points. In other words, when comparing only to the data points which are at the origin of the EMC-BCDMS discrepancy, the reanalysed SLAC data are in good agreement with the steeper x dependence of the BCDMS results.

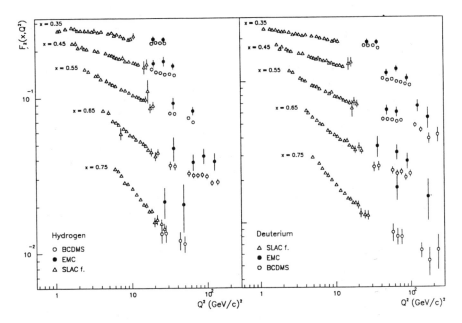

Figure 5: EMC and BCDMS disagreeing F_2 data points compared to the SLAC measurements.

So, the combination of the results from BCDMS and from the SLAC reanalysis provides a consistent set of high precision data on the proton and deuteron stucture function F_2. As far as the Q^2 range is concerned, these data are nicely complementary; the wide Q^2 region that they cover (from 0.5 to 250 GeV^2) allows to perform on them a very fruitful common QCD analysis (see [8]).

2. Experimental future of charged lepton deep inelastic scattering

In this second part, I will briefly say why the continuation of a program in this field (including fixed-target experiments) is necessary and what this program should be.

2.1 Motivations

The essential motivations to maintain an experimental effort for deep inelastic scattering (d.i.s.) studies are twofold.

i) D.i.s. is by far the best way of measuring parton distributions. These distributions are fundamental quantities of matter. They are not predicted by perturbative QCD (and apparently still far from being reliably determined from non-perturbative techniques) but they are needed as input to all hadronic studies and will be essential to interpret the data of the future hadron super-colliders (LHC, SSC). They will have to be known with precision over a wide x range because, even if most of the LHC or SSC events will correspond to very low x partons, the larger x region will play an important role in discovering or putting limits on new high energy phenomena. In addition, for non-hydrogen target, the very large x region (up to $x > 1$) is very interesting for nucleus studies.

ii) D.i.s. is, in principle as well as in practice, one of the "cleanest" way to perform significant and quantitative tests of perturbative QCD. It is characteristic for example that the most precise measurement of α_s (or equivalently Λ_{QCD}) has been obtained from d.i.s. [9]. This domain is privileged for QCD studies essentially because, on the theoretical side, there exist unambiguous predictions computed at NLO of perturbative QCD with a well determined energy scale (Q^2) and, on the experimental side, no fragmentation phenomena are involved (only the incoming and outgoing leptons are measured). The predictions concern the absolute value of the structure function R (once F_2 and the gluon distribution are measured) and the Q^2 evolution of any structure function F; the magnitude of the logarithmic slope $d\ln F/d\ln Q^2$ is directly related to the value of α_s (see e.g. [10]).

These motivations are valid for d.i.s. studies in general. I shall not discuss here the relative merits of neutrino, electron or muon scattering but I will concentrate on charged lepton d.i.s. since it concerns most of the future programs in this field and seems to me the easiest way to significantly reduce the experimental errors (both statistical and systematic).

2.2 Limitations of present results

In the first section, we have seen that the recent SLAC reanalysis has contributed to clarify an experimental situation which, in view of the EMC-BCDMS discrepancy, was somewhat obscure. It nevertheless remains that the errors on present results (at high Q^2) are dominated by systematics. As an example, one can cite the most probable dominant sources of such systematic uncertainties for the two muon CERN experiments: spectrometer magnetic field absolute calibration for BCDMS and hardware and reconstruction-software efficiencies for EMC. These dominant errors are different for the two experiments because they correspond to radically different types of apparatus (based on a magnetised iron toroid for BCDMS and on an air gap dipole magnet for EMC). When added to the numerous other sources of systematic uncertainties (beam energy calibration, detectors alignement and efficiencies, relative normalisation of data sets corresponding to different beam energies, energy losses and multiple scattering simulation, hadronic shower simulation, ...) these systematic errors are, for both results, larger than the statistical ones in most of the (x-Q^2) domain and up to 4 times larger in certain regions.

Together with the reduction of the systematic errors, an increase of the measured x range toward lower and larger values (down to 10^{-4} and up to 1, or more for non-hydrogen targets) is desirable. Moreover, although statistics is not the present limiting factor, a 10-fold increase is necessary to reach the precision allowing new types of QCD tests like, for example, to establish unambiguously the running of α_s (i.e. to disentangle the logarithmic Q^2 variation of α_s from any possible "higher twist" effect). In connection to these types of studies, measurements at higher Q^2, that are potentially more free of non-perturbative effects, are also highly desirable.

2.3 Future results

2.3.1 Fixed-target approved program

The only approved fixed target experiments that are expected to provide new data in charged lepton d.i.s. are the NMC and SMC collaborations at CERN and the E665 experiment at Fermilab ; in neutrino d.i.s., promising new data will come soon from the CCFR collaboration. The NMC experiment has been especially designed to be able to measure ratios of structure functions (like F_2^n/F_2^p or $F_2^A/F_2^{D_2}$) with very small systematic errors and down to x regions as low as 0.005. Absolute values of F_2 on hydrogen an deuterium targets are also to be extracted. This will further clarify the experimental situation and extend the measured kinematical domain to lower x values but, in the already measured region, the F_2 measurements are not expected to be more accurate than the present ones. The SMC experiment will measure spin structure functions with comparable errors. The E665 collaboration has chosen to concentrate on hadronic and nuclear studies.

2.3.1 HERA

Indisputably, the main experimental effort in d.i.s. will be pursued in the forthcoming years at the big electron-proton collider HERA which is scheduled to be operational in the middle of 1991. HERA will collide electrons of 30 GeV against protons of 820 GeV and two big experiments (H1 and ZEUS) will study the resulting d.i.s. interactions at Q^2 values up to 40000 GeV^2. For QCD studies, one of the main interests of this collider is that, at such high values of Q^2, one can be confident that all non-perturbative effects are totally negligible. More generally, the high center of mass energy may also allow the discovery of new phenomena like leptoquarks or a quark substructure. On the other hand, at these high Q^2, the cross sections are rather small and, as collider luminosities are limited compared to fixed-target ones, the structure function measurements will be limited by statistical errors in part of the accessible kinematical range. This is illustrated in Figure 6a in which we show a simulation of the expected HERA F_2 measurements after two years of running with 100% data-taking efficiency (luminosity of 200 pb^{-1}). This prediction is based on a QCD-inspired phenomenological parametrisation of $F_2(x,Q^2)$ which is also represented in the Figure together with the SLAC and BCDMS data on which it has been fitted[1]. For the HERA simulated measurement, the errors shown are purely statistical and the kinematical range corresponds to the region where the systematic errors are expected to remain below 10% ($0.1 < y < 0.9$). It is clear from Figure 6 that F_2 measurements at HERA will only be accurate in the low and very low x-region (say from 0.15 down to 0.0001). In this region, QCD studies are complicated because the Q^2-dependence of F_2 results from an interplay of many features: α_s, the R structure function (which can be rather large), the gluon distribution (which is essentially unknown), the radiative corrections (that can be as large as 100%). Moreover, at very low x values, the validity of perturbative QCD is questionable, but some asymptotic behaviours are predicted. Actually, the sensitivity to all these effects proves that this region is very rich and as it is, for $x < 0.005$, an uncharted domain, the HERA data will be of the greatest interest.

---

[1] In this fit the BCDMS data were allowed to vary according to the main systematic error and the errors used in the χ^2 expression (and represented on the Figure) are the pseudostatistical ones (see section 1.5).

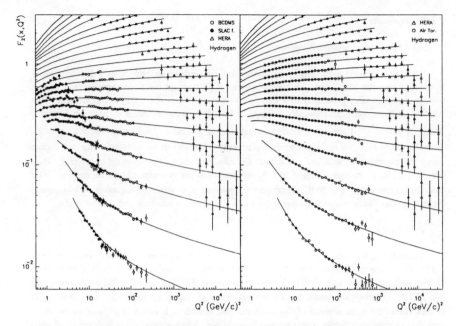

Figure 6: Simulated HERA measurements. Together with a) SLAC and BCDMS data, b) simulated measurements from a new fixed-target experiment (air-core toroid) in the FNAL muon beam. From the upper part of the figure downwards, the x-bins correspond to values of .0003, .0005, .0008, .0015, .003, .005, .008, .015, .03, .05, .07, .10, .14, .18, .225, .275, .35, .45, .55, .65, .75 .

2.3.2 Desirable extension

Complementary to HERA, it would also be of great interest to use the high energy FNAL muon beam to measure the nucleon structure functions in the x range from ~ 0.01 to 1 and at Q^2 values up to 700GeV^2, with a new muon fixed-target experiment, especially designed to lower the systematic uncertainties by one order of magnitude compared to the accuracy of present data. Up to now, two possibilities have been considered: a modification of the presently running E665 experiment (based on an air-core double dipole magnet) [11] and a new experiment based on a superconducting air-core toroidal magnet [12]. The two possible detectors as well as the physics that can be addressed with them at the upgraded Tevatron muon beam have been studied in the workshop "Physics at Fermilab in the 90's" [13]. The conclusions of the working group on this subject has been that «the air-core toroid is clearly the superior apparatus and is the experiment that is recommended to address the physics in question».

The principal advantages of the air-core toroid compared to the double dipole (E665 upgrade) concern the luminosity (it can use a 25 m long target compared to a 10 m one for the dipole), the crucial high Q^2 acceptance (its acceptance remains flat along Q^2 up to ~ 700 GeV^2 against

~250 GeV² for the dipole) and the control of systematics. In connection with this last point, I would like to stress that such a new muon experiment, which involves a large effort lasting several years, only makes sense if one can be sure that the final errors will be significantly lower than the present ones. So, a very good control of all sources of systematics *is crucial* and the aire-core toroid, which combines the advantages and avoids the drawbacks of air dipoles and iron toroids, definitely allows such a control. In particular :

• The pattern recognition is easy because of the azimuthal (phi) symmetry (scattered muon tracks stay in a radial plane), phi segmentation, and the magnetic sweeping and shielding of particles from the hadronic shower.

• The magnetic field calibration, the required precision of which is 10^{-4}, is relatively easy to get and reliable because the absence of ferromagnetic material makes an accurate computation possible. The computed field can be checked by direct measurements and is constrained by Ampère's law.

• The resolution on the scattered muon energy measurement is good (0.5%) and depends little on the energy (the more energetic the scattered muon, the longer its path through the magnetic field).

• The alignment of the detectors (wire chambers) is less crucial than for dipole experiments because of the phi symmetry.

• The measurements are highly redundant and there are many possibilities of cross-checks because the experiment is actually equivalent to 8 identical sub-experiments (one for each phi-octant) running in parallel.

For these reasons, and though the air-toroid may be four times more expensive than the E665 upgrade, it is the experiment to be run in the FNAL muon beam in the 90's. To give an idea of the potentialities of such an experiment, we show on Figure 6b a simulation of the F_2 data that it can provide in one year of running with a hydrogen target. Compared to the expected HERA data, it is noticeable that the covered kinematical range is complementary in x and roughly comparable in Q^2 if one restricts the HERA range to where the bulk of the statistics is ($x < 0.10$). The already impressive accuracy of the data over the whole x range can be six times better if one uses, instead of hydrogen, a carbon target and, even in this case which corresponds to a huge statistical power, the apparatus is designed to keep the systematic errors below the statistical ones. This high precision would allow a series of very significant QCD tests (see *e.g.* [12]).

In conclusion, together with the HERA data, such an experiment would yield (for ~3% of the HERA cost) a very accurate measurement of the nucleon structure functions in an extremely wide kinematical range (x from 0.0001 to 1 and Q^2 up to 1000-5000 GeV²). This provides an unmatched opportunity of detailed studies of QCD.

3. Conclusions

Charged lepton deep inelastic scattering is still a very fruitful field ; the accessible physics is far from being fully explored, even in the presently accessible kinematical domain. The recent reanalysis of SLAC data has contributed to a clarification of the experimental situation.

In the future, the new domain of very low x will be explored at HERA. A new generation high precision muon scattering fixed-target experiment is highly desirable and would be nicely complementary to HERA. An air-gap superconducting toroid is the safest way to reach the required accuracy.

References

1. See *e.g.* S.R. Mishra and F. Sciulli, Annu. Rev. Nucl. Part. Sci. 1989.39: 259-310, or J. Feltesse, Proceedings of the XIV International Symposium on Lepton and Photon Interactions, Stanford, August 6-12, 1989.

2. EMC collaboration, J.J. Aubert et al., Nucl. Phys. B259 (1985) 189.

3. EMC collaboration, J.J. Aubert et al., Nucl. Phys. B293 (1987) 740.

4. BCDMS collaboration, A.C. Benvenuti et al., Phys. Lett. 223B (1989) 485.

5. BCDMS collaboration, A.C. Benvenuti et al., Phys. Lett. 237B (1990) 592.

6. L. Whitlow, Ph.D. Thesis, SLAC-REPORT-357 (1990), and A. Bodek, contribution in these proceedings.

7. A. Milsztajn et al., CERN-EP preprint in preparation (1990).

8. A. Milsztajn, contribution in these proceedings.

9. BCDMS collaboration, A.C. Benvenuti et al., Phys. Lett. 223B (1989) 490.

10. M. Virchaux, Thèse, Université Paris-7 (1988).

11. F.M.Pipkin et al., *Letter of Intent for a precise and accurate measurement of nucleon structure functions (inelastic form factors) using the E665 apparatus*, Presented to FERMILAB, March 5, 1990.

12. C. Guyot et al., *A New Fixed Target Experiment for Precise Tests of QCD (updated version)*, Unnumbered Saclay Preprint, March 14, 1989.

13. K.Bazizi et al., *High Luminosity Muon Scattering at FNAL − Report of the Study Group on Future Muon Scattering Experiments from "Physics at Fermilab in the 90's"*, Breckenridge, Colorado, August 15-24, 1989, FERMILAB-Conf-90/39,

STRUCTURE FUNCTIONS, PARTON DISTRIBUTIONS AND QCD-TESTS AT HERA

JOHANNES BLUEMLEIN

Institute for High Energy Physics, Academy of Sciences

Platanenallee 6, Zeuthen, 1615, GDR

and

GERHARD A. SCHULER

II. Institute for Theoretical Physics

University of Hamburg

Luruper Chaussee 149, D-2000 Hamburg 50, FRG

ABSTRACT

The possibilities are reviewed to measure the structure functions and quark and gluon distributions in the HERA energy range. Conditions are discussed to measure Λ_{QCD} and $\alpha_s(Q^2)$ from the scaling violations of structure functions.

1. Introduction

In early 1991 two experiments ZEUS an H1 [1] will be set into operation at the ep-collider HERA [2]. One of the main subjects to be studied by these experiments is the measurement of the proton structurefunctions, parton-distributions and the scaling violations of these quantities. HERA will operate at a cms energy $s = 4 \cdot E_e \cdot E_p = 4 \times 30 \times 820$ GeV^2 and a nominal luminosity $L = 1.5 \cdot 10^{34}$ cm^{-2} sec^{-1}. The kinematical range which has been explored by lepton-nucleon scattering experiments so far [3] $1 < Q^2 < 300$ GeV^2 and $0.01 < x < 0.9$ can be extended by roughly two orders of magnitude both in x and Q^2 by these experiments.

A precise measurement of the structurefunction $F_2(x, Q^2)$ will be possible down to very low x ($x \sim 10^{-4}$) and up to $Q^2 \sim O(10^4$ GeV), which allows to test QCD in new kinematical ranges [4,5]. Particularly, it will be interesting to investigate the low-x range of F_2 since yet untested QCD contributions [6-8] could influence its x- and Q^2-dependence in this range. At high Q^2 different possibilities exist to combine $e^{\pm}$p-neutral and charged current cross sections to unfold individual quark distributions and combinations of them [9,10]. Further, the gluon distribution can be determined using different quantities as $R = \sigma_L/\sigma_T$ [11], $J/\psi -$ [12] and open charm production [13] and the singlet component in QCD-fits [4] of F_2.

The measurement of all these quantities presumes detailed investigations of a series of systematic effects and theoretical conditions as the smearing functions, the calibration uncertainties of the calorimeters, the electroweak radiative corrections and further theoretical corrections in some of the measurements. Due to the size of these effects only a limited part of the available phase space can be used for the measurement of the structure functions. Because most of the issues which we will discuss are rather dependent on these conditions they will be discussed in section 2 briefly. In section 3 possible measurements of structure functions are considered. Different ways to unfold quark distributions from the deep inelastic scattering cross sections are discussed in section 4. Estimates on the potential of HERA to measure Λ_{QCD} and $\alpha_s(Q^2)$ including estimates on systematic effects are given in section 5. In section 6 a survey on different possibilities to determine the gluon distribution at HERA is given and section 7 contains the conclusions.

2. Systematic Effects and Theoretical Corrections

2.1. Smearing Corrections and Systematic Effects

These effects have been considered in detail in ref. [14] both for the measurement of the final state electron and the current jet. Due to the finite resolution of the detectors one has to account for smearing and systematic effects to derive the structurefunctions from measured event distributions. The dominant contributions due to the smearing functions are implied by the resolutions of both the electromagnetic and hadronic calorimeters. They are given by $\delta E_H/E_H = 0.35 ... 1.0 /E_H^{1/2} + 0.02$ and $\delta E_e/E_e = 0.13 ... 0.17/E_e^{1/4}$. Further, one has to consider the precision of the absolute calibration of the energy measurement also, for which 1 % for the electron measurement and 2 % for the jet measurement might be obtained. Using a Monte Carlo calculation contours in x and Q^2 were derived in [14]. They are depicted in fig.1 for the cms energy $\sqrt{s}$ = 314 GeV and a possible low energy option at HERA with E_e = 15 GeV and E_p = 300 GeV, $\sqrt{s}$ = 134 GeV [15].

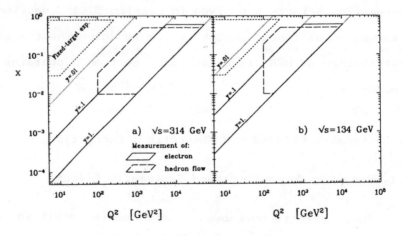

Fig. 1 Kinematical ranges

The electron measurement is bounded by 0.1 < y but ranges to very low x. Due to trigger conditions the hadron-flow measurement is bounded by 100 GeV^2 < Q^2 but the resolution allows to measure down to y ~ 0.03. Comparing this range, which we will refer to as the accessible kinematical range at HERA subsequently, with the kinematical range explored so far one observes a gap of about one order of magnitude in Q^2 for the nominal HERA energy. This could be narrowed by measuring the structure functions at lower energies (see fig.1b) and/or seeking for a further improvement of the resolution at low y. Note, that the closest contact to the kinematical range of the fixed target experiments comes from the range of jet measurement in both cases.

Among the systematic effects the calibration uncertainty of the calorimeters is one of the important contributions. Because this effect is particularly large e.g. in the case of the measurement of Λ_{QCD} [5] we discuss it in more detail here.

A shift in the measured electron or hadronic energy $\delta E_{e,H}/E_{e,H} = \pm \varepsilon_{e,H}$ will induce a systematic shift in the kinematical variables x, y and Q^2 which are given for $\varepsilon_{e,H} \ll 1$ by

$$\hat{x}/x \sim 1 + \varepsilon_e/y \qquad \hat{x}/x \sim 1 + \varepsilon_H/(1-y) \qquad (1a)$$

$$\hat{y}/y \sim 1 + \varepsilon_e - \varepsilon_e/y \qquad \hat{y}/y \sim 1 + \varepsilon_H \qquad (1b)$$

$$\hat{Q}^2/Q^2 \sim 1 + \varepsilon_e \qquad \hat{Q}^2/Q^2 \sim 1 + \varepsilon_H(2-y)/(1-y) \qquad (1c)$$

From these relations it follows, that one can get a nearly

unperturbed Q^2 measurement from the final state electron and a nearly unperturbed y measurement from the final state hadrons as long as $\varepsilon_{e,H}$ are small. Otherwise the x and y (or x and Q^2) measurements fail at low (high) y for the electron (hadron) measurement due to the finite miscalibration. Note, that the relations (1) are derived using the kinematics of the Born process only. Modified results would be obtained for the $O(\alpha)$ photon Bremsstrahlung contributions.

2.2. Radiative Corrections

The electroweak radiative corrections to the inclusive neutral and charged current cross sections are the major theoretical corrections which have to be considered for the extraction of structure functions in the HERA energy range. They have been calculated by different groups during the last years [16-21] in complete [16,17,20] and leading log calculations [18,19,21]. Recently, numerical agreement has been obtained among all approaches at the per cent level. Because of the high value of $s \sim 10^5$ GeV^2 both the corrections to the neutral and charged current cross section are widely dominated by the QED-corrections, even by their leading log contributions. This was demonstrated in [18,19] for the neutral current and in [19] for the charged current reactions. A comparison for the relative correction $\delta_i =$

$d^2\sigma(O(\alpha))/d^2\sigma_0$ to neutral and charged current e^- p-scattering for the complete [17] and approximate calculation [19] is given in fig. 2 defining the variables x and y at the leptonic vertex.

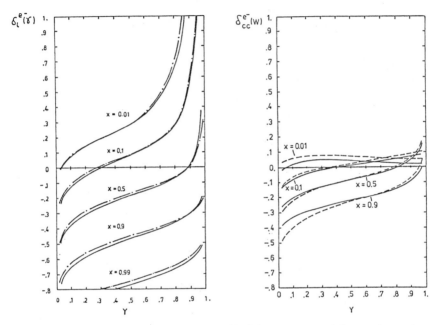

Fig. 2: Comparison of the radiative corrections to deep inelastic neutral and charged current e^- p-reactions. Full lines: leading log approximation [19];—·—·and ——— line: complete calculations [17].

The neutral current radiative corrections become rather large at low x and high y while in the case of charged current reactions the corrections are of O(10%) in this range. At low y soft photon bremsstrahlung causes large negative corrections.

As outlined in sect. 2.1 in wide ranges of x and Q^2

the kinematical variables can only be determined via the jet measurement. It is shown in [21] that the radiative corrections can be calculated analytically for this case and yield new y-dependent factors in front of the structure functions of the Born cross sections. The corrections are shown in fig. 3 for different values in x in the leading log approximation.

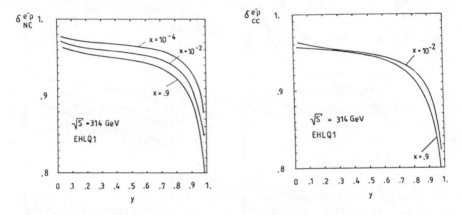

Fig. 3: Radiative corrections to deep inelastic neutral and charged current $e^{\pm}p$ reactions. The kinematical variables are defined from jet measurement [21].

The variation in x is only logarithmic and the corrections are rather small compared to the case discussed before. At high y a large negative contribution is caused due to soft photon emission. Since there is a range in which both the electron and jet measurement can be used to measure the kinematical variables the inclusive radiative corrections can be also controlled experimentally. Subsequently we will discuss the measurement of structure functions and parton distributions assuming that the radiative corrections are carried out already and consider the Born cross sections only.

3. Measurement of Structure Functions

The Born cross sections for deep inelastic neutral and charged current reactions are determined by 14 structure functions. They can be written in the form

$$d^2\sigma(e^{\pm}p \to \overset{e^{\pm}}{\underset{\nu_e}{\varepsilon}_i} X)/dx\,dQ^2 = \frac{2\pi\alpha^2}{xQ^4} \hat{\sigma}_i^{\pm} \quad (2)$$

with

$$\hat{\sigma}_{NC}^{\pm} = y^2 2x F_1^{\pm}(x,Q^2) + 2(1-y) F_2^{\pm}(x,Q^2) + Y_- x F_3^{\pm}(x,Q^2) \quad (3)$$

and

$$\hat{\sigma}_{CC}^{\pm} = \tfrac{1}{2} x_W^2(Q^2)\left[y^2 2x W_1^{\pm} + 2(1-y)W_2^{\pm} \mp Y_- x W_3^{\pm}\right] \quad (4)$$

for neutral and charged current reactions. In leading order the Callen-Gross relations $2x F_1^{\pm} = F_2^{\pm}$ and $2x W_1^{\pm} = W_2^{\pm}$ hold. The functions $F_i^{\pm}$ are not the generic structure functions but contain also propagator terms. They are given by

$$2x F_1^{\pm} = 2x F_1 - v_e \mathscr{X}_Z(Q^2) 2x G_1 + \mathscr{X}_Z^2(Q^2)(v_e^2 + a_e^2) 2x H_1 \quad (5a)$$

$$F_2^{\pm} = F_2 - v_e \mathscr{X}_Z(Q^2) G_2 + \mathscr{X}_Z^2(Q^2)(v_e^2 + a_e^2) H_2 \quad (5b)$$

$$x F_3^{\pm} = \pm a_e x_E(Q^2) \times G_3 \mp x_Z^2(Q^2) 2 v_e a_e \times H_3 \quad (5c)$$

The neutral and charged current structure functions are related to the quarkdistributions by

$$2x(F_1, G_1, H_1) = x \sum_q (Q_q^2, 2Q_q v_q, v_q^2 + a_q^2)(q + \bar{q}) \quad (6a)$$

$$x(G_3, H_3) = 2x \sum_q (Q_q a_q, v_q a_q)(q - \bar{q}) \quad (6b)$$

$$W_2^{\pm} = 2x \sum_i q_{d(u)} + \bar{q}_{u(d)} \quad (6c)$$

$$x W_3^{\pm} = 2x \sum_i q_{d(u)} - \bar{q}_{u(d)} \quad (6d)$$

and $f_2 = 2x f_1 + f_L$, $f_i = F_i$, G_i, H_i, where f_L can be rewritten in terms of the according longitudinal parton densities. Here $Y_- = 1-(1-y)^2$, $x_Z(Q^2) = Q^2/(Q^2+M_Z^2)/4 s_\theta^2 c_\theta^2$, $x_W(Q^2) = Q^2/(Q^2+M_W^2)/4 s_\theta^2$, and Q_f, $a_f = I_f^3$, $v_f = I_f^3 - 2Q_f \sin^2\theta_W$ are the electroweak couplings.

To illustrate the event rates which can be measured at HERA the integrated event rates for $Q^2 > Q_o^2$ are depicted for the different cross sections, $x > 0.01, y > 0.03$ and an integrated luminosity of 1oo pb^{-1} in fig. 4. At Q^2 lower than $\sim 10^3$ GeV^2 both neutral current cross sections are nearly identical and nearly no charged current events are found in this range. Thus, the only structure functions which can be measured in this range are F_2 and F_L.

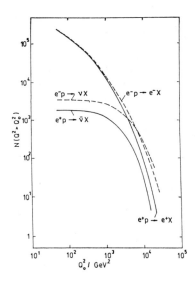

Fig. 4: Neutral and charged current event rates for $Q^2 > Q_0^2$ [9].

The statistical precision which can be obtained for the measurement of F_2 is illustrated in fig. 5. At low x the slopes in $\ln(Q^2)$ are rather large. For $x > 0.01$ at high y the contributions due to the electroweak terms become visible. Due to the smaller statistics in this range F_2 remains to be a good approximation up to rather high Q^2.

For the measurement of the longitudinal structure function F_L the measurement of the neutral current cross section at at least two different beam energies is required to determine F_L from the y dependence in equal (x, Q^2)-bins. Choosing $E_p = 820$ and $E_p = 300$ GeV in [9] the precision of $R = F_L / F_2$ was estimated assuming a luminosity of

100 pb^{-1} for each option.

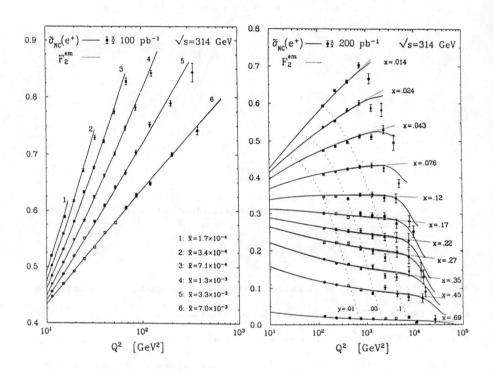

Fig. 5: Statistical precision for F_2

As shown in fig. 6 R can be precisely measured at x < 0.01 which is important for the determination of F_2 and a possibility to measure the gluon distribution at low x [11], cf. sect. 6.

From the difference of the cross sections $\hat{\sigma}_{NC}^{\pm}$ the structure function xG_3 arising in the γ-Z interference term can be measured.

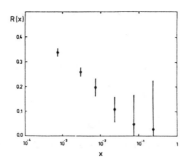

Fig. 6: Statistical precision for a R-measurement using E_p= 820 and E_p= 300 GeV.

For this distribution only the x-shape can be measured with sufficient precision. As seen from fig. 4 , this distribution is only measurable for $Q^2 > 10^3$ GeV^2 . Because it is a valence distribution its Q^2 dependence is rather weak only and can not be resolved within the statistical errors.

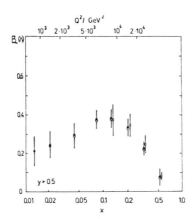

Fig. 7: Determination of xG_3 from $\hat{\sigma}_{NC}^-$ and $\hat{\sigma}_{NC}^+$. The luminosities used are 100 pb^{-1} per beam.

The charged current cross sections contain six different structurefunctions. Even if one assumes the validity of the Callen-Gross relation an unfolding of the structure functions is not possible. However, at low y the Y_- terms become small and both from the cross sections $\hat{\sigma}_{cc}^+$ and $\hat{\sigma}_{cc}^-$ the functions $W_2^\pm$ can be determined in an approximate way. This is illustrated in fig. 8 for the structurefunction W_2^-.

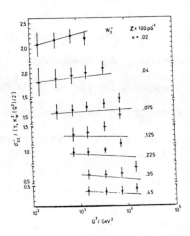

Fig. 8: Determination of W_2^- from $\hat{\sigma}_{cc}^-$ at low y.

4. Unfolding of Quark Distributions

Since four different deep inelastic scattering cross sections can be measured at HERA at most four combinations of quark distributions can be unfolded from the cross sections. If a suitable basis of those functions is choosen one has to solve the linear problem [9,10]

$$\begin{pmatrix} \hat{\sigma}^+_{NC} \\ \hat{\sigma}^-_{NC} \\ \hat{\sigma}^+_{CC} \\ \hat{\sigma}^-_{CC} \end{pmatrix} = (A_{ij}(y,Q^2)) \begin{pmatrix} xq_1 \\ xq_2 \\ xq_3 \\ xq_4 \end{pmatrix} \quad (7)$$

with $\hat{\sigma}_i = xQ^4 d^2\sigma_i/dxdQ^2/2\pi\alpha^2$

In principle (7) can be solved in the whole available kinematical range but practically one has to guarantee that i) $(A_{ij}(y,Q^2))$ is not degenerate and ii) the errors of $\hat{\sigma}_i$ are sufficiently small to yield a meaningful measurement for the distributions $xq_j(x,Q^2)$. At low Q^2 ($Q^2 < 10^3$ GeV^2) both conditions are not met: because of the dominance of γ-exchange both neutral current cross sections are nearly equal and $(A_{ij}(y,Q^2))$ tends to degenerate. The charged current statistics is rather poor in this range which would induce large errors for xq_j. Thus, this method can be applied best at high $Q^2 \gtrsim O(10^3 \text{ GeV}^2)$ where all $\hat{\sigma}_i$'s are of comparable magnitude. Because the overall statistics both for neutral and charged current events is of a few $10^{3..4}$ only the determination of the x-shapes of the distributions xq_j is possible. However, due to the high Q^2 still differences in shape might be found comparing with measurements at lower Q^2 in fixed target experiments [10].

Apart from the exact unfolding method (7) one can extract some of the combinations xq_j also approximately in some parts of the kinematical range. This is discussed in ref. [9,10]. An example for this method is the extraction of parton distributions in the valence range $x > 0.25$. The cross sections $\hat{\sigma}^{\pm}_{cc}$ are proportional to xu_v and xd_v in this range resp. From $\hat{\sigma}^{\pm}_{NC}$ one can measure $\frac{4}{9}xu_v + \frac{1}{9}xd_v$ for $Q^2 \lesssim 10^3$ GeV^2. More complicated examples are discussed in ref. [10].

To illustrate the statistical precision of these methods fig. 9 shows possible measurements of the x-shape of $F_s = \sum_i x(q_i + \bar{q}_i)$. The statistics used both for the approximate representation $\hat{\sigma}^-_{cc} + \hat{\sigma}^+_{cc}/(1-y)^2$ (fig.9a) and the exact unfolding (fig.9b) using the basis (xu_v, xd_v, $x\sum_i u_i + \bar{u}_i$, $x\sum_i d_i + \bar{d}_i$) amounts to 400 pb^{-1}.

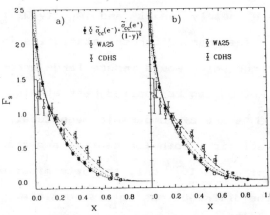

Fig. 9: Possible determination of $F_s(x)$ via an approximate representation and exact unfolding [10]

Although in principle 14 different x-distributions can be determined only 7 can be measured with a sufficient precision. Particularly it will be difficult to measure $x\sum_i d_i$, $x\sum_i \bar{d}_i$, and at lower x also of xd_v [10].

5. QCD Tests

The analysis of the Q^2 dependence of the measured structure functions at constant x allows to determine the QCD scale Λ and $\alpha_s(Q^2)$ [22]. In a systematic analysis [9,5,6] a variety of structure functions and combinations of parton distributions which can be unfolded from the neutral and charged current cross sections was investigated. It turns out that only $F_2(x,Q^2)$ can be measured at a sufficient precision required for a QCD-test.

F_2 can be measured from $\hat{\sigma}_{NC}^{\pm}$ directly at $Q^2 \lesssim O(10^3 \text{GeV}^2)$. At higher values of Q^2 electroweak terms contribute but also the statistical errors become larger. One can define a cut

$$|1 - \hat{\sigma}_{NC}^{\pm}/F_2 Y_+|/\partial F_2 \lesssim \varepsilon \qquad (8)$$

for the data used in the analysis. For $\varepsilon \lesssim 1$ stable fits are obtained yielding the Λ value used for the Monte Carlo generation of the data.

The statistical precisions for Λ obtained in the

analysis using different cms-energies, different ranges in x and comparing the results for the full phase space and the range given in fig.1 are summarized in table 1.

Table 1   Statistical precision on Λ and α_s from QCD fits to $\tilde{\sigma}_{NC}(e^+) \approx F_2^{em}$

x-range	type of fit	Λ [MeV]	restricted range		
			Λ [MeV]	α_s	$\langle Q^2 \rangle$ [GeV]
a) $\sqrt{s}$ = 314 GeV, $Q^2 \geq 100$ GeV^2, $\int \mathcal{L} dt = 200$ pb^{-1}					
$x \geq 0.25$	non-singlet	145 ± 48	175 ± 176	0.132 ± 0.023	2770
$x \geq 10^{-2}$	" + singlet	297 ± 76	177 ± 135	0.159 ± 0.026	400
$x \geq 10^{-2}$	— " — $xG(x,Q_0^2)$ fixed	215 ± 16	201 ± 25	0.164 ± 0.005	400
b) $\sqrt{s}$ = 314 GeV, $Q^2 \geq 10$ GeV^2, $\int \mathcal{L} dt = 100$ pb^{-1}					
$x \geq 10^{-4}$	non-singlet + singlet	196 ± 5	225 ± 25	0.204 ± 0.006	80
c) $\sqrt{s}$ = 134 GeV, $Q^2 \geq 18$ GeV^2, $\int \mathcal{L} dt = 100$ pb^{-1}					
$x \geq 0.25$	non-singlet	200 ± 47	460 ± 263	0.193 ± 0.028	530
$x \geq 10^{-2}$	— " — + singlet	211 ± 27	227 ± 58	0.188 ± 0.012	160

Using the range fig.1 for $x > 10^{-2}$ and L = 200 pb^{-1} the error for Λ amounts $\delta\Lambda \sim$ 140 MeV. This can be improved further using constraints on xG from other measurements and including data at lower y. The extension of the kinematical range to lower x would yield a significant improvement. In the analysis [5] only the leading-log Altarelli-Parisi evolution was considered. For this case a precision $\delta\Lambda \sim$ 25 MeV is estimated for $x > 10^{-4}$ and L = 100 pb^{-1}. Since new dynamical effects are expected to influence the data at very low x [6-8] the analysis of the low-x HERA data requires the inclusion of these effects which has not yet been considered. For a low-energy option at HERA with $\sqrt{s}$ = 134 GeV the precision of Λ is estimated by $\delta\Lambda \sim$ 60 MeV for $x > 10^{-2}$ and L =

100 pb^{-1}. Pure non-singlet fits of F_2, i.e. for $F_2^{val} \simeq F_2$ at $x > 0.25$ yield rather large errors even for L=200 pb^{-1} since most of the data are located at smaller x.

A major systematical effect in the measurement of Λ are systematic shifts of the cross section due to miscalibrations of the hadronic and electromagnetic calorimeters, cf. section 2. To illustrate this we considered 1% shifts in the measured hadronic or electromagnetic energy E_H and E_e resp. For $Q^2 > 100$ GeV^2, $x > 10^{-2}$ and $\delta E_H/E_H = \pm .01$ $\Delta\Lambda = \pm 70$ MeV is obtained. Since $\delta E_H/E_H$ is expected to be of 2% this error is of the size of the statistical error for L=200 pb^{-1}. For $Q^2 <$ 100 GeV^2, $10^{-4} < x < 10^{-2}$ and $\delta E_e/E_e = \pm .01$ one finds $\Delta\Lambda = \mp 40$ MeV which also is comparable to the according statistical error.

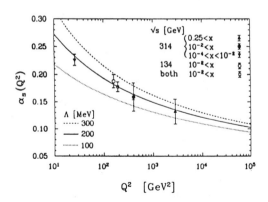

Fig. 10: Statistical precision of the Q^2 dependence of $\alpha_s(Q^2)$

In fig. 10 different measurements for Λ (cf. table 1) are represented in terms of $\alpha_s(Q^2)$ using measurements at the two beam energies $\sqrt{s}$ = 134 GeV (L=100 pb^{-1}) and 314 GeV (L=200 pb^{-1}) and different ranges of x. The inner error bars correspond to the case were the gluon distribution is assumed to be known completely. A variation of $\alpha_s(<Q^2>)$ can be seen comparing the values at high and low Q^2. Together with measurements at intermediate values of Q^2, this could give evidence for the running of the strong coupling.

6. The Gluon Distribution

For many processes in $p\bar{p}$-collisions at high s the detailed knowledge of the gluon distribution at small x is required. HERA offers different possibilities to measure or to constrain this distribution at small x in different ranges of Q^2.

From the simultaneous solution of the Altarelli-Parisi evolution equations of F_2 and xG the gluon distribution can be measured at a given Q_0^2. The statistical precision of this measurement was estimated in [4] using the range $x > 10^{-3}$ and $Q^2 > 10$ GeV^2. In fig. 11 a 1σ error contours are given for $xG(x,Q_0^2)$ [23], Q_0^2 = 4 GeV fixing Λ = 200 MeV. The hatched range corresponds to the statistical error using an integrated luminosity of L=100pb^{-1} the dotted line illustrates the effect of additional overall systematic errors of 2 %. Results on a joint fit of $xG(x,Q_0^2)$ and Λ are shown in fig.11 b for different input distributions: xG(DO) and $0.3xG(DO)/\sqrt{x}$ which can be distinguished clearly in the low x range.

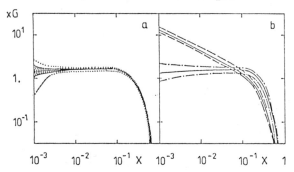

Fig. 11 Extraction of $xG(x,Q_0^2)$ in a QCD analysis for $x > 10^{-3}$

Due to the simplicity of the splitting functions P_{qq}^L and P_{qg}^L for the longitudinal structure function for γ-exchange an approximate representation for $xG(x,Q^2)$ can be obtained [11] at low x.

$$xG(x,Q^2) \approx \frac{3}{5} 5.9 \left[\frac{3\pi}{4\alpha_s} F_L(0.4x,Q^2) - \frac{1}{2} F_2(0.8x,Q^2) \right] \quad (9)$$

To measure xG(x) at x a measurement of F_L at 0.4 x and of F_2 at 2*0.4x is required. As lined out in [11] xG can be determined in the range $Q^2 \sim 50...100$ GeV^2 for $x \sim 0.003...0.02$. In fig. 11 two estimates are given on the precision to measure xG using (9). The statistical errors correspond to L = 100 pb^{-1}. Again the input distributions $xG(x,Q_o^2) \sim$ const. and $xG(x,Q_o^2) \sim 1/\sqrt{x}$ can be distinguished for $x \to 0$.

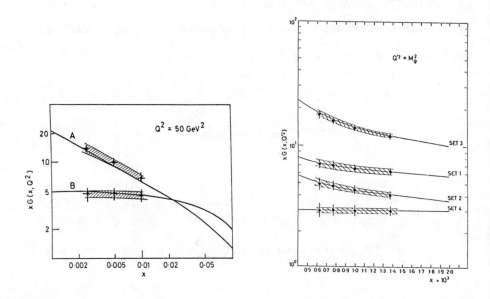

Fig.12 Determination of $xG(x,Q^2)$ from $F_L(x,Q^2)$ and $F_2(x,Q^2)$

Fig.13: Determination of $xG(x,M_\psi^2)$ from $d\sigma/dy(\gamma g \to J/\psi x)$

The gluon distribution can be measured from the of the production cross section $\gamma g \to J/\psi X$ also [12]. In this process Q^2 is fixed by $Q^2 \sim M_\psi^2$. The y dependence of the cross section is given by

$$\frac{d\sigma}{dy} = 1.5 \frac{\alpha}{\pi} \frac{1+(1-y)^2}{y} \bar{x} G(\bar{x},M_\psi^2) \ln \frac{Q^2_{max}}{Q^2_{min}} nb \qquad (10)$$

with $\bar{x} \approx 1.5 x$. Estimates on the precision $xG(x,M_\psi^2)$ using different parametrizations (SET 1-3 [24] and SET 4: $xG(x) = 3(1-x)^5$) are illustrated in fig. 13 [12]. The statistical errors correspond to $L=100pb^{-1}$ and the error due to overall normalization.

As discussed in [13] also cross section measurements of open charm production $e^{\pm} p \to e^{\pm} c\bar{c} X$ can be used to constrain the gluon distributions. To apply this method detailed Monte Carlo studies are required to simulate the background reactions correctly.

7. Conclusions

Structure function measurements at HERA will extend the kinematic range of present day lepton scattering experiments up to $Q^2 \sim$ few 10^4 GeV^2 and down to $x \sim 10^{-4}$. The first time charged current reactions will be measured in $e^{\pm}$ p-scattering at a statistics of a few 10^3 events. Aside the measurement of the kinematical variables from the scattered electron their measurement from the current jet becomes rather important to control systematic effects.

The $O(\alpha)$ electroweak radiative corrections to deep inelastic neutral and charged current reactions are well understood now [16-21] and are controlled at the per cent level.

Among the various structure functions and combinations of parton distributions to be measured only $F_2(x,Q^2)$ can be determined with sufficient precision both in x and Q^2 [9]. Other distributions can be measured as shapes in x by unfolding from four measured cross sections $\hat{\sigma}^{\pm}_{NC}$, $\hat{\sigma}^{\pm}_{CC}$ or approximations of combinations of them which are valid in certain kinematical ranges [9,10].

Different ways exist to measure or to constrain the gluon distribution [4,11-13], particularly at small x. Statistically one can distinguish the behaviour $xG(x,Q^2) \sim$ const. and $xG(x,Q^2) \sim 1/\sqrt{x}$.

The structure function $F_2(x,Q^2)$ can be precisely measured up to very low x and could be a frist testing ground for new contributions influencing the evolution and shape of F_2 at low x [6-8].

The statistical precision on Λ for $\sqrt{s}$ = 314 GeV, $x > 10^{-2}$, $y > .03$, L = 200 pb^{-1} is $\sim$100 MeV. It could be improved by: i) using constraints on $xG(x,Q^2)$ from other measurements in the analysis, ii) including data at lower y, $y \sim 0.01$, iii) running also at lower $\sqrt{s} \simeq$ 134 GeV, iv) extending the analysis down to $x = 10^{-3}...10^{-4}$. It is effected by miscalibrations of the electromagnetic and hadronic calorimeters.

All measurements of the above quantities are long term tasks and require the control of the systematics at the per cent level.

References

[1] ZEUS Collaboration, Technical proposal, DESY,1986; H1 Collaboration, Technical proposal, DESY,1986.

[2] HERA - a proposal for a large electron-proton colliding beam facility, DESY/HERA 81/10 (1981).

[3] For reviews on deep inelastic scattering experiments,cf.:
F. Eisele, Rep. Progr. Phys. 49 (1986) 233;
M. Diemoz, F. Ferroni, E. Longo, Phys. Rep. 130 (1986) 293;
T. Sloan, G. Smadja, R.Voss, Phys.Rep. 162 (1988) 45.

[4] J. Bluemlein, M. Klein, T. Naumann, Proc 'New Theories in Physics', Kazimierz, Poland, 1988, Eds. Z. Ajduk et al., World Scientific, 1989, p.228.

[5] J. Bluemlein, G.Ingelman, M.Klein, R.Rueckl, Z. Phys. C45(1990)501.

[6] J. Kwiecinski, Z. Phys. C29 (1985) 561.[7] L.V. Gribov, E.M. Levin, M.G. Ryskin, Phys. Rep. 100(1983) 1.

[8] J. Bartels, J. Bluemlein, G. Schuler, in preparation.

[9] J. Bluemlein, M. Klein, T. Naumann, T. Riemannn Proc. of the HERA Workshop, Hamburg 1987, Ed. R.D. Peccei, (DESY Hamburg,1988) Vol.1, p.67.

[10] G. Ingelman, R.Rueckl, Phys. Lett.B 201 (1988) 369; Z.Phys. C44 (1989) 291.

[11] A.M. Cooper-Sarkar, G. Ingelman, K.R. Long, R. G. Roberts, D.H. Saxon, Z. Phys. C39 (1988) 281.

[12] S.M. Tkaczyk, W.J. Stirling, D.H. Saxon, Proc. of the HERA Workshop, Hamburg 1987, Ed. R.D. Peccei, (DESY Hamburg,1988) Vol.1, p.265; A.D.Martin, C.K.Ng and W.J.Stirling, Phys.Lett. B191 (1987) 200.

[13] G.Barbagli, G.D'Agostini, Proc.of the HERA Workshop, Hamburg 1987, Ed. R.D.Peccei, (DESY Hamburg, 1988), Vol.1, p.135.

[14] J. Feltesse, Proc. of the HERA Workshop, Hamburg 1987, Ed. R.D.Peccei, (DESY Hamburg,1988) Vol.1, p.33.

[15] R. Brinkmann, F. Willeke, private communication.

[16] M. Boehm, H. Spiesberger, Nucl. Phys. B294(1987) 1081; B303(1988) 749.

[17] D. Y. Bardin, C. Burdik, P.C. Christova, T. Riemann, Z. Phys. C42 (1989) 679; C44 (1989) 149.

[18] W. Beenakker, F. A. Berends, W.L. van Neerven, Proc. 'Radiative corrections for e e collisions', Ringberg, Bavaria, 1989, Ed. J.H. Kuehn, (Springer, Berlin, 1989) p.3.

[19] J. Bluemlein, Berlin preprint PHE 89-08;89-12 and Z. Phys. C in press.

[20] H. Spiesberger, in preparation.

[21] J. Bluemlein, Berlin preprint PHE 90-06.

[22] W. Furmanski, R. Pretronzio, Nuck. Phys. B195 (1982) 237;
L.F. Abbott, W.B. Atwood, R.M. Barnett, Phys. Rev. D22 (1982) 582.

[23] D.W. Duke, J.F. Owens, Phys. Rev. D30 (1984) 49.

[24] A.D. Martin, R.G. Roberts, W.J. Stirling, Phys. Rev. D37 (1988) 1161.

Comments on Precision Measurements of Nucleon Structure Functions

Richard Wilson
Harvard University

A controversy about the accuracy of the nucleon structure functions has occupied much of this session of the meeting. It appears there are differences between the two set of measurements at CERN (EMC and BCDMS) about 15% in cross section and 8% in the value of the structure function. In this brief note I go back over (a part of) the historical record to throw light upon the discrepancy and to suggest what features are important in any future experiment and analysis.

The first extensive experiments on scattering of leptons by protons, and the extraction of the proton elastic form factor was by Bumiller et al.[1] at Stanford. The accuracy was about 10% in cross section and 5% in the form factor. A major step in accuracy was taken at Orsay by Lehmann et al.[2]. They made a limited number of cross section measurements, but carefully examined the systematic errors. They claimed an accuracy of ± 2% in the cross section or ± 1% in the form factor. The detector had an efficiency of 99.5 ±0.5 %. More recent data from Mainz suggests that these measurements gave too large a cross section by about the quoted error, and the radius of the nucleon is 6% larger than these measurements suggest. Better radiative corrections seem to be the major improvements.

The major set of experiments at SLAC on elastic and inelastic electron proton scattering, (giving elastic form factors and structure functions) had a similar absolute accuracy[3]. The reanalysis of these data presented earlier in this session seems to reduce the systematic errors by another factor of 2.

Experiment E98 at Fermilab was the first important muon scattering experiment, with the aim of precise measurements of nucleon structure function[4]. E98 was designed to have an absolute accuracy of 1% but due to shortage of beam time, and lower intensity than planned, there were severe limitations in the statistical accuracy. The actual absolute accuracy achieved was somewhat better than 5% in the cross section or 2% in the structure function.

The experiments of EMC at CERN were first presented 2 years later at a seminar at CERN. The beam intensity and beam availability were both high enough that the statistical accuracy was vastly superior to E98, and it became usual to ignore E98 from then on. But there were a few puzzling aspects. A simple plot showed that at high ν the cross section was systematically 15% lower than measured by E98. A search was made, by the residue of the E98 collaboration for the "mistake" but none was found. At this ν the scattered muon was identified and measured at a claimed 100% efficiency; and if this was an overestimate the discrepancy would be worse. The radiative correction was large, but the correction was made using the Mo-Tsai[5] program, adjusted for the muon mass and with the 3 usual corrections made. EMC also used the Mo-Tsai procedure, although using a different computer program in the same (Rutherford Lab) computer. In retrospect this discrepancy in cross sections should have been considered as a clue to possible problems in EMC.

The magnitude of this discrepancy between E98 and EMC is about the difference between the EMC and the BCDMS data sets discussed earlier in this session. It is now clear that the efficiency of trackfinding and measurement in the EMC was <u>not</u> close to 100%, and it therefore becomes likely that the <u>inefficiency</u> was underestimated.

If the EMC data have to be modified because of a correction for inefficiency, it is far from clear that this can be handled by a simple scale factor. All global fits to form factors discussed in this session attempt this unjustified procedure. The inefficiency is likely to vary considerably over the parameters of the experiment.

The Experience of E98 and BCDMS suggests that close to 100% efficiency for the scattered muon (within the acceptance) is possible. In E98 this was achieved by using two groups of 4 spark chambers (with double-sided read out) each with efficiency 97-99% per plane, close together to create the "seed" for identifying and tracking the scattered particle. This was a location where the muons had been deflected or scattered away from the beam. Although spark chambers are now obsolete, it is noteworthy that modern drift chambers and proportional chambers have lower efficiencies (typical 95% or less) and many more planes close together would therefore be needed.

Earlier in this session Virchaux implied that the problem with the EMC experiments was due to using a dipole magnet configuration, and that toroidal arrangement of the BCDMS, or the new air-cored toroid proposed by Virchaux and collaborators will avoid these problems. Since E98 and EMC had similar magnet configurations the discussion above suggests that the difference has another cause. Indeed Virchaux also emphasizes the importance of a high efficiency in identifying and measuring tracks. The uncertainty created by an <u>inefficiency</u> is usually proportional to the inefficiency itself. If the inefficiency is calculated by a complex Monte-Carlo program to be 90%, it does not seem unlikely that neglected effects could really make it 80%. Whereas if the calculated efficiency is 99% an efficiency of less than 98% is improbable. Striving for 100% efficiency is therefore especially important. There is an exception; all experiments have a well-defined geometrical acceptance. In the early ones this was a solid aperture, now it is the area of the wire detector at a crucial location.

Virchaux emphasizes the importance of measurement of magnetic field. The accuracy must be considerably better than 0.1%. There are <u>two</u> measurements. The incident muon and the scattered muon. The two spectrometers, may easily be intercalibrated by the muon beam to the desired accuracy. In E98, although the claimed accuracy of $\int B\, dl$ for each magnet was 0.1%, there was a measured difference of 1%. The middle value was taken with an assigned accuracy of ±1/2%. No source for this discrepancy was found in 1979, but in 1985, when the magnet was remeasured with the same equipment, an error was found in the computer controlled stepping of the measuring probe which explained the whole discrepancy.

In the future it would be wise to check the magnet plot made by stepping a probe by an integral measurement of $\int B\, dl$ using a long coil. This can easily be done to better than 0.1% precision. The ease in doing this is far greater for a dipole magnet configuration than for the toroidal geometry proposed by Virchaux, and this simplicity may lead to a greater believability of the accuracy.

There is a considerable advantage in a complex experiment, to being able to check crucial alignment parameters by symmetry. In a dipole arrangement, such as E98, EMC or E665, the rear detector where the muon is measured is one single device. For example, a small error in knowledge of beam position can give an error in the angle when the muon scatters upwards, but this is compensated by an opposite error when the muon scatters downwards. Virchaux pointed out that the same is true of a toroidal geometry, and indeed he proposes 6 detectors. In his case, however, the proposed detectors will be mechanically separate, and the distance between them must be precisely known and controlled. This is, of course, possible but it is not automatic.

The history of EMC shows that physicists have used unnecessarily inaccurate structure functions for 10 years. There is now a theoretical demand for an accuracy of 1% over a wide range, and a need for accurate structure function by those using hadron colliders. Virchaux has proposed in this session, an experiment to do so using an air-cored toroid. There is also a letter of intent to Fermilab to do so with a dipole. History suggests that more than one experiment, each with differing systematic errors, are needed to be sure of these important quantities.

Because it is likely to be cheaper and therefore could be done sooner, an experiment with a dipole may be the first choice.

References

1. F. Bumiller, M. Croissiaux, E. Dally and R. Hofstadter, Phys. Rev. 124, 1623 (1961).

2. P. Lehmann, R. Taylor and Richard Wilson, Phys. Rev. 126, 1183 (1962).

3. e.g. J.S. Poucher et al. Phys. Rev. Lett. 32, 118 (1974); W.B. Atwood et al., Phys. Lett. 64B, 479 (1976).

4. B.A. Gordon et al., "Measurement of the nucleon structure function" Phys. Rev. D20, 2645 (1979).

5. L.W. Mo and Y.S. Tsai, Rev. Mod. Phys. 4, 205 (1969).

UNCERTAINTIES IN QCD ANALYSIS OF STRUCTURE FUNCTIONS

R. LEDNICKY

Joint Institute for Nuclear Research, Dubna, USSR

Abstract

The influence of various approximations and theoretical uncertainties due to preasymptotic corrections on the determination of the QCD mass-scale parameter Λ, and, - on the QCD tests is studied with the help of the BCDMS hydrogen data. A small (negative) contribution of higher twists in the transverse proton structure function is obtained in the region of moderate x and $Q^2 > 10$ GeV^2, and, in agreement with theoretical models, it is well described in terms of one parameter $k \approx 0.2$ GeV/c - the parton transverse momentum generated dynamically. The dominant theoretical uncertainty in the $\Lambda(\overline{MS})$ due to the higher-order corrections is estimated as $+120/-30$ MeV. At the same time, this uncertainty appears to be of minor importance for QCD tests based on the x-dependence of scaling violations.

1. Introduction

The deep-inelastic scattering represents one of the best places for testing quantum chromodynamics (QCD) and determination of the fundamental QCD mass-scale parameter Λ or the strong coupling constant α_s. Present deep-inelastic scattering data are not accurate enough to check the most specific QCD prediction: running of α_s. However, the precision and the ranges of the Bjorken x and the momentum transfer squared Q^2 of the recent BCDMS data [1-3] appear to be sufficient to test quantitatively the QCD predictions for scaling violations and to determine Λ reliably. The question however arises concerning the influence of various approximations and theoretical uncertainties on the QCD analysis of experimental data. In the present paper we study this question with the help of the BCDMS hydrogen data [2,3]. More details can be found in ref. [4].

2. Leading-twist QCD fits

According to the QCD factorization theorem, the nucleon structure functions $F_k(x,Q^2)$, describing the lepton deep-inelastic scattering, are given as the convolution of quark and gluon densities with the coefficient functions $C_k(x,Q^2)$ (C_k is proportional to the corresponding cross section of the hard process of the intermediate boson absorption by a parton). The Q^2-dependence of the parton densities is governed by solutions of the generalized Altarelli-Parisi-Lipatov evolution equations. The one of the coefficient functions is determined by an expansion in powers of the running coupling constant $\alpha_s(Q^2)$. In the following we use the next-to-leading order (NLO) expression [5]:

$$\alpha_s^{(1)}(Q^2) = [4\pi/\beta_0 \ln(Q^2/\Lambda^2)] \cdot [1-(\beta_1/\beta_0^2)\ln\ln(Q^2/\Lambda^2)/\ln(Q^2/\Lambda^2)], \quad (1)$$

where $\beta_0 = 11-2f/3$, $\beta_1 = 102-38f/3$; f is the number of active flavours. The evolution equations allow one to calculate the parton densities at any Q^2 provided they are given at a certain reference point Q_0^2. The x-dependence of the nonsinglet-, singlet-quark and gluon densities is usually parametrized on the basis of plausible theoretical assumptions concerning the behaviour near the end points $x = 0, 1$, e.g.:

$$xq^{NS}(x,Q_0^2) = a_{NS} x^{\mu_{NS}} (1-x)^{\nu_{NS}} (1+\gamma_{NS} x),$$
$$xq^{SI}(x,Q_0^2) = a_{SI} [x^{\mu_{SI}} (1-x)^{\nu_{SI}} + a_{SEA} x^{\mu_{SEA}} (1-x)^{\nu_{SEA}}],$$
$$xG(x,Q_0^2) = a_G x^{\mu_G} (1-x)^{\nu_G}. \quad (2)$$

A sufficient flexibility of these parametrizations should be provided in order not to bias the comparison of the QCD predictions with data.

To perform the QCD fits we have used the computer code developed within the BCDMS collaboration [4,6,7] based on analytical solution of the evolution equations with the help of Jacobi polynomial expansion. The parameters in Eqs. (2) are determined together with the QCD mass-scale parameter Λ by fitting the QCD predictions to the BCDMS cross section data. Both the relevant structure functions F_2 and $F_L=(1+\varepsilon)F_2 - 2xF_1$ ($\varepsilon = 4M^2x^2/Q^2$) are calculated in the $\overline{MS}$ scheme [5].

An excellent agreement of the QCD predictions with the data is demonstrated in fig. 1 and Table 1. It indicates a sufficient flexibility of the quark x-parametrizations in Eqs. (2). We have confirmed this with the help of polynomial modifications of the parametrizations and found that the subsequent change of Λ is negligible (< 2 MeV).

In the fits we have constrained the size of the gluon density with the help of the momentum sum rule. This may be questionable as it requires an interpolation of the singlet-quark and gluon densities into the unmeasured region of $x < 0.06$. It appears, however, when treating both $<x_S>$ and $<x_G>$ as free parameters and assuming $\mu_{SEA} = \mu_G = 0$, that the results of Table 1 remain practically unchanged (except for 50% increase of the error in ν_G), and, that the sum rule is well satisfied: $1 = 1.05 \pm 0.13$.

Table 1. The results of NLO leading-twist QCD fits to the BCDMS hydrogen data [2] (Λ in MeV), assuming 4 massless flavours and parametrizing the parton densities according to Eqs.(2) at $Q_0^2 = 5\ GeV^2$; $\mu_{SEA} = \mu_G = 0$. The kinematic cuts of ref. [3] are applied; in particular, $x > 0.06\ (0.25)$ in a SI+NS (NS) fit.

Fit	μ_{NS}	ν_{NS}	γ_{NS}	a_{NS}	μ_{SI}	ν_{SI}	$\langle x_q \rangle$	ν_{SEA}	a_{SEA}	ν_G	$\Lambda_{\overline{MS}}$	$\frac{\chi^2}{DOF}$
SI+NS	0.5 ±0.2	3.5 ±0.2	10 ±2	1.1 ±0.2	0.8 ±0.1	4.5 ±0.6	0.45 ±0.08	13 ±4	0.17 ±0.05	9.0 ±1.5	207 ±21	$\frac{258}{270}$
NS	0.6 ±0.2	3.5 ±0.3	0.1 ±0.8	2.2 ±0.7	–	–	–	–	–	–	198 ±20	$\frac{178}{198}$

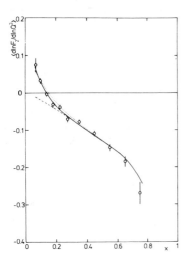

Fig. 1. The comparison of the logarithmic slopes of the BCDMS proton structure function $F_2(x,Q^2)$ [2] with the result of a complete SI+NS QCD fit (full curve) described in the caption of Table 1; dotted and dashed curves correspond to $\nu_G = 5$ and $\langle x_G \rangle = 0$, respectively.

The softness of the gluon distribution makes it possible to neglect its contribution in the evolution equations at sufficiently large x-values and to determine Λ with the help of a more constrained nonsinglet analysis. We have found [4] that a NS fit at $x > 0.25$ underestimates $\Lambda_{\overline{MS}}$ by 5 +15/-5 MeV.

3. Higher-order corrections (HOC)

The problem of HOC is closely related to the one of the renormalization scheme (RS) dependence of a truncated perturbation series (see, e.g., [8-10]). Thus the QCD predictions obtained to a finite order $O(\alpha_s^m)$ differ in various schemes at $(m + 1)$-th order when expanded in powers of the coupling constant in some reference RS. At m-th order, the RS is specified by m conditions. The corresponding unphysical parameters may be identified as the renormalization point μ and the beta function coefficients β_j, $j = 2,...$ m [8]. Thus the RS-dependence of the

NLO predictions is entirely equivalent to the problem of the best choice of the renormalization point within one particular scheme. Changing the renormalization point μ by another one $\tilde{\mu} = \mu/æ$, the NLO expansion coefficient $B_{k,n}^{(1)}$ of the Mellin transformed coefficient function $C_k(n,Q^2)$ and the mass-scale parameter Λ become $æ$-dependent. E.g. [5,11]

$$\tilde{B}_{2,n}^{NS(1)} = B_{2,n}^{NS(1)} - \gamma_+^{(0)}(n) \cdot ln æ, \qquad \tilde{\Lambda} = æ \cdot \Lambda. \qquad (3)$$

It is implied that $\Lambda/\mu \ll æ \ll \mu/\Lambda$, so the $æ$-rescaling of the parameter Λ in Eq. (3) is approximately equivalent to the rescaling of α_s:

$$\tilde{\alpha}_s = \alpha_s \cdot (1 + \frac{\alpha_s}{4\pi} \beta_0 ln æ^2). \qquad (3')$$

Choosing the $\overline{MS}$ scheme as the reference RS and assuming $f = 4$, we have, e.g. $æ = 2.17$ for the MOM scheme [10], or, $æ = 0.377$ for the MS scheme [5]. The complete NLO contribution to the predicted scaling violations is, of course, independent of the RS, as would be a full-all-order calculation. However, since we truncate the calculation in the NLO, different choices of $æ$ yield different estimates of the higher-order terms.

Fortunately, within the accuracy of present data the higher-order uncertainty (RS-dependence) of the predicted x-dependence of the scaling violations is rather small. In particular, fig. 2 indicates only a weak $æ$-dependence of the χ^2 of a NS fit to the BCDMS hydrogen data within the range of approximate validity of the NLO expansion (in which the $æ$-rescalings in Eqs. (3) and (3') yield near-by results). At the same time, substantial dependence of the fitted value of $\Lambda_{\overline{MS}}$ on $æ$ is observed.

It follows that the check of perturbative QCD with the help of measured scaling violations and the determination of the mass-scale parameter Λ are quite different tasks - the uncertainty due to higher

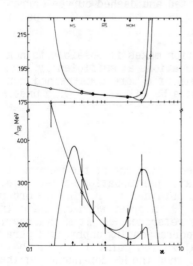

Fig. 2. The results of nonsinglet NLO QCD fits (described in the caption of Table 1) to the BCDMS hydrogen data [2] as functions of the parameter $æ$ specifying the renormalization scheme. The closed and open circles correspond to the $æ$-rescalings in Eqs. (3) and (3'), respectively.

orders being probably small in former but not in latter case. Based on fig. 2, we may estimate that a NLO fit at $x = 1$ yields $\Lambda_{\overline{MS}}$ with an uncertainty of $+120/-30$ MeV. Similar uncertainty is also indicated by a recent calculation [12] of the third-order correction to $R(e^+e^-)$ in the $\overline{MS}$ scheme. This correction leads to ~10% (40%) decrease of α_s (Λ).

As compared with $R(e^+e^-)$, there is an additional problem due to the higher-order contributions containing terms proportional to powers of $\alpha_s ln(1-z)$ and $\alpha_s lnz$ ($x \leq z \leq 1$) which destroy the validity of the perturbative QCD predictions for the inclusive leptoproduction at high and low values of x. Various procedures for resummation of these terms have been suggested. Concerning the low-x region, the perturbative QCD predictions seem to be valid until x as small as 10^{-3}-10^{-4} [13,14]. The resummation (exponentiation) in the high-x region is quite ambiguous but, fortunately, not very important in the BCDMS kinematic range. E.g., exponentiating the terms containing powers of $ln(1-z)$ according to the recipe of ref. [15], a NLO fit of BCDMS hydrogen data yields, in the nonsinglet approximation, Λ by 5 MeV higher and χ^2 by 1 unit better than the standard fit.

Related to the high-x problem there is a question of the HOC starting at three loops. Thus a consideration of the evolution in terms of the timelike variable $W^2 = m^2 + Q^2(1-z)/z$ (the effective mass-squared of a final state in the photon-parton subprocess) instead of $q^2 = -Q^2$ leads to the effective replacement in Eq. (1) [16]: $ln(Q^2/\Lambda^2) \to [ln^2(Q^2/\Lambda^2) + \pi^2]^{1/2}$, indicating that the contribution of the $O(\alpha_s^3)$ terms may be substantial. Neglecting other possible HOC, it would mean less steep running of $\alpha_s(Q^2)$ as compared with the prediction of Eq. (1), i.e. the NLO Λ-value increasing with Q^2. Thus to the BCDMS NLO value of $\Lambda = 210$ MeV would correspond 230 (240) MeV at UNK (HERA) and 300 MeV at asymptotic energies. This effect should be trackable in the experiments planned to measure running of α_s at SPS CERN and UNK with the statistical error of a few MeV in Λ [17].

We may conclude that at present experimental errors the HOC-uncertainties in the predicted x-dependence of the scaling violations seem to be practically negligible. At the same time, the HOC may lead to an uncertainty as large as ~10% in α_s or ~40% in Λ.

4. Flavour threshold corrections (FTC)

The corrections due to a flavour excitation treshold are of the order $O(m_i^2/Q^2)$ at Q^2 much larger than a heavy quark mass squared m_i^2. They become large at $Q^2 < m_i^2$, since, according to the intuitive decoupling theorem [18], the contribution of a heavy flavour vanishes at $Q^2 \ll m_i^2$. Generally, the FTC arise from the mass dependence of $\alpha_s(Q^2)$, coefficient functions and splitting functions (anomalous dimensions). The $\overline{MS}$ (MS) scheme does not satisfy the decoupling theorem since its beta function (α_s) and anomalous dimensions are mass independent by definition. There was some hope [19,20] that the FTC can be calculated within the MOM renormalization scheme. In this scheme, the mass-dependence of α_s and the gluon-gluon splitting function is merely connected with the replacement of the number of flavours by an effective one. It appears that the usual ansatz of four massless flavours with appropriately rescaled Λ-value provides a reasonable approximation to the mass-dependent $\alpha_s(Q^2)$ in rather wide Q^2 range. Even better approximation is achieved when calculating the coupling with the number of massless flavours appropriate for a given Q^2-range and requiring its continuity at flavour excitation thresholds. The corresponding parameters Λ_f are only weakly dependent on the threshold positions. Choosing them at $f(1-5)m_f^2$, we have in the NLO: $\Lambda_3:\Lambda_4:\Lambda_5 \approx 1.3:1:0.65$. Since the mean approximation accuracy is much better than the accuracy of present measurements of $\langle\alpha_s\rangle$ (2.5 % in the case of the BCDMS hydrogen data), we may conclude that the mass-dependence of $\alpha_s(Q^2)$ is of minor practical importance.

The situation with the mass-dependence of the coefficient and splitting functions is less clear [21]. Therefore, similar to refs. [22,23], we estimate the FTC with the help of a phenomenological approach based on the perturbative calculation of the γ-gluon fusion $\gamma g \to \bar{q}_i q_i$. To retain the simplicity of the $\overline{MS}$ renormalization scheme, we follow ref. [22] (see also [24]) and treat the mass-dependence of the anomalous dimensions in a simplified way assuming them mass-independent in a Q^2-range between the neighbouring thresholds $Q_i^2 \approx 5m_i^2$, and, include the explicit mass-dependence entirely into the coefficient functions. The uncertienty in Λ due to various prescriptions for the mass-dependent coefficient functions is estimated to be less than 1 MeV in the BCDMS kinematic range [4].

It appeares that the χ^2 of the fits to the BCDMS hydrogen data as well as the fitted gluon density are practically insensitive to the number of flavours used to calculate the evolution. At the same time, the parameter Λ is strongly correlated to this number: changing $f = 4$ to $f = 5$ decreases $\Lambda_{\overline{MS}}$ by 60 MeV. Assuming the bottom threshold at $Q^2 = (5 \pm 2) m_b^2$ we get for the shift of $\Lambda_{\overline{MS}}(f=4)$ due to the FTC a value of $-5 +3/-10$ MeV. We may conclude that the influence of the FTC on the QCD tests and the determination of α_s is negligible at present.

5. Target mass- and higher twist- corrections (TMC and HTC)

In the simple model of free massless partons the TMC arises from the target mass M (through the Nachtmann variable ξ) and the intrinsic transverse parton momentum $\tilde{k}_\perp$ [25-28]. It may be seen from fig. 3 that the TMC to $F_2(x,Q^2)$ is practically negligible at $x < 0.55$ ($<10\%$ GeV^2/Q^2 at $Q^2 > 10$ GeV^2), but rapidly increases at higher values of x. The TMC to $F_L(x,Q^2)$ is relatively very large at moderate values of x and Q^2 ($\sim 100\%$ at $x = 0.55$, $Q^2 = 10$ GeV^2), however, at lower x, where F_L essentially deviates from zero, the TMC is much less important.

There are however problems in the above approach. First, the TMC do not vanish for unphysical values of $1 < x \leq 1/(1-M^2/Q^2)$. This effect is of nonperturbative origin and cannot be cured by switching on the interaction with the spectator quarks through a truncated series of the HTC [29,30]. Fortunately, this overestimation rapidly vanishes with Q^2, and, in fact, can be avoided in a finite order expansion of the TMC in inverse powers of Q^2. Second, vanishing of $\tilde{k}_\perp$ in the boundary regions $x \to 0, 1$ or in the case of a small target mass^1 (following from the free parton approximation) is in apparent contradiction with the uncertainty principle. To solve this problem the simple picture of on-shell partons

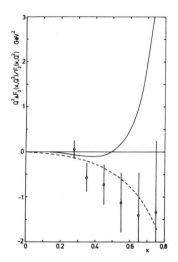

Fig. 3. Power corrections to the proton structure function $F_2(x,Q^2)$ at $Q^2 = 60$ GeV^2. The TMC (full curve) is calculated according to refs. [25-28]. The HTC_0 result from NS-fits to the BCDMS hydrogen data [2] using Eq. (4) ($k = 0.2$ GeV – dashed curve) and Eq. (5) (points) in the region $x > 0.25$, $Q^2 > 10$ GeV^2; y-cuts of ref. [3] are applied.

with intrinsic transverse momentum should be improved and generalized in the presence of interaction. It is well-known that the complete QCD result can be formally written as: $HTC = TMC + HTC_0$, where the twist-4 corrections HTC_0 calculated in the limit of zero target mass depend on certain parton correlation functions which measure the transverse momentum generated by the interaction together with the transverse components of the gluon field [29]. At sufficiently high x-values the integration over the gluon momenta, implicit in their definition, is dominated by the soft gluon region. Assuming the soft gluon approximation valid also at lower x-values, the x-dependence of the correlation functions is essentially given by the shape of the quark distribution function $\tilde{F}_2(x)/x$ [31]. For the twist-4 corrections HTC_0 we then have (see also [32]):

$$\Delta F_2(x,Q^2) = 4(k^2/Q^2)x\frac{\partial}{\partial x}\tilde{F}_2(x,Q^2), \quad \Delta F_L(x,Q^2) = 4(k^2/Q^2)\tilde{F}_2(x,Q^2). \quad (4)$$

The parameter k^2 satisfies the positivity constraint $k^2 + \langle k_\perp^2 \rangle > 0$ and may be interpreted as the mean transverse momentum squared generated dynamically. From the twist-4 corrections to the Gross-Llewellyn Smith sum rule calculated in ref. [33] within three different approaches (the QCD sum rules, the vector dominance approximation and the nonrelativistic quark model) we may estimate $k^2 = 0.013$–0.033 GeV^2.

The twist-4 corrections in Eqs. (4) contain only the two-fermion contributions. The four-fermion ones vanish for the longitudinal structure function but not for $F_2(x,Q^2)$. Their x-dependence is expected [34] to be similar to the one in Eq. (4). Based on the bag model calculations in ref. [35] we may estimate the corresponding k^2 as 0.004 GeV^2.

The results of NS fits to the BCDMS hydrogen data, taking into account the HTC according to Eqs. (4), are given in Table 2 (Fit 1-3). The values of the parameter k^2 fitted in various kinematic regions agree within the errors with the theoretical expectation of 0.02–0.04 GeV^2. They should be compared with previous determinations of this

Table 2. The results of NLO nonsinglet QCD fits to the BCDMS hydrogen data [2] as described in the caption of Table 1 but taking into account the HTC according to Eqs. (4) or (5).

Fit	Q^2-cut GeV^2	y-cuts [10]	HTC_0 Eq.	$\Lambda_{\overline{MS}}$ MeV	k^2 $(GeV/c)^2$	χ^2/DOF
1	20	yes	(4)	204 ± 53	0.02 ± 0.03	174/197
2	10	yes		218 ± 46	0.04 ± 0.03	201/228
3	10	no		247 ± 36	0.06 ± 0.02	222/250
4	20	yes	(5)	238 +130/−100	−	171/192
5	10	yes		273 +80/70	−	192/223
6	10	no		297 ± 70	−	212/245

parameter: $k^2 = 0.10 \pm 0.16\ GeV^2$ [32], $0.12-0.20\ GeV^2$ [30] and $0.04-0.12\ GeV^2$ [36], based on EMC-, EMC + neutrino- and SLAC-data, respectively. A systematic decrease of the fitted k^2-value when cutting low-y (Q^2) points may indicate the sensitivity of the fits to systematic errors.

Large Λ-errors in these fits result from a competition of the scaling violations in $F_2(x,Q^2)$ predicted by the leading twist QCD evolution and the HTC in Eq. (4), which have similar x-dependences in the moderate-x region. This leads to a flat χ^2 curve as a function of k^2 and to a substantial correlation between Λ and k^2, as it is demonstrated in fig. 4. The correlation weakens with Q^2, e.g. its slope at $Q^2 \geq 10\ GeV^2$ decreases by 30% at $Q^2 > 40\ GeV^2$. Taking $k^2 = 0.04 \pm 0.04\ GeV^2$ as a combined result of Table 2 (roughly including systematic errors) we may conclude from fig. 4 that the standard NS fit of the BCDMS data (Table 1) underestimates $\Lambda_{\overline{MS}}$ by $35\ +60/-50\ MeV$.

The error in Λ becomes still larger if we do not fix the x-depen-

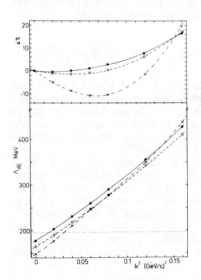

Fig. 4. The results of NS fits to the BCDMS hydrogen data as functions of the transverse momentum squared generated dynamically (see Eq. (4)); $\Delta\chi^2 = \chi^2 - \chi^2(k^2=0)$. Closed circles, open circles and crosses correspond to Fits 1,2 and 3 in Table 2. The dotted line indicates $\Lambda_{\overline{MS}}$ fitted with $HTC = 0$.

dence of the HTC in F_2 according to Eq. (4) and instead use the ansatz:

$$\Delta F_2(x_j, Q^2) = (a_j/Q^2) \, F_2(x_j, Q^2), \qquad (5)$$

where a_j is a free parameter in each x-interval (see Fit 4-6 in Table 2). The fitted values of the parameters a_j (Fit 5) are compared in fig. 3 with the prediction of Eq. (4). It may be seen that this prediction does not contradict with the BCDMS data, and, that the global power corrections are slightly negative in the moderate-x region whereas at large x they are dominated by the positive TMC. The small negative power corrections at moderate x qualitatively agree with the results of neutrino experiments [37]. However, they are in disagreement with rather large positive HTC required to tail the low-Q^2 SLAC data to the deep-inelastic muon-proton scattering data of EMC and BCDMS [38,39]. This discrepancy may be due to the twist-6 contribution $\propto 1/Q^4$ to the charged lepton deep-inelastic scattering, being dominant in the SLAC kinematic region but getting negligible in the CERN one. Note that the twist-6 contribution to neutrino scattering is expected to be negligible [43,40].

6. Conclusions

The influence of various approximations and theoretical uncertainties on the determination of the QCD mass-scale parameter $\Lambda_{\overline{MS}}$ in a non-singlet NLO fit (using Eq. (1) for $\alpha_s(Q^2)$ and assuming four massless flavours) to the BCDMS hydrogen data in the region: $x = 0.25-0.8$, $Q^2 = 20-260$ GeV^2 (y-cuts of ref. [3] are applied) is summarized in Table 3. All the sources of the uncertainties, except the last one, appear to be of minor importance for the QCD test based on the x-dependence of scaling violations.

Table 3. Corrections and uncertainties for the parameter $\Lambda_{\overline{MS}}$ fitted in a leading-twist NS approximation from the BCDMS hydrogen data as described in the caption of Table 1.

Source	$\Delta\Lambda_{\overline{MS}}$ MeV
NS approximation (gluon x-parametrization)	5 +15/-5
Quark x-parametrizations	± 2
HOC	+120/-30
FTC	-5 +3/-10
HTC Eq. (4)	35 +60/-50
Eq. (5)	42 +130/-100

The contribution of higher twists in the proton structure function $F_2(x,Q^2)$ is found small (and negative) in the moderate-x region and $Q^2 > 10$ GeV^2, and, in agreement with theoretical models, it is well described in terms of one parameter $k \approx 0.2$ GeV/c - the transverse momentum generated dynamically. This circumstance allows one to use the

BCDMS data for a stringent QCD test and a reliable determination of the mass-scale parameter Λ. Combining the results of Tables 1-3 and ref. [3], we get:

$$\Lambda_{\overline{MS}} = 204 \pm 53(stat.) \pm 60(syst.) \pm {}^{120}_{\;30}(theor.) \text{ MeV},$$

or, including the HTC uncertainty into the theoretical error:

$$\Lambda_{\overline{MS}} = 230 \pm 20(stat.) \pm 60(syst.) \pm {}^{140}_{\;60}(theor.) \text{ MeV}.$$

The statistical error in former case is dominated by the effect of the HTC. It would be twice as large if the theoretical ansatz in Eq. (4) is replaced by the phenomenological one in Eq. (5). The theoretical error is dominated by the higher-order uncertainty.

The next generation experiments may decrease the statistical and systematical errors in $\Lambda_{\overline{MS}}$ at least 5 times [17] and thus make it possible to perform fine QCD tests, i.e. to clarify running of α_s and the influence of the higher-order corrections. These measurements could be also of great importance for lattice computations and Grand Unified Theories provided the theoretical uncertainty in Λ will be diminished by the next-to-next-to-leading order calculations.

References

1. BCDMS, A.C.Benvenuti et al., Phys. Lett. 195B (1987) 91, 97.
2. BCDMS, A.C.Benvenuti et al., Phys. Lett. 223B (1989) 485.
3. BCDMS, A.C.Benvenuti et al., Phys. Lett. 223B (1989) 490.
4. R.Lednicky, CERN/EP/NA4 Note 89-03, 30 October 1989; JINR B2-2-89-791, Dubna 1989.
5. W.A.Bardeen et al., Phys. Rev. D19 (1978) 3998.
6. V.G.Krivokhizhin et al., Z. Phys. C36 (1987) 51.
7. V.G.Krivokhizhin et al., JINR E1-90-330, Dubna 1990, submitted to Z. Phys. C.
8. P.M.Stevenson, Phys. Rev. D23 (1981) 2916; D27 (1983) 1968; P.M.Stevenson, H.D.Politzer, Nucl. Phys. B277 (1986)
9. D.W.Duke, R.G.Roberts, Phys. Rep. 120 (1985) 275.
10. W.Celmaster, R.J.Gonsalves, Phys. Rev. Lett. 42 (1979) 1435; Phys. Rev. D21 (1980) 3112.
11. A.Devoto et al., Phys. Rev. D30 (1984) 541; D.W.Duke et al., Phys. Rev. D25 (1982) 71.
12. S.G.Gorishny et al., Phys. Lett. B212 (1988) 238.
13. L.V.Gribov et al., Phys. Rep. C100 (1983) 1.
14. J.Kwiecinski, Z. Phys. C29 (1985) 147; 561.
15. G.Curci, et al., Nucl. Phys. B175 (1980) 27.
16. M.R.Pennington et al., Phys. Lett. 120B (1983) 204.
17. C.Guyot et al.: "A new fixed-target experiment for precise tests of QCD", Proposal, Saclay, 14 March 1989.
18. T.Applequist, J.Carrazone, Phys. Rev. D11 (1975) 2856.
19. A.DeRujula, H.Georgi, Phys. Rev. D13 (1976) 1296; H.Georgi, H.D.Politzer, Phys. Rev. D14 (1976) 1829.
20. B.J.Edwards, T.D.Gottschalk, Nucl. Phys. B196 (1982) 328.

21. S.Wada, Phys. Lett. 92B (1980) 163.
22. S.P.Luttrell, S.Wada, Nucl. Phys. B182 (1981) 381.
23. M.Gluck et al., Z. Phys. C13 (1982) 119.
24. J.C.Collins, Wu-Ki Tung, Nucl. Phys. B278 (1986) 934.
25. R.Barbieri et al., Nucl. Phys. B117 (1976) 50.
26. A.DeRujula et al., Ann. Phys. 103 (1977) 315.
27. M.Gluck, E.Reya, Nucl. Phys. B145 (1978) 24.
28. A.Buras et al., Nucl. Phys. B131 (1977) 308.
29. R.K.Ellis et al., Nucl. Phys. B212 (1983) 29.
30. J.L.Miramontes, J.Sanchez Guillen, Z. Phys. C41 (1988) 247.
31. R.K.Ellis et al., Nucl. Phys. B207 (1982) 1.
32. V.A.Bednyakov et al., Yad. Fiz. 40 (1984) 770.
33. V.M.Braun, A.V.Kolesnichenko, Nucl. Phys. B283 (1987) 723.
34. S.P.Luttrell et al., Nucl. Phys. B188 (1981) 219.
35. R.L.Jaffe, M.Soldate, Phys. Lett. 105B (1981) 467.
36. J.L.Miramontes et al., Phys. Rev. 40 (1989) 2184.
37. BEBC-WA59, K.Varvell et al., Z. Phys. C36 (1987) 1.
38. R.Voss, Proceedings of the 1987 Int. Symp. on Lepton and Photon Int. at High Energies, Hamburg, 1987; CERN/EP 87-223.
39. EMC, J.J.Aubert et al., Nucl. Phys. B259 (1985) 189.
40. E.V.Shuryak, A.I.Vainshtein, Nucl. Phys. B199 (1982) 451; B201 (1982) 144.

Derivation of Weak Neutral Current Structure Functions from Distributions in a Hadronic Scaling Variable

Harald P. Borner

The Queen's College

OXFORD OX1 4AW, U.K.

Recently, we outlined [1] a new method to determine (anti)neutrino neutral current (NC) x_{Bj} distributions by using the variable $w = \frac{p_{T,Had}^2}{2p_{L,Had}M_N}$, constructed from the hadronic system X of an inclusive NC reaction like $\nu + N \to \nu + X$ with $p_X = (\nu + M_N, p_{L,Had}, \mathbf{p}_{T,Had})$, and shown to be equal to $x_{Bj}(1-y_{Bj})$ in the Bjorken scaling limit [2]. The basic idea was to concentrate on that part w of $x_{Bj} = w + \frac{p_L - \nu}{M}$ that can be measured sufficiently accurate, and to exploit the fact that we do not have to know x_{Bj} for each event individually if we are interested in $\frac{d\sigma}{dx}$ only.

The use of this variable is advantageous from the point of view of experimental analysis for various other reasons. For instance, discrimination against CC events faking NC events, due to an outgoing lepton having been misidentified as a hadron, is easy since those events cluster at $w = 0$. The apparent Q^2 variation of the distributions in the scaling variables, due to a cut in the hadronic energy $E_H = \nu + M_N$, is much weaker for $\frac{d\sigma}{dw}|_{Q^2 fix}$ than that of $\frac{d\sigma}{dx}|_{Q^2 fix}$. Furthermore, the mean Q^2 for a fixed w bin, $<Q^2>_w$, is significantly less dependent on w than $<Q^2>_x$ is on x_{Bj}, where a strong correlation is found.

To apply the method we had to incorporate some knowledge about the cross section dependence on y_{Bj}. However, we found that the influence of the input marginal distribution in y_{Bj} on the result $\frac{d\sigma}{dx}$ is weak, pointing to a rather direct dependence of $\frac{d\sigma}{dx}$ on the shape of the w distribution.

Here we intend to show that all the information about the weak NC structure functions $F_i^{\nu NC}, i = 1, 2, 3$ is actually contained in $\frac{d\sigma}{dw}^{\nu NC}$, in the Bjorken scaling limit. No (parton or other) model assumptions are necessary to obtain the results, we only assume the general structure of the cross section which follows from Lorentz covariance and the V,A nature of the NC interaction [3].

Our method rests on two main facts: firstly, ν and $\bar{\nu}$ NC structure functions are identical since the hadronic neutral current is hermitian [4]. Secondly, thinking of them as operators $F_i(*, Q_0^2)$ for fixed Q_0^2, to determine their behaviour it is not necessary to have them act on x_{Bj}, the variable which gives them their physical interpretation. If we can establish a relation between functions, like $F_i(*) = \mathcal{F}_i\{f_j(*)\}$, the f_j tell us all about the F_i, no matter whether we use x_{Bj} as an argument or not. Let us start by writing the ν (and $\bar{\nu}$) NC cross section as a function of x, w:

$$\frac{d^2\sigma^{NC}}{dwdx} = \frac{G^2 s}{2\pi}\frac{1}{x}\{\frac{1}{2}(1-\frac{w}{x})^2\, 2xF_1(x) + (\frac{w}{x}(1+\epsilon) - \epsilon)\, F_2(x) \pm \frac{1}{2}(1-\frac{w^2}{x^2})\, xF_3(x)\} \quad (1)$$

where we did not write out the dependence of the F_i on Q_0^2, and we denote $\epsilon = \frac{xM}{2E}$. The lower sign applies to the antineutrino case. Introducing

$$\eta(x, Q_0^2) = \frac{1 + R(x, Q_0^2)}{1 + (2Mx/Q_0)^2}, \quad (2)$$

where $R \equiv \sigma_L/\sigma_T$ so that $F_2 = 2\eta x F_1$, and neglecting the target mass term $y\epsilon$, we obtain

$$\frac{4\pi}{G^2 s}\frac{d^2\sigma^{NC}}{dwdx} = \{1 + 2(\eta(x) - 1)\frac{w}{x} + (\frac{w}{x})^2\}\, 2F_1(x) \pm (1 - (\frac{w}{x})^2)\, F_3(x) \quad (3)$$

We make the approximation that the structure functions scale (the time for extracting scaling violations in νNC reactions has not yet come) and assume the Bjorken limit $\nu, Q^2 \to \infty, Q^2/\nu = cst$. Thus $\eta = 1 + R(x)$. The helicity structure of the weak interaction, giving rise to terms up to order w^2 only, now allows the following procedure: we integrate over all possible x_{Bj} values for fixed w:

$$\frac{4\pi}{G^2 s}\frac{d\sigma^{NC}}{dw} = \int_w^1 dx(2F_1 \pm F_3) + w^2 \int_w^1 dx \frac{2F_1 \mp F_3}{x^2} + 2w \int_w^1 dx(\eta - 1)\frac{2F_1}{x} \quad (4)$$

where we suppressed the dependence of the F_i and η on x_{Bj} for brevity. Taking the derivative of this w distribution $\sigma(w) \equiv \frac{d\sigma}{dw}$ then gives

$$\frac{2\pi}{G^2 s}\frac{d^2\sigma^{NC}}{dw^2} = -2F_1(w) + w \int_w^1 dx \frac{2F_1 \mp F_3}{x^2} + \frac{d}{dw}\{w \int_w^1 dx(\eta - 1)\frac{2F_1}{x}\}. \quad (5)$$

This holds seperately for neutrino and antineutrino, so that we can take linear combinations, keeping in mind the precious property $F_i^{\nu NC} = F_i^{\bar{\nu} NC}$:

$$\frac{2\pi}{G^2 s}(\frac{d^2\sigma^\nu}{dw^2} - \frac{d^2\sigma^{\bar{\nu}}}{dw^2}) = -2w \int_w^1 dx \frac{F_3(x)}{x^2} \quad (6)$$

independent of $\eta(x)$, as expected. Defining

$$\Gamma_\pm(w) = \frac{2\pi}{G^2 s}(\frac{d\sigma^\nu}{dw} \pm \frac{d\sigma^{\bar{\nu}}}{dw}) \quad (7)$$

as the scaled sum and difference of the w distributions, we obtain the simple result

$$F_3(w) = \frac{1}{2}(w\Gamma''_-(w) - \Gamma'_-(w)) \quad (8)$$

where the primes indicate derivatives w.r.t. w. The solutions we present are unique after imposing $\Gamma_\pm(1) = \Gamma'_\pm(1) = 0$.

Since the w distributions can be parametrised in a simple way [1], the determination of the derivatives does not pose as difficult a problem as it may appear at first sight. In addition, the $\frac{d\sigma^{NC}}{dw}$ fall so rapidly that a point to point calculation of the derivatives, using the various discretised numerical approximations [5], can be performed.

Note that we end directly with F_3 rather than with xF_3. This is seen to be useful, for instance, if we want to extract the integral over F_3, as suggested by the quark parton model. We finally arrive at

$$\int_0^1 dx F_3(x) = \Gamma_-(0) \tag{9}$$

From (5) it follows that F_1 will be given from Γ_+ alone. This requires the solution of an integral equation that is straightforward for the case $\eta - 1 = 0$ and a little more involved for the general case [6]. In the former case, ie. when the Callan-Gross relation is supposed to hold, one can easily show that

$$2F_1'(w) = \frac{1}{2}(-\Gamma_+'' + \frac{1}{w}\Gamma_+'(w)) \tag{10}$$

which can be integrated to give $2F_1$ in terms of Γ_+' alone. This result partially answers the question why the shape of the w distribution is rather similar to that of F_1 (see [6] for a discussion). It should be noted that by using eqs (8,10) we shall learn about the functions $F_{1,3}$ in the domain of the measured w values, which can extend to much smaller arguments than the accessible x_{Bj} values. In this way we profit maximally from the scaling approximation which leaves us with two dimensionless variables for $\sigma(s)/s$ that we can in principle freely combine into two new independent variables, resulting in a new parametrisation of the cross section. We showed that the particular combination x, $w = x(1-y)$ leads to powerful relations and can be motivated experimentally.

We conclude that the special structure of the neutrino NC cross sections allows to establish a direct link between the NC structure functions and distributions in the scaling variable w. It leads to simple results that justify the investigation of $\frac{d\sigma}{dw}$ in its own right.

References

[1] H. P. Borner, Oxford preprint OUNP-89-20, to appear in NIM A, (1990)

[2] J. D. Bjorken, Phys. Rev. 179, p. 1547, (1969)

[3] F. E. Close, An Introduction to Quarks and Partons, p. 236, London: Academic Press (1979)

[4] E. Leader, E. Predazzi, An Introduction to Gauge Theories and the New Physics, p. 278, Cambridge University Press, (1982)

[5] S. D. Conte, C. de Boor, Elementary Numerical Analysis, Mc Graw-Hill, (1981)

[6] H. P. Borner, in preparation

W/Z AND LEPTON-PAIR PRODUCTION AND PARTON DISTRIBUTIONS

PROTON STRUCTURE FROM $p\bar{p}$ COLLIDERS

F. HALZEN and S. KELLER
*Department of Physics, University of Wisconsin
Madison, WI 53706, U.S.A.*

ABSTRACT

With improved statistics, ambiguities in the quark structure functions of the proton will be the dominant error in the measurement of the W mass and width in $p\bar{p}$ collider experiments. We discuss how these ambiguities can be controlled with minimal reliance on other experiments by measurements of the asymmetry $A(y) = [\sigma_{W^+}(y) - \sigma_{W^-}(y)]/[\sigma_{W^+}(y) + \sigma_{W^-}(y)]$ and cross section ratio $R_\sigma(y) = [\sigma_{W^+}(y) + \sigma_{W^-}(y)]/\sigma_Z(y)$ over a sufficient range of rapidity y. We illustrate our proposal with an analysis of the preliminary CDF data on the W asymmetry.

1. Introduction

When extracting physics from W, Z events one is inevitably confronted with the fact that $p\bar{p}$ machines "collide quark structure functions." For our discussion it is important to realize[1] that at $\sqrt{s} = 630$ GeV weak bosons are produced by valence quarks. At $\sqrt{s} = 1.8$ TeV annihilations of valence with sea quarks dominate the interactions because of the reduced value of $x = M_W/\sqrt{s}$; see Fig. 1. Precision experiments require precise knowledge of quark structure functions. Two crucial measurements are the determination of M_W and Γ_W. With the Z-boson parameters accurately measured by e^+e^- colliders, the $p\bar{p}$ colliders are still our leading source of information on $W^\pm$ properties. The pivotal nature of the measurement of the W-mass is illustrated in Fig. 2. It tabulates the most probable values for the top mass derived by solving the relation[2]

$$\Gamma(\mu \to e\bar{\nu}_e \nu_\mu) = F(M_W, M_Z, M_{\text{top}}) . \tag{1}$$

Here F represents the one-loop corrected calculation of the precisely measured muon decay width in the standard model. The top mass enters the calculations via $t\bar{b}$ and $t\bar{t}$ loop corrections to W and γ, Z propagators. We obtain from Fig. 2

$$M_{\text{top}} = 143 \pm 38 \text{ GeV} \tag{2}$$

with

$$M_{\text{top}} < 198 \text{ GeV at 90\% C.L.} \tag{3}$$

One can dream and reduce the measurement error ΔM_W on the W-mass, which is the dominant ambiguity in (1), from 500 MeV to 150 MeV. For the same central

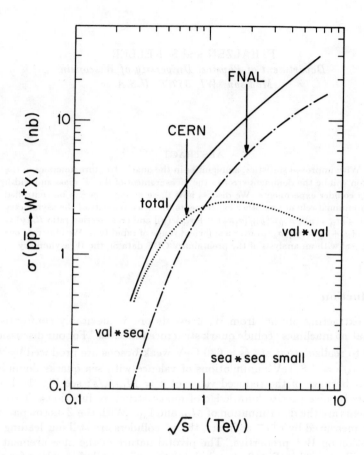

Figure 1. W-production cross section for $p\bar{p}$ collisions as a function of $\sqrt{s}$. Also shown are the separate contributions from valence-valence and valence-sea annihilations. The contribution from sea-sea quarks interactions is small at present energies. Parton distributions of MRS 1 were used.

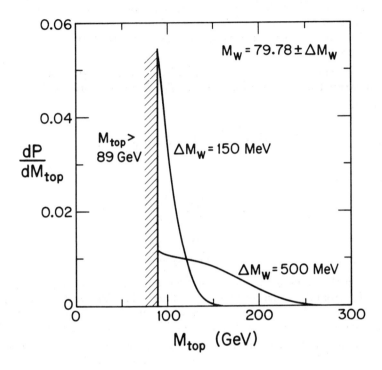

Figure 2. Probability distribution for the top mass from a one-loop calculation of the muon lifetime. The calculation is shown for the present and future error on the value of M_W.

value[3] of $M_W = 79.78$ GeV, the value of M_{top} obtained from (1) is 104 ± 12 GeV. In this dream the top mass is known and its discovery imminent. In reality however $\Delta M_W \gtrsim 100$ MeV as a result of our imprecise knowledge of structure functions.[4] Structure functions affect the shape of the jacobian peak of the decay electron from which the value of M_W is derived. In the near future quark structure functions will represent the dominant error, larger than errors related to statistics, the underlying event structure, energy scale, *etc.*

The importance of improving our knowledge of quark structure functions is further illustrated by the measurement of Γ_W from the ratio of $W \to e\nu$ and $Z \to ee$ events[5]

$$\frac{\#W \to e\nu}{\#Z \to ee} \equiv R = R_\sigma R_\Gamma \frac{\Gamma_Z}{\Gamma_W} \,. \tag{4}$$

In Eq. (4)

$$R_\sigma \equiv \frac{\sigma(p\bar{p} \to W)}{\sigma(p\bar{p} \to Z)} \,, \tag{5}$$

$$R_\Gamma \equiv \frac{\Gamma(W \to e\nu)}{\Gamma(Z \to ee)} \,. \tag{6}$$

R_Γ is fixed in the standard model by $\sin^2 \theta_W$ and α_s. Measuring R allows us to determine Γ_W from Γ_Z measured by e^+e^- colliders. The experimental ambiguity on $\Delta\Gamma_W$ from imprecise knowledge of the quark structure functions is of order 5%, with 3% resulting directly from the calculation of R_σ and 2.5% from corrections for the finite acceptance of the experiment[6] requiring a calculation of the y dependence of the weak bosons. The two ambiguities overlap and the overall error can be reduced by evaluating R_σ within experimental cuts, rather than correcting data on R for geometrical acceptance. At present the statistical errors on R are of the order of 10%. As in the example of the W-mass measurement, the structure functions will become the dominant error in the near future.

2. Improving Structure Function at $Q^2 = M_W^2$?

We will argue that the $p\bar{p}$ colliders can to a large extent control their own destiny in cornering the quark structure of the proton with improved precision. The dominant ambiguities are the ratio $d(x)/u(x)$ and the charm structure function $c(x)$ which significantly contributes to R at the higher energy of the Tevatron.[1,7] Our main point is that from the measurements of $\sigma_{W^+}(y)$, $\sigma_{W^-}(y)$ and $\sigma_Z(y)$ one can determine two ratios of cross sections with high precision:

$$A(y) = \frac{\sigma_{W^+}(y) - \sigma_{W^-}(y)}{\sigma_{W^+}(y) + \sigma_{W^-}(y)} \,, \tag{7}$$

and

$$R_\sigma(y) = \frac{\sigma_{W^+}(y) + \sigma_{W^-}(y)}{\sigma_Z(y)} . \qquad (8)$$

A and R are free of normalization errors, theoretical (K-factors) as well as experimental errors (luminosity). We will illustrate how $A(y)$ corners $d(x)/u(x)$ while $R(y)$ determines $c(x)$.

Why $A(y)$ is sensitive to d/u is illustrated in Fig. 3. As $u(x) > d(x)$ in the vicinity of $x = M_W/\sqrt{s}$, W^+ bosons produced via $u\bar{d}$ are boosted towards positive rapidity relative to the proton beam, see Fig. 3. Structure functions with smaller d/u ratio give the larger asymmetry $A_{W^+}(y)$. The measurement can also be performed with W^- as $A_{W^-}(y) = A_{W^+}(-y)$. Also the sensitivity of $R(y)$ to $c(x)$ is easy to understand. Although the production of $W^\pm$, Z via charm is small, the rate for $c\bar{c} \to Z$ is much smaller than $c\bar{s} + \bar{c}s \to W$ thus significantly affecting the cross section ratio. An illustrative calculation is shown in Fig. 4.

It has not escaped our attention that a measurement of $R(y)$ cannot be used to determine both $c(x)$ and Γ_W from Eq. (4). In order to achieve a determination of Γ_W it is important to pursue alternative ways to determine the charm structure of the proton, e.g., by the process[8]

$$g + c \to \gamma + c(\to \mu) . \qquad (9)$$

Production of prompt photons by gq Compton scattering proceeds preferentially via charge $\frac{2}{3}$ u and c quarks as the cross section is proportional to αe_q^2. One can tag c-quarks by their leptonic decay as suggested in Eq. (9). Events with prompt photons produced in association with prompt muons thus measure the charm content of the nucleon. Another possibility is to measure the process[1]

$$c\bar{s} + \bar{c}s \to W + c(\to \mu) , \qquad (10)$$

where the charm in the final state is the spectator c-quark associated with the active quark forming the W. This charm quark could be tagged by muon decay as before, but its detection is likely to be more difficult as it will tend to be part of the beam jet.

3. Structure Functions at $Q^2 = M_W^2$: Some Numerical Games

We will now illustrate the feasibility of our proposal using the preliminary CDF data[9] on the W asymmetry shown in Fig. 5. We will parametrize quark structure functions in the usual form

$$q(x, Q^2 = M_W^2) = a(1-x)^n/x^J \qquad (11)$$

with

$$J \equiv \left(\frac{1}{2} + \epsilon\right) \text{ or } (1+\epsilon) \qquad (12)$$

for valence or sea distributions. The d_v/u_v ratio is modelled with a sea-type parametrization. For u_v and d_v/u_v, the J-parameter will be fixed by the sum rule

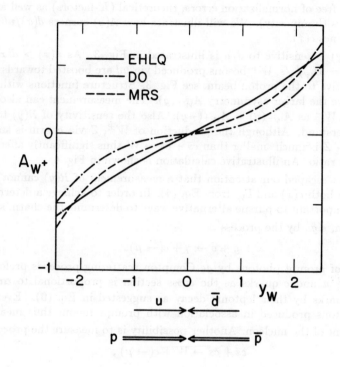

Figure 3. Asymmetry $A(y)$ for W^+ production. Because $u > d$ the W^+ is boosted in the direction of the proton beam.

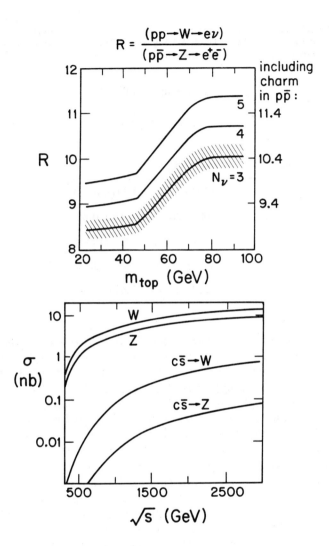

Figure 4.
a) The ratio R is plotted as a function of the top mass M_{top} assuming 3, 4, and 5 light neutrinos. The calculation is repeated including (right-hand scale), and excluding (left-hand scale) charm quark contributions to W, Z production.
b) The charm quark contribution to W and Z production cross section as a function of $\sqrt{s}$ is compared to the total cross section. The parton distributions of EHLQ 1 were used.

Table 1. The values for a, n, $J(\epsilon)$ obtained by fitting the generic parametrization of Eq. (11) to the parton distributions EHLQ 1, DO 1, MRSB′ and MRSE′. The "errors" represent the range for each structure function covered by the different sets.

$$q(x) = a(1-x)^n / x^J$$

$$J = \left(1, \frac{1}{2}\right) + \epsilon$$

$q(x)$	$a\,(\times 100)$	n	$\epsilon\,(\times 100)$
u_v	192 (±50)	3.39 (±0.33)	sum rule
d_v/u_v	68 ± 28	1.47 ± 0.62	sum rule
$u_s = \bar{u}_s = \ldots$	10.7 (±0.82)	7.98 (±0.65)	30.0 (±3.2)
$s = \bar{s}$	8.02 (±2.67)	7.98 (±0.65)	31.7 (±4.9)
$c = \bar{c}$	3.86 (±1.5)	9.67 (±1.9)	42.6 (±10)

counting 2 u's and 1 d in the proton. In order to illustrate our strategy for determining $d_v(x)/u_v(x)$ and $c(x)$, we need some information about the other structure functions. We construct a set of structure functions of the form (11) by fitting to off-the-shelf sets [10-12] EHLQ 1, DO 1, MRSB′, and MRSE′. The parameters, along with their variation implied by the different sets, are tabulated in Table 1. It is of course very difficult to evaluate how realistic the quoted errors are, but they are likely to reflect the ambiguities in the input low Q^2 data. EHLQ 1 is the only set derived without assuming SU(3) symmetry for sea quarks. In order to reproduce the CCFR measurement [13] of the suppression of the strange sea at low Q^2, and to allow at the same time evolution to $Q^2 = M_W^2$, we restricted a_s to be $(\frac{1}{2}-1)\,a_{\bar{u}}$. Asymptotically of course, $s = \bar{u}$. Unfortunately, at present the 10% statistical error on R does not allow a meaningful extraction of $c(x)$. We will conclude in the end that the W, Z statistics have to be increased by at least a factor of 10. We therefore use $c(x)$ as given in Table 1. The range of the parameters might not include the true answer in this case.

This sets the stage for a determination of d_v/u_v from the asymmetry data in Fig. 5. Given the limited rapidity range of the present data we only fit the normalization of the d_v/u_v ratio assuming the slope from Table 1. The result of our fit to the electron asymmetry is given in Table 2 and turns out not to be very sensitive to the assumed slope. The "errors" represent the range shown in Fig. 5, along with the best fit. Notice that the experiment measures the electron, not the W-asymmetry. The W-rapidity cannot be reconstructed because the missing-energy

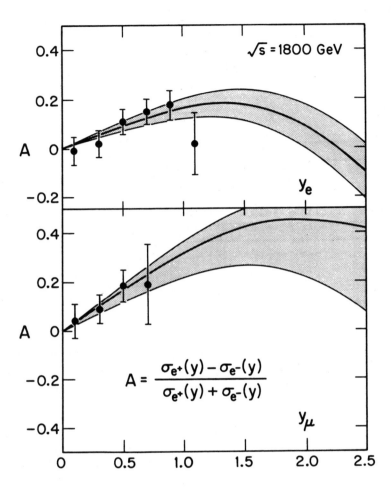

Figure 5.
a) Preliminary CDF result for the asymmetry $A(y_e)$ of the electron rapidity from W production. The solid curve corresponds to our best fit and the dashed area shows the variation of $a_{d/u}$ given in Table 2. Data and fits include a cut on the transverse mass of the W of bigger than 50 GeV.
b) Same as a) for the muon from W production.

Table 2. The d_v/u_v parameters extracted from the electron asymmetry (ASY), the F_{2n}/F_{2p} ratio (DIS), and neutrino data (BEBC). For comparison, we have also shown the d_v/u_v parameters obtained from structure functions, as in Table 1.

	$a_{d/u}\,(\times 100)$	$n_{d/u}$
ASY	• 32^{+14}_{-10}	1.47 ± 0.62
BEBC	• 39 ± 8	0.85 ± 0.28
DIS	• 99 ± 12	2.11 ± 0.25
SETS	• 68 ± 28	1.47 ± 0.62

measurement does not determine the longitudinal momentum of the neutrino from $W \to e\nu$. This does not represent a problem. Inclusion of the decay of the W is straightforward and the electron and W-distributions essentially contain the same information. The d_v/u_v ratio derived from the muon asymmetry, also shown in Fig. 5, is smaller. Because of the poor statistics we will not use this result in subsequent calculations. The exercise illustrates that the preliminary CDF data are relevant to the determination of the quark structure functions, even after allowing for the present errors.

Traditionally, the d_v/u_v ratio is determined from deep inelastic scattering data.[14,15] The most direct method determines u_v, d_v, u_s and d_s from the four measurements $\nu(\bar{\nu}) + p(n) \to \ell + X$. Fitting the BEBC data,[16] shown in Fig. 6, we obtain an independent determination of d_v/u_v which is in nice agreement with the asymmetry result. The fit is shown in Fig. 6 and the parameters listed in Table 2. Yet another probe of the d_v/u_v ratio is provided by comparing the lepton scattering cross section on protons and neutrons. We performed a global fit to measurements[17-19] of the quantity F_{2n}/F_{2p}. The resulting value of d_v/u_v tends to be larger for all data sets. The fit to the BCDMS data is shown for illustration in Fig. 7. The parameters are listed in Table 2. The conclusion is not surprising. Off-the-shelf structure functions underestimate the large-y asymmetry measurements. Inspection of Table 2 verifies that their d_v/u_v is larger than the value we obtained by inverting the $A(y)$ data. This inevitably results in a smaller asymmetry as was qualitatively demonstrated in Fig. 3.

4. How Well Do We Know the Structure Functions?

One of the frustrating aspects of structure functions is that they have no errors. It is extremely dangerous to estimate errors by comparing different sets. *E.g.* EHLQ 1 and DO 1 agree on the value of R calculated from Eqs. (4)–(6). Closer inspection reveals that this agreement is the result of cancellations between the contributions from individual components for which they significantly disagree. This is illustrated in Fig. 8 using fits of the form (11) to the two sets of structure

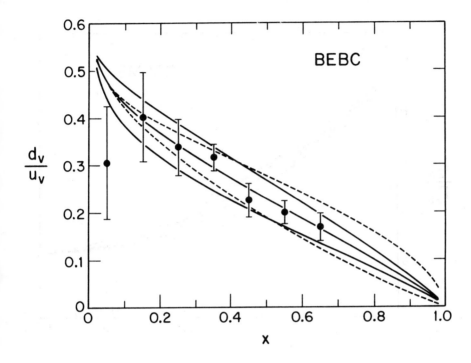

Figure 6. BEBC data for the d_v/u_v ratio as a function of Bjorken-x. The central solid curve corresponds to our best fit listed in Table 2. The other solid and dashed curves correspond to the variation of $a_{d/u}$ and $n_{d/u}$, respectively.

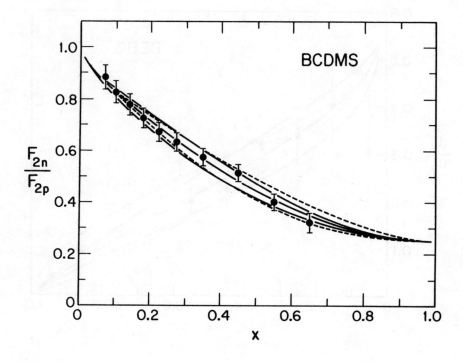

Figure 7. BCDMS data for F_{2n}/F_{2p} as a function of Bjorken-x. Curves as in Figure 6.

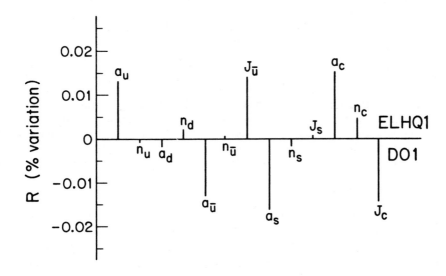

Figure 8. Calculation of R using EHLQ1 and DO1 structure functions. The percentage variation of DO1 relative to EHLQ1 is shown for each parameter. Despite sizeable variations with individual components, the final value for R agrees as a result of cancellations.

functions. It would be equally naive to promote our own results in Tables 1 and 2 as structure functions with errors. Some errors are conservative, others such as those on d_v/u_v are certainly underestimated because we neglected the correlations between the errors. The d_v/u_v errors reflect experimental errors on A, F_{2n}/F_{2p} and not the ambiguities associated with the other structure functions entering the extraction of d_v/u_v. Therefore the different results for this quantity tabulated in Table 2 do not represent a serious discrepancy.

It is nevertheless instructive to tabulate the percentage variation of A and R at $\sqrt{s} = 1.8$ TeV corresponding to the "errors" of Tables 1 and 2. Figure 9 supports our assertion that a measurement of R is sensitive to the charm structure function although the correlation is by no means direct and ambiguities in other components of the structure functions cannot be ignored. A is mostly sensitive to u, d and we should emphasize that this is also the case at $\sqrt{s} = 630$ GeV. The $Sp\bar{p}S$ could make a significant contribution here. Determining $c(x)$ at this energy from R is however very difficult because of the dominance of valence quark annihilation; see Fig. 1.

5. Determination of Γ_W

Using the structure functions of Tables 1 and 2 we evaluate R_σ defined in Eq. (5). The results for the two differing values of d_v/u_v obtained from $A(y_e)$ and F_{2n}/F_{2p} are tabulated in Table 3. Also shown in the table are the experimental values[6,20,21] of R and the values of Γ_W/Γ_Z obtained from Eqs. (4)-(6). For the standard model parameters, we used $M_Z = 91.169$ GeV,[22] $\sin^2\theta_W = 0.220 \pm 0.0095$,[21] and $\Lambda_{QCD} = 200$ MeV. As we are computing ratios, only the error on the $\sin^2\theta_W$ matters. The resulting ratio of the W and Z widths for the two values of d_v/u_v are compared to the standard model prediction in Fig. 10. The values shown were obtained after averaging the results at $\sqrt{s} = 630$ and 1800 GeV. The R-value obtained from the CDF asymmetry agrees with the standard model prediction with $M_{top} > M_W - M_b$. The result calculated from F_{2n}/F_{2p} agrees at the 1σ level.

6. Conclusions

(i) Even with the present statistics W asymmetry measurements probe quark structure functions at $Q^2 = M_W^2$. They are sensitive to d_v/u_v, a quantity otherwise extracted from F_{2n}/F_{2p} data with a history of controversy and disagreements. As is seen in Table 3 precise knowledge of d_v/u_v is less critical at the higher energy.

(ii) In order to determine $c(x)$ from R, a measurement at the 3% level or better is required. The experimental error on R is now of order 10%. It has to be reduced to the 3% level which is the present error on R_σ, an error which is dominated by the charm structure function as shown in Fig. 9. With a factor of 10 increase in W, Z data, R will become sensitive to the charm content of the proton.

Table 3. Results for the ratio $\sigma(W)/\sigma(Z)$, Γ_W/Γ_Z and Γ_W taking $\Gamma_Z = 2.537 \pm 0.027$ GeV from the LEP results.[22] The calculations are repeated at $\sqrt{s} = 630$ GeV and $\sqrt{s} = 1800$ GeV for two different values of d_v/u_v (ASY and DIS) in Table 2. Other structure functions are taken from Table 1.

	$\dfrac{\#W}{\#Z}$	$\dfrac{\sigma(W)}{\sigma(Z)}$	$\dfrac{\Gamma_W}{\Gamma_Z}$
630 GeV	$9.33 ^{+0.77}_{-0.64}$		
DIS		$3.22 ^{+0.14}_{-0.13}$	$0.938 ^{+0.073}_{-0.084}$
ASY		$2.77 ^{+0.26}_{-0.28}$	$0.805 ^{+0.093}_{-0.104}$
1800 GeV	10.2 ± 0.9		
DIS		$3.23 ^{+0.11}_{-0.12}$	0.859 ± 0.080
ASY		$3.21 ^{+0.12}_{-0.13}$	0.854 ± 0.081

Combined:

$$\Gamma_W = 2.28 ^{+0.14}_{-0.15} \quad \text{DIS}$$
$$2.11 \pm 0.16 \quad \text{ASY}$$

7. Acknowledgements

Discussions with A. D. Martin and R. G. Roberts are gratefully acknowledged. This research was supported in part by the University of Wisconsin Research Committee with funds granted by the Wisconsin Alumni Research Foundation, and in part by the U. S. Department of Energy under contract DE-AC02-76ER00881.

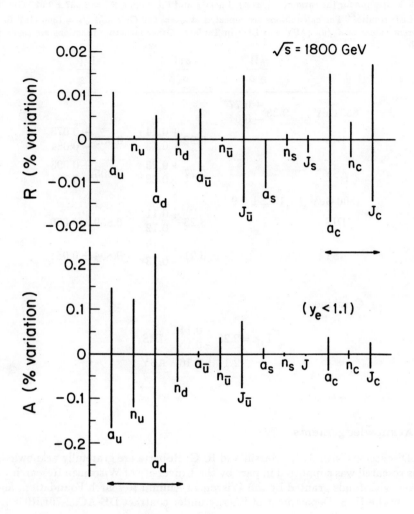

Figure 9. Percentage variation of the ratio R and the asymmetry $A(y_e)$ with each of the parameters listed in Tables 1 and 2. The ASY parameters for d_v/u_v were used. Cuts on the transverse mass of the W as in Fig. 5.

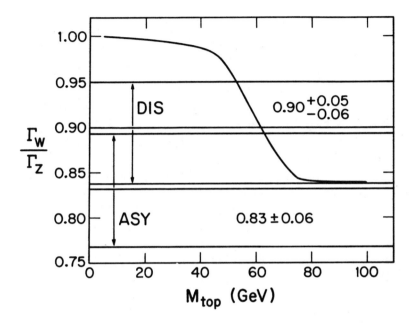

Figure 10. The ratio of the total width Γ_W/Γ_Z as a function of the top mass M_{top}. Also shown are the two results obtained from the ratio of W to Z events in $p\bar{p}$ collisions using the structure functions of Tables 1 and 2 with the ASY and DIS parameters. Some results are listed in Table 3.

8. References

1. E. L. Berger, F. Halzen, C. S. Kim, and S. Willenbrock, *Phys. Rev.* **D40** (1989) 83; Erratum, *ibid*, **D40** (1989) 3789.
2. See *e.g.* F. Halzen and D. A. Morris, *Phys. Lett.* **B237** (1990) 107.
3. L. Pondrom, talk presented at the *Z Phenomenology Symposium*, Madison, Wisconsin, April 1990.
4. W. J. Stirling and A. D. Martin, *Phys. Lett.* **B237** (1990) 551.
5. F. Halzen and M. Mursula, *Phys. Rev. Lett.* **51** (1983) 857.
6. CDF Collaboration, F. Abe *et al.*, *Phys. Rev. Lett.* **64** (1990) 152.
7. K. Hagiwara, F. Halzen, and C. S. Kim, *Phys. Rev.* **D41** (1989) 1471.
8. R. S. Fletcher, F. Halzen, and E. Zas, *Phys. Lett.* **221B** (1989) 403.
9. J. Hauser, these proceedings.
10. E. Eichten, I. Hinchliffe, K. Lane, and C. Quigg, *Rev. Mod. Phys.* **56** (1984) 579; Errata, Fermilab-Pub-86/75-T, 1986.
11. D. W. Duke and J. F. Owens, *Phys. Rev.* **D30** (1984) 49.
12. A. D. Martin, R. G. Roberts, and W. J. Stirling, *Mod. Phys. Lett.* **A4** (1989) 1135.
13. CCFR Collaboration: C. Foudas *et al.*, *Phys. Rev. Lett.* **64** (1990) 1207.
14. A. D. Martin, R. G. Roberts, and W. J. Stirling, *Phys. Lett.* **B189** (1987) 220.
15. F. Halzen, C. S. Kim, and S. Willenbrock, *Phys. Rev.* **D37** (1988) 229.
16. G. T. Jones *et al.*, *Z. Phys.* **C44** (1989) 379.
17. EMC Collaboration, J. J. Aubert *et al.*, *Nucl. Phys.* **B293** (1987) 740.
18. BCDMS Collaboration, A. C. Benvenuti *et al.*, CERN-EP/89-171, submitted to Phys. Lett. B.
19. NMC Collaboration: presented by J. Nassalski at the *Europhysics Conference on High Energy Physics*, Madrid, 1989.
20. UA1 Collaboration, C. Albajar *et al.*, *Z. Phys.* **C44** (1989) 15.
21. UA2 Collaboration, J. Alitti *et al.*, CERN-EP/90-20, submitted to Z. Phys. C.
22. R. D. Peccei, summary talk presented at the *Z Phenomenology Symposium*, Madison, Wisconsin, April 1990.

W/Z and Lepton Pair Production at the Tevatron

JAY HAUSER
Fermilab

ABSTRACT

We look in the CDF data from our 1988-1989 run at $\sqrt{s}=1.8$ TeV for those aspects of W/Z and lepton pair (Drell-Yan) production at the Tevatron which appear most sensitive to parton distribution functions, namely, the W^+ and W^- rapidity distributions and the low mass region of Drell-Yan production. The W rapidity distributions are sampled through the lepton asymmetries in the $e\nu$ and $\mu\nu$ decay channels. The analysis on low-mass Drell-Yan production is just beginning and future prospects are briefly discussed.

1. Introduction

CDF data at $\sqrt{s}=1.8$ TeV allows us to probe the parton distribution functions at high values of q^2 and generally low values of x. Data from deep-inelastic scattering (DIS) experiments constrain the quark and anti-quark distributions fairly well down to $x \approx 0.01$. Since the leading diagrams for W/Z and lepton pair production depend on the quark and anti-quark distributions, the production cross-sections for these processes are well predicted by existing data, with a couple of *important* exceptions: the rapidity distribution of W production, and the production of low-mass and/or high-rapidity Drell-Yan pairs.

The rapidity distribution of W particles depends sensitively on the ratio $d(x)/u(x)$ in the region of $x<0.2$ or so. At $\sqrt{s}=1.8$ TeV, more than 85% of the W's are created by a valence-valence or valence-sea interaction (at all rapidities). Thus, a W^+ will be produced primarily by a u quark from the proton striking a d quark from the antiproton. Because the u quarks have, on average, higher momentum than the d quarks, the W^+ will tend to be boosted along the proton direction, while a W^- will tend to be boosted along the anti-proton direction. By measuring the W rapidity distributions, we can reduce the systematic uncertainties in two important measurements. First, the W mass is measured by fitting the W transverse mass spectrum for different W mass hypotheses. This spectrum is distorted by the finite acceptance in rapidity of our detector. Corrections for this acceptance introduce a systematic uncertainty in the W mass measurement. The size of this uncertainty, on the order of 60 MeV at present[1], will become more significant in future collider runs in which we expect much smaller statistical errors on the W mass measurement. In addition, the measurement of the cross-section ratio $R \equiv \sigma(p\bar{p} \to W \to e\nu)/\sigma(p\bar{p} \to Z \to ee)$ contains systematic uncertainties from the low-x behavior of the u and d quark distributions. This ratio, which has given us our best knowledge of the total W decay width, suffers systematic uncertainties both in the theoretical prediction of the W and Z production cross-sections[2], and in the calculated CDF detector acceptances for the two final states[3]. Both sources of uncertainty can be reduced by an accurate measurement of the W rapidity distributions.

The measurement of low-mass and/or high-rapidity Drell-Yan cross-sections is sensitive to the general behavior of the parton distribution functions at low x. Because of the high center-of-mass energy, very low values of x (down to 10^{-3} or less) can be probed with pairs of moderate mass (>10 GeV/c^2). The large uncertainty in SSC cross-sections for W/Z and Drell-Yan production because of the unknown small-x behavior of the parton distributions has been previously pointed out[4] in the 1988 Snowmass workshop. Measurements of Drell-Yan production at the Tevatron should help reduce this uncertainty.

II. W Rapidity Distributions (Asymmetry)

The data sample which we are using comes from 4.3 pb^{-1} collected in the '88-'89 CDF data run. We have isolated W samples in the electron and muon final states by requiring the lepton to have transverse energy $E_T>20$ GeV, as well as by imposing tight cuts on the identification of the lepton candidates. Because it is important to know the charge of the leptons, we use only leptons detected in the central region, $|\eta|<1.0$ for electrons and $|\eta|<0.7$ for muons. In the electron channel, we also require missing-$E_T>20$ GeV. We then impose a transverse mass cut, $M_T>50$ GeV/c^2, and require that no jets with observed $E_T>10$ GeV be found in the W candidate events. The no-jet requirement not only reduces background, but should also reduce the effect of higher-order W production diagrams, making it easier to compare the data with leading-order calculations. We find 923 $W^+\to e^+\nu$, 994 $W^-\to e^-\nu$, 411 $W^+\to\mu^+\nu$, and 386 $W^-\to\mu^-\nu$ candidate events in our data sample.

We actually do not observe the rapidity of the W particles themselves, but rather the rapidity distributions of the charged leptons from the $W\to\ell\nu$ decay. Finding the rapidity of the parent W particle from the lepton energy and direction and the missing transverse energy involves an equation having two solutions. At CERN collider energies, the rapidity distribution of W production is sufficiently peaked near zero that the ambiguity between these solutions can usually be resolved by taking that solution which minimizes $|Y_W|$. At our energies, the rapidity distribution of W's is sufficiently broad so that this "trick" works very poorly for leptons detected in the central region of our detector. Therefore, we base our comparisons to the predictions from different parton distribution functions on the charged lepton rapidity distributions alone.

In order to reconstruct the lepton rapidity distributions, one must know the reconstruction efficiencies quite precisely as a function of rapidity. It is much easier to just measure the lepton asymmetry defined by the ratio:

$$A(\eta) = \frac{\sigma_+(\eta) - \sigma_-(\eta)}{\sigma_+(\eta) + \sigma_-(\eta)} . \qquad (1)$$

In this ratio, the reconstruction efficiency drops out if it is the same for +/- leptons. We must demonstrate that this is true, and also that the backgrounds in the W sample are negligible.

A number of studies have been performed using our W samples, Monte Carlo events, cosmic ray events, and inclusive muon events to test our assumption that the reconstruction efficiencies are the same for both lepton charges. These studies test this assumption down to the level of 1%. No discrepancies are observed.

The backgrounds from normal QCD events are estimated at less than $(1.0\pm0.5)\%$ in the electron sample and $(0.7\pm0.7)\%$ in the muon sample. The effect of this source of background is to dilute any observed asymmetry (which, as will be shown, is typically less than 15%). Backgrounds from $W \to \tau\nu \to e\nu\nu\nu$ sequential decays are estimated at 3% in both samples. This source of background has the same asymmetry as the signal except for a very small correction due to the transverse mass cut. The net effect of all sources of background is to change the observed asymmetry by less than 1%, and can thus be ignored.

The CDF asymmetry data are shown in Fig. 1 together with curves from a leading-order calculation. This calculation numerically integrates the W rapidity spectrum with the expected angular distribution $\propto (1-\cos\theta^*)^2$ of the leptons in the W rest frame. In the CDF convention, θ is defined from the proton beam direction. We have plotted the lepton asymmetry as a function of $|\eta|$, having combined the asymmetry measurements from $\eta<0$ with the opposite sign (assuming CP conservation) to the measurements from $\eta>0$. The curves represent predictions of EHLQ[5], DO[6], DLFM[7], and HMRS[8] parton distribution function sets. The order of the curves, from largest to smallest asymmetries, is: HMRSB, EHLQ1, EHLQ2, DFLM3, DFLM2, HMRSE, DFLM1, DO2, and DO1. It is apparent that the parton distribution sets which predict the largest asymmetries also agree best with the CDF data.

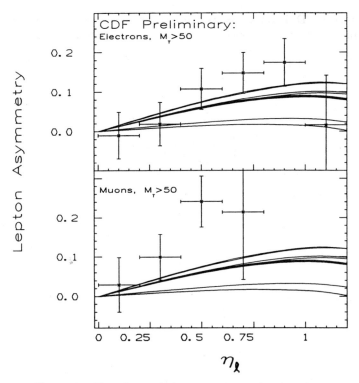

Figure 1: The observed lepton asymmetries in electron and muon event samples, compared to the predictions of a leading-order calculation for various parton distributions.

The effect of higher-order corrections on the observed asymmetry has been investigated with the Papageno Monte Carlo. W+1-jet events were produced requiring $E_T(\text{jet}) > 5$ GeV. Fig. 2 shows the resulting variation of the predicted asymmetry (integrated over $|\eta|<1.0$) with $P_T(W)$. From the lines drawn to fit the Monte Carlo data, we estimate that the mean observed $P_T(W)$ of 4.7 GeV/c in our data sample raises the observed asymmetry by approximately 0.5% from that predicted by the leading-order calculation. Thus, we have limited the corrections due to higher-order effects to a negligible level.

The results of fitting the curves shown in Fig. 1 to the observed data are listed in Table 1. The large χ^2 observed for the DO1 and DO2 distribution functions appear to rule them out. The low asymmetry predicted from Duke&Owens is directly related to a particularly slow fall-off in the $d(x)/u(x)$ ratio in the range $0.01<x<0.2$.

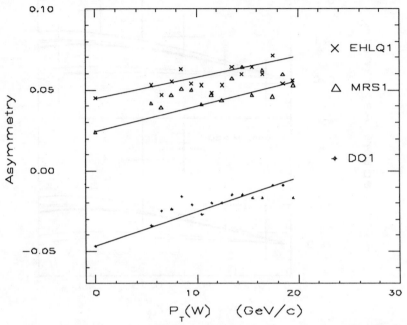

Figure 2: The mean asymmetry in the central region predicted by Papageno vs. $P_T(W)$.

Table I

Function Set	Electrons χ^2, 6 d.o.f.	Muons χ^2, 4 d.o.f.	Combined χ^2, 10 d.o.f.	Combined $P(\chi^2)$
HMRSB	3.5	7.8	11.3	34%
EHLQ1	3.6	8.1	11.8	30%
EHLQ2	4.9	9.3	14.2	16%
DFLM3	5.1	9.3	14.5	15%
DFLM2	5.4	9.5	15.0	13%
HMRSE	5.4	9.8	15.2	12%
DFLM1	5.7	9.8	15.4	11%
DO2	13.6	14.8	28.4	0.16%
DO1	16.6	16.1	32.6	0.03%

III. Drell-Yan Production

We now turn to the case of Drell-Yan production at $\sqrt{s}=1.8$ TeV. In general, one can access di-lepton masses down to about twice the lepton P_T threshold. In our di-electron and di-muon triggers, we ran with nominal P_T thresholds of 5 and 3 GeV/c, respectively. In addition, the di-lepton rapidities (Y^*) which can be accessed are roughly the same as the pseudo-rapidity coverage of the detector systems: $|\eta|<1.0$ for central electrons and $|\eta|<0.7$ for central muons. The relevant equations in the Drell-Yan process are:

$$m_{\ell+\ell-}^2 = s x_1 x_2, \quad x_1 = \sqrt{\frac{m^2}{s}} e^{-Y_{\ell\ell}}$$

Central Drell-Yan production ($Y_{\ell\ell} \approx 0$) accesses x ranges $x_1 \approx x_2 \approx \sqrt{\frac{m^2}{s}}$, whereas forward Drell-Yan (x_2 close to 1) can probe x regions all the way down to $x_1(\min) = \frac{m^2}{s}$. At the Tevatron, di-lepton masses of 10 GeV/c^2 are in principle sensitive down to $x \approx 6 \cdot 10^{-3}$ in the central region and $x \approx 3 \cdot 10^{-5}$ in the most forward region. These x values are low enough to be applicable to calculations of SSC cross-sections for W/Z and Drell-Yan production.

We have analyzed di-electron candidate events in a preliminary analysis which is geared more toward exploring high di-lepton masses for signals from additional Z^0's and quark-lepton compositeness than for measuring the Drell-Yan cross-section down to the lowest possible masses. Nonetheless, the analysis retains reasonable efficiency down to masses of about 30 GeV/c^2. Fig. 3 demonstrates this with an acceptance curve derived from Isajet v6.10. In Fig. 4 is shown the differential cross-section for lepton pairs as a function of pair mass after acceptance corrections. The curve plotted is also from Isajet v6.10, setting parameters $q_T(W)=0$ and using EHLQ1 parton distributions. The data and the Monte Carlo predictions agree down to a mass of 30 GeV/c^2.

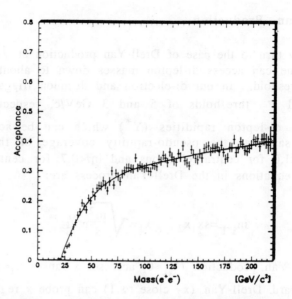

Figure 3: Acceptance curve (from Isajet) for the di-electron analysis.

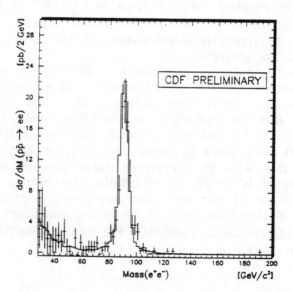

Figure 4: Differential cross-section for $p\bar{p} \to X \to e^+e^-$.

We have also begun work on an analysis of the central di-muon data from the '88-'89 data run. Again using Isajet to simulate Drell-

Yan production, we find that our central di-muon analysis becomes quite efficient when pair masses exceed 10 GeV/c².

In principle, we could use also analyze di-muon events from our forward muon toroid system, which covers pseudo-rapidities between 1.9 and 3.6. However, during our '88-'89 data run, we collected only a very small integrated luminosity while running with a forward di-muon trigger. Unfortunately, the high-rapidity data which is accessible to the forward muon system probes the lowest values of x. This is shown in Fig. 5, where the minimum x probed is plotted versus di-lepton rapidity ($Y_{\ell\ell}$) for several different cases: the present electron analysis, reaching down to pair masses of 30 GeV/c²; a di-electron analysis in progress which should reach down to about 16 GeV/c²; the central muon analysis reaching down to 10 GeV/c², and a hypothesized analysis based on forward di-muon data from our next ('91) CDF data run which might reach down to 16 GeV/c² as well. Given the lack of reliable data from DIS experiments below x≈0.01, it is clear that the CDF Drell-Yan data can probe the parton distribution functions in a previously unexplored region of x.

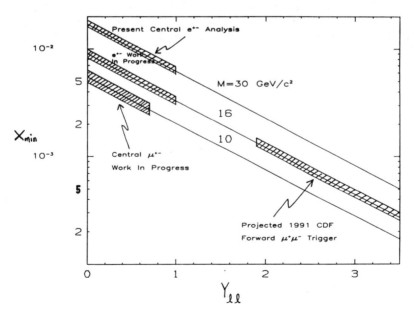

Figure 5: The minimum x probed by Drell-Yan at several pair masses relevant to CDF data analysis.

IV. Conclusions

We have shown that the CDF data can supply some useful information about the parton distribution functions in the low x region responsible for W/Z production and below. The lepton asymmetry in the W sample is on the high side of expectations, and appears to rule out Duke&Owens structure functions. Drell-Yan data down to pair masses of 30 GeV/c^2 agrees with expectations. Further analyses are in progress attempting to measure Drell-Yan production down to lower masses. Finally, with the improvement in luminosity and our increasing knowledge of our detector's capabilities, we expect to measure these processes with much better precision in future CDF data runs.

---

[1] W. Trischuk, ANL-HEP-TR-90-30 section 6.2 (1990, doctoral thesis).
[2] E.L. Berger, F. Halzen, C.S. Kim, and S. Willenbrock, Phys. Rev. D40: 83 (1989).
[3] F. Abe et al., Phys. Rev. Lett. 64:152 (1990).
[4] W.K. Tung et al., Proceedings of the (1988) Summer Study on High Energy Physics in the 1990s: 305 (1989).
[5] E. Eichten, I. Hinchliffe, K. Lane, and C. Quigg, Rev. Mod. Phys. 56: 579 (1984).
[6] D. Duke and J.F. Owens, Phys. Rev. D30:49 (1984).
[7] M. Diemoz, F. Ferroni, E. Longo, and G. Martinelli, Z. Phys. C39:21 (1988).
[8] P.N. Harriman, A.D. Martin, W.J. Stirling, and R.G. Roberts, DTP/90/04 and RAL/90/007 (1990).

MEASUREMENT OF THE RATIO R = $\sigma(W \to \mu\nu)/\sigma(Z \to \mu\mu)$ IN UA1 AT $\sqrt{s} = 0.63$ TeV

M.W. van de Guchte
Nikhef-H, P.O. Box 41882
Amsterdam, 1009 DB, The Netherlands

ABSTRACT

A preliminary analysis of W and Z boson production at UA1, using 4.7 pb^{-1} of new data from the 1988/1989 CERN p$\bar{\text{p}}$ Collider runs at $\sqrt{s} = 0.63$ TeV, yields $\sigma(W \to \mu\nu)$ = 672 ± 50(stat) ± 98(sys) pb, $\sigma(Z \to \mu\mu)$ = 55.0 ± 7.1(stat) ± 7.7(sys) pb and for their ratio R = 12.22 $^{+2.06}_{-1.60}$(stat) ± 1.38(sys). In the framework of the standard model, the measured value of R, combined with an earlier measurement, is used to extract a value for the total width of the W boson, $\Gamma(W) = 2.19 ^{+.27}_{-.30}$ GeV, and to derive a lower limit on the top quark mass of 43 GeV/c^2 (95% CL) independent of the top decay mode.

1. Introduction

Data taken at the CERN p$\bar{\text{p}}$ Collider in the period 1982-1985, in the following referred to as the old data, has been used by the UA1 experiment to determine the production cross sections of the W and Z bosons[1].

A new data sample 7 times larger, corresponding to an integrated luminosity of 4.7 pb^{-1}, has been collected in the 1988 and 1989 runs. During these runs the UA1 detector was operated without an electromagnetic calorimeter. Therefore W and Z detection was only possible in the muon channel, in contrast to the earlier runs where both electrons and muons were used. In spite of the increased luminosity the number of W's and Z's obtained from the new data has only slightly increased with respect to the old data. This is due to the fact that the acceptance and detection efficiency for W→eν is almost 6 times larger than for W→$\mu\nu$. The new sample is used to calculate values of $\sigma_W^\mu \equiv \sigma(\text{p}\bar{\text{p}} \to W+X)\cdot BR(W \to \mu\nu)$ and $\sigma_Z^\mu \equiv \sigma(\text{p}\bar{\text{p}} \to Z+X)\cdot BR(Z \to \mu\mu)$ at $\sqrt{s} = 0.63$ TeV. The constraints on theoretical predictions are not as powerful as the ones obtained from the old data due to larger systematic uncertainties.

Many of the systematic uncertainties (both experimental and theoretical) which contribute to the individual cross sections drop out in the ratio. This ratio can be expressed in the following Standard Model relation[2]:

$$R \equiv \frac{\sigma_W^\mu}{\sigma_Z^\mu} = \frac{\sigma_W}{\sigma_Z} \cdot \frac{\Gamma(W \to \mu\nu)}{\Gamma(W)} \cdot \frac{\Gamma(Z)}{\Gamma(Z \to \mu\mu)} = R_\sigma \cdot R_\Gamma \quad (1)$$

where $\sigma_W(\sigma_Z)$ is the total W(Z) production cross section and the Γ's are the total and partial widths for boson decays. The main uncertainty on the right hand side comes from R_σ. The W and Z total widths are obtained by summing over all decay modes and are therefore sensitive to additional decays of the bosons in other channels. The experimental error on R (~16%) is still larger than the uncertainty of the parameters on the right hand side of Eq. 1 (~3%). It is dominated by the statistical error on the number of Z's. By adding old and new data slightly more than a factor 2 is gained in Z statistics.

Traditionally Eq. 1 has been used for neutrino counting at hadron colliders by determining $\Gamma(Z)$ and taking $\Gamma(W)$ as given by the Standard Model[3]. At present, however, $\Gamma(Z)$ has been precisely measured at the SLC and at LEP[4]. Therefore, it is now possible to extract a value for $\Gamma(W)$ by making use of the complementary results from hadron colliders and e^+e^- machines[5]. $\Gamma(W)$ is sensitive to m_t as long as $m_t < M_W - m_b$, where m_t, m_b and M_W are the masses of the t quark, the b quark and the W respectively, in contrast to $\Gamma(Z)$, which is sensitive to m_t for $m_t < M_Z/2$, where M_Z is the mass of the Z. This m_t dependence is essentially model independent (except for QCD corrections to $W \to t\bar{b}$) in contrast to direct top quark searches at $p\bar{p}$ colliders where always a value for the leptonic branching ratio $BR(t \to bl\nu)$ of about 10% is assumed.

2. UA1 Apparatus

After 1985 several important changes where made to the UA1 detector. The central electromagnetic calorimeters (Gondolas and Bouchons) and the forward calorimeters were removed in preparation for the installation of a new Uranium-TMP calorimeter[6]. The calorimetric measurements now rely on the hadron calorimeters alone, which have an hadronic energy resolution of $\Delta E/E = 80\%/\sqrt{E}$.

The Central Driftchamber (CD) was operated in a new mode in order to cope with the higher currents induced by the increased Collider luminosity. The wire gain was lowered by a factor four, which was partially compensated for by increasing the electronics gain by a factor three. The net effect was a momentum resolution of about $\Delta p/p = 0.02*p$ corresponding to an error of 100% at p = 50 GeV (before 1988 it was $\Delta p/p = 0.005*p$).

The muon detection system was improved by the addition of 820 ton of iron shielding in the forward region. The external muon chambers allow muon detection in the rapidity range $|\eta| < 2.3$, the 1μ trigger however, is limited to $|\eta| < 1.5$.

3. W sample

The effect of the limited CD resolution is illustrated in fig. 1, which shows a MonteCarlo (ISAJET) prediction for the transverse mass of µν pairs, $M_T(\mu\nu)$, from W's before and after detector simulation (with an M_T cut at 30 GeV). The usual Jacobian peak collapses after detector smearing into a distribution where many W's are reconstructed at low M_T, which is a region where a large background is expected from muons coming from semileptonic heavy flavor decays and from pions and kaons decaying in flight.

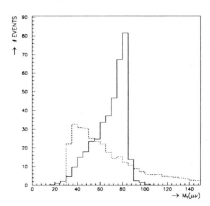

Fig. 1: ISAJET prediction for the W transverse mass before (solid) and after (dashed) detector simulation.

3.1. Event selection and expectation from MonteCarlo

Most of the heavy flavor and decay backgrounds can be removed by the cuts described below:
- p_T(muon) > 15 GeV/c
- good quality track
- muon isolation: $\sum_{\Delta R<0.4} P_T < 1$ GeV/c and $\sum_{\Delta R<0.7} E_T < 5$ GeV, where $\Delta R = \sqrt{(\Delta\phi^2+\Delta\eta^2)}$ is a cone around the muon and where the muon itself is excluded from the momentum and energy sums.
- veto events where jets are nearby or back to back with the muon
- $M_T(\mu\nu) > 30$ GeV/c^2

With this set of cuts 526 W candidates were obtained. After scanning 392 events were good, in 84 events the muon track showed a kink in the CD and in 50 events the scanner could not decide whether the event was good or bad.

The ISAJET MonteCarlo expectation for W's and background processes together with a separate prediction for the decay background[7] is shown in table 1. The only background contributing to kinks comes from π, K decays.

Table 1: The left column displays the observed number of W candidates with the scanning results, the right one shows the MC prediction.

# Events observed	# Events predicted
392 good	351 W
84 kink	9 Z^0
50 doubt	16 DrellYan
526 total	7 $c\bar{c}/b\bar{b}$
	90 π, K decays
	473 total

3.2. Cross section

In order to determine the cross section a 2 parameter maximum likelihood fit of signal and background is done to the data based on the shape of the M_T distribution. The normalization of the W and decay background contribution are left free. Fig. 2 shows the M_T distributions for the data, W, decay background, other backgrounds and the overall fit. It shows that one is sensitive to the different shapes, because above 60 GeV/c^2 one is dominated by W's, while below this value the shape of the background is very different from the signal. Note also that the shapes of all backgrounds are very similar. Therefore, some uncertainty in the normalization of the other backgrounds, especially $c\bar{c}/b\bar{b}$, is taken care of while renormalizing the decay background.

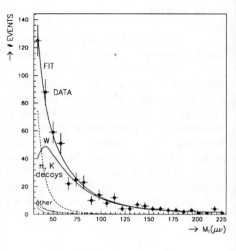

Fig. 2: Transverse mass of data, W, π/K decays, other backgrounds (τ, DY/Z0, $c\bar{c}/b\bar{b}$) and fit.

The result from the fit is that one has to renormalize the W signal and decay background by W-norm and D-norm respectively, where: W-norm = 1.05 ± .08 (stat) ± .05 (shape of background) and D-norm = 1.43 ± .22. The 43% upwards rescaling of the decay background is just allowed by the systematic error on the absolute prediction. After renormalization the amount of decay background predicted is 129 events, which is consistent with scanning results (84 kinks + 50 doubtfuls).

With the W normalization now being fixed, the following value for the cross section has been obtained:

$$\sigma_W^\mu = 672 \pm 50(\text{stat}) \pm 98(\text{sys}) \text{ pb}.$$

The contributions to the systematic error are: 8% from the luminosity, 5% from the uncertainty on the shape of the background, 9% from the CD reconstruction uncertainty and some smaller ones. Adding them in quadrature gives in total: 14.5%.

4. Z sample

The situation for Z's is much simpler, because the only background in this case comes from DrellYan pairs. Event selection criteria:
1st track: similar cuts as the ones applied in the W selection.
2nd track:
- p_T(muon) > 8 GeV/c
- muon isolation: $\sum_{\Delta R<0.4} P_T < 10$ GeV/c and $\sum_{\Delta R<0.7} E_T < 10$ GeV
- Mass(μμ) > 50 GeV/c^2

With these cuts 60 Z^0 candidates have been obtained. The expectation from MonteCarlo is 58 Z^0 and 2 DrellYan events. The requirement of a second high p_T track effectively removes all decay background. Fig. 3 displays the dimuon mass spectrum starting at 30 GeV/c^2 for data and MonteCarlo, where the DrellYan background has been subtracted from the data. The following value for the cross section has been obtained:

$\sigma_Z^\mu = 55 \pm 7.1(\text{stat}) \pm 7.7(\text{sys})$ pb.

The systematic errors are similar to ones mentioned earlier in section 3.2, in total 14%.

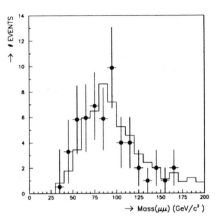

Fig. 3: Dimuon mass of data after background subtraction (points with error bars) and MC (histogram).

5. Comparison with the Standard Model

The different inputs to Eq. 1 will be described with the associated uncertainties. Only the combined UA1 result willl be used for comparison with theory.

Table 2: Measurements of R for the different experiments and associated Z^0 statistics on which each measurement is based.

Experiment	R	# Z^0's
UA1(88/89)	$12.22 ^{+2.06}_{-1.60}(\text{stat}) \pm 1.38(\text{sys})$	60
UA1($\leq$85)	$9.1 ^{+1.7}_{-1.2}$	52
UA1(combined)	$9.99 ^{+1.35}_{-1.18}$	112
UA2[8]	$9.38 ^{+0.82}_{-0.72}(\text{stat}) \pm 0.25(\text{sys})$	169
CDF[9]	$10.2 \pm 0.8(\text{stat}) \pm 0.4(\text{sys})$	193

5.1. R

The results for σ_W^μ and σ_Z^μ were used to calculate a value for the cross section ratio, where the errors were propagated with a simple MonteCarlo technique taking into account partial correlations between systematic errors. The result is displayed in table 2 together with the old[3] and combined UA1, UA2 and CDF results. A maximum likelihood method was used to combine the old and new measurements from UA1. For each experiment the number of Z^0's on which the result is based is indicated, because the statistics in the Z^0 sample is still the dominating uncertainty on R. From the table it follows that the combined UA1 result has an uncertainty of 12.5%, while for the other two experiments it is about 9%. With more integrated luminosity from future Collider runs at CERN and FNAL the uncertainty on R can eventually come down to about 3%, which then is dominated by systematics.

5.2. $\Gamma(W \to \mu\nu)/\Gamma(Z \to \mu\mu)$ and $\Gamma(Z)$

The partial widths can be expressed in the following Standard Model relations:

$$\Gamma(W \to l\nu) = \frac{G_F}{6\pi\sqrt{2}} M_W^3 \quad ; \quad \Gamma(Z \to ll) = \frac{G_F}{12\pi\sqrt{2}} M_Z^3 (1 - 4\sin^2\theta_W + 8\sin^4\theta_W) \quad (2)$$

Recently the measurements of M_Z from CDF, SLC and LEP have dramatically improved. From the four LEP experiments alone one now has: $M_Z = 91.164 \pm 0.033$ GeV[10] (better than 1‰). M_W is known at the 1% level from measurements of UA2[11] and CDF[12], a weighted average gives: $M_W = 80.26 \pm 0.51$ GeV. It is cleaner however, to use measurements of the ratio M_W/M_Z, because in that case the systematic error due to the energy scale drops out. The following value is used, which is again a weighted average of UA2[11] and CDF[12] results:

$$M_W/M_Z = 0.883 \pm 0.006 \quad (3)$$

Also for $\sin^2\theta_W$ results from UA2 and CDF are used[10]. A weighted average gives:

$$\sin^2\theta_W = 0.223 \pm 0.008 \quad (4)$$

With these inputs the ratio of the partial widths is known up to 1%.

SLC and LEP have greatly improved the measurement of $\Gamma(Z)$ with respect to earlier results from UA1, UA2 and CDF. Combining the LEP results gives[10]:

$$\Gamma(Z) = 2.531 \pm 0.027 \text{ GeV} \quad (5)$$

5.3. $R_\sigma = \sigma_W/\sigma_Z$

The main uncertainty on the right hand side of Eq. 1 comes from R_σ. The ratio of the production cross sections of W and Z at parton level, $\sigma(q\bar{q} \to W)/\sigma(q\bar{q} \to Z)$, can be reliably calculated in QCD[2]. However, what is needed is: $\sigma(p\bar{p} \to W)/\sigma(p\bar{p} \to Z)$. The uncertainty comes from the parton flux factors relating $p\bar{p}$ and $q\bar{q}$ cross sections. This factor depends largely on the ratio of structure functions u/d, which can be extracted from data on F_2^p and F_2^n measured in leptoproduction experiments. In fig. 4 recent results on the ratio F_2^n/F_2^p from the BCDMS and NMC collaborations are compared with theoretical expectations for some of the various structure function parametrizations.

Clearly, in the x range applicable to CERN and FNAL, the set E and B parametrizations of the HMRS group[13] describe the data best. R_σ was calculated in 0th order[14] at $\sqrt{s}$ = 630 GeV and 1800 GeV using the HMRSE and HMRSB parametrizations. Taking the mean of the results obtained with the 2 sets gives:

$R_\sigma = 3.18 \pm .08$ at $\sqrt{s}$ = 630 GeV,
$R_\sigma = 3.17 \pm .08$ at $\sqrt{s}$ = 1800 GeV   (6)

The .08 is due to uncertainties in $\sin^2\theta_W$ and M_W, systematics present in the BCDMS and NMC data and the fact that higher order corrections need to be applied to R_σ. The uncertainty in R_σ has now been constrained to within 3%.

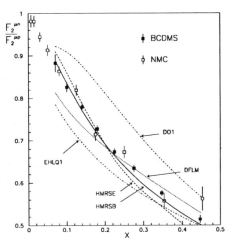

Fig. 4: Ratio of F_2^n/F_2^p as measured by BCDMS and NMC and theoretical predictions using different structure function parametrizations. The error bars are statistical only. For BCDMS (NMC) there is an additional systematic error σ_{sys} = .035 (.010) over the full x range.

5.4. Direct comparison of R with theory

With the input from Eq. 2, 3, 4 and 6 and by summing over all partial decay widths of the W and Z[15], we are now ready to compare the experimental results on R with the right hand side of Eq. 1. R is exhibited as a function of m_{top} in fig. 5 for 3 and 5 neutrino families, which in principle should be displayed separately for $\sqrt{s}$ = 630 GeV and 1800 GeV. As the value of R_σ is almost the same however, advantage is taken to merge them into one plot, so that the Fermilab result can be compared directly to the ones from CERN. The 1σ band originates from δR_σ. The rise in R above about $M_Z/2$ is the result of switching off the decay $Z \to t\bar{t}$, while just below M_W also the decay $W \to t\bar{b}$ is no longer kinematically allowed, so that R becomes independent of m_{top}.

Also indicated in fig. 5 are the measurements of R from UA1, UA2 and CDF. The three results were combined and can be compared to the combination of the UA2 and CDF result alone, in order to show the impact of the UA1 result. Recently, CDF has reported a lower limit on the top mass of 89 GeV/c^2 coming from a direct top search[16]. Therefore, knowing that the top is heavy, that is, $m_{top} > M_W$, from this measurement of R one can first deduce an upper limit on the number of neutrino families from R_{exp}. At 90% confidence level the limits for the 3 experiments are: $N_v < 4.8$ (UA1), $N_v < 3.3$ (UA2) and $N_v < 4.6$ (CDF). The UA1, UA2 and CDF results combined give: $N_v = 2.24 \pm .85$, which is no longer competitive with the recent measurements coming from SLC and LEP ($N_v = 2.96 \pm .14$)[10].

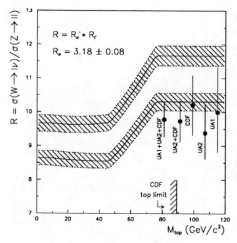 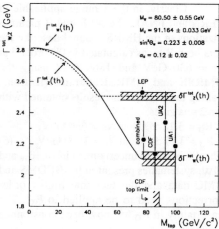

Fig. 5: Comparison of experimental results on R with the Standard Model prediction as a function of m_{top} for 3 and 5 neutrino families.

Fig. 6: Comparison of experimental results on $\Gamma(W)$ and $\Gamma(Z)$ with the Standard Model prediction as a function of m_{top}. $\delta\Gamma^{tot}_W$(th) mainly comes from the uncertainty in M_W, while $\delta\Gamma^{tot}_Z$(th) is due to the uncertainty in $\sin^2\theta_W$. The parameters used in the theoretical calculation are indicated in the upper right hand corner.

5.5. $\Gamma(W)$ and a lower limit on m_{top}

Instead of taking the Standard Model value for $\Gamma(Z)$, one can use the experimental result given in Eq. 5, so that a value for $\Gamma(W)$ can be extracted from Eq. 1. Fig. 6 displays the Standard Model prediction for $\Gamma(W)$ as a function of m_{top}, together with the measurements from UA1, UA2 and CDF and the combined result. From $\Gamma(W)$ alone the three experiments do not yet exclude the possibility of a light top. In table 3 numerical values for $\Gamma(W)$ and lower limits on m_{top} (at 95% CL) are listed. As the mass limits are independent of the top decay mode, the results have to be compared with results from LEP, which give $m_{top} > M_Z/2 \simeq 45$ GeV/c^2. Note that the combined result on $\Gamma(W)$ has an uncertainty of about 5%, which in future Collider runs will eventually come down to about 3% probably before the start of LEP II.

Table 3: Measurements of $\Gamma(W)$ and lower limits on m_{top} (95% CL) for the different experiments.

Experiment	$\Gamma(W)$ (GeV)	top lim. (GeV/c^2)
UA1	$2.19^{+.27}_{-.30}$	43
UA2	$2.34^{+.20}_{-.22}$	33
CDF	$2.14^{+.20}_{-.20}$	51
combined	$2.22^{+.12}_{-.14}$	53

6. Conclusion

With new data from the 1988/1989 runs our measurement of R has improved with respect to an earlier result due to an increase in Z^0 statistics. With additional recent results from e^+e^- machines just available it was possible to derive a value for the total width of the W boson with an experimental error of 12.5%. From this a lower limit on the mass of the top quark was extracted of 43 GeV/c^2 at 95% CL independent of the top decay mode. The experimental error on $\Gamma(W)$ is reduced to 5% if one combines the measurements of R by UA1, UA2 and CDF, which gives a lower limit on the top mass of 53 GeV/c^2 at 95% CL.

REFERENCES

[1] UA1 Collaboration, C. Albajar et al., Z. Phys. **C44** (1989) 15.
[2] F. Halzen, *Top Search*, MAD PH/372 (1987).
[3] C. Stubenrauch, *Thèse de Doctorat d'Etat*, Université Paris-Sud, note CEA-N-2399 (1987)
  UA1 Collaboration, C. Albajar et al., *Phys. Lett.* **B198** (1987) 271.
  UA2 Collaboration, R. Ansari et al., *Phys. Lett.* **B186** (1987) 440.
  P. Colas, D. Denegri and C. Stubenrauch, Z. Phys. **C40** (1988) 527.
[4] MARK II Collaboration, G.S. Abrams et al., *Phys. Rev. Lett.* **63** (1989) 724.
  G. Altarelli, *Summary on the Z^0 Line Shape at LEP*, CERN seminar, 1-3-1990.
[5] D. Denegri and C. Stubenrauch, *The Total Width of the W and a Lower Limit on the t-Quark Mass*, CERN-EP-90-28 (1990).
[6] M. Albrow et al., *Nucl. Instr. Methods* **A265** (1988) 303.
  A. Gonidec et al., *Ionization chambers with room-temperature liquids for calorimetry*, CERN-EP-88-36 (1988).
[7] M.P. Jimack, *A Study of Heavy Flavour Production in Muon Evenets at the CERN $p\bar{p}$ Collider*, RAL/T/053 (1987).
[8] UA2 Collaboration, J. Alitti et al., *Measurement of W and Z Production Cross Sections at the CERN $p\bar{p}$ Collider*, CERN-EP-90-20 (1990).
[9] CDF Collaboration, F. Abe et al., *Phys. Rev. Lett.* **64-2** (1990) 152.
[10] G. Altarelli, *Summary on the Z^0 Line Shape at LEP*, CERN seminar, 1-3-1990.
[11] UA2 Collaboration, J. Alitti et al., *A Precise Determination of the W and Z Masses at the CERN $p\bar{p}$ Collider*, CERN-EP-90-22 (1990).
[12] CDF Collaboration, T.J. Phillips et al., *W and Z Masses and Standard Model Parameters*, to appear in Proc. 8th Topical Workshop on $p\bar{p}$ Collider Physics, Castiglione della Pescaia, 1989.
[13] P.N. Harriman et al, *Parton Distributions Extracted from Data on Deep-Inelastic Lepton Scattering, Promt Photon Production and the Drell-Yan Process*, DTP/90/04 and RAL/90/007 (1990).
[14] F.A. Berends et al., *Phys. Lett.* **B224** (1989) 237.
[15] P. Colas et al., *Z. Phys. C - Particles and Fields* **40** (1988) 527.
[16] CDF Collaboration, K. Sliwa et al., *Top Search at CDF*, to appear in the Proc. of the Moriond Conf. (1990).

RECENT RESULTS FROM UA2 WITH RELEVANCE TO HADRON STRUCTURE FUNCTION DETERMINATION

The UA2 Collaboration
Bern - Cambridge - CERN - Heidelberg - Milano - Orsay (LAL) -
Pavia - Perugia - Pisa - Saclay (CEN)

presented by
Gary F. Egan
University of Melbourne, Australia

ABSTRACT

The W and Z production cross sections and P_T distributions determined from 7.4 pb^{-1} of data collected during 1988 and 1989 with the UA2 detector are presented. Comparisons are made to theoretical predictions at three different perturbative orders and utilising three different structure function sets. Preliminary results for the inclusive jet P_T distribution are also given.

1. INTRODUCTION

During the operation of the CERN $\bar{p}p$ collider in 1988 and 1989 at $\sqrt{s} = 630$ GeV the upgraded UA2 detector collected an integrated luminosity of 7.8 pb^{-1}. The construction of the detector has been described elsewhere in these proceedings[1].

Measurement of the W and Z production cross sections, their ratio and their p_T distributions provide an opportunity of testing the Standard Model (SM) of electroweak interactions. One of the theoretical uncertainties in the SM predictions arises from the parton distribution functions. However, since this systematic uncertainty together with other experimental and theoretical uncertainties is common to both cross section measurements, the ratio of the cross sections is least affected and provides a precision test of the SM.

This report firstly describes the measurement of the W production cross section, including electron and neutrino identification in UA2, followed by the Z production cross section measurement. The p_T distributions of the W and Z are presented. Finally the status of the inclusive jet cross section analysis is presented.

2. σ_W^e MEASUREMENT

Electron Identification

An electron candidate must have a track reconstructed in the tracking detectors which points to an electromagnetic cluster in the calorimeter. The track must originate from a reconstructed vertex which is displaced less than 250 mm along the beam direction from the center of the detector. In addition, a preshower cluster must be reconstructed which is consistent with the position of the electron candidate track.

Energy corrections are applied according to the precise electron direction and impact point in the calorimeter based on data obtained from 40 GeV test beam electrons. The corrected energy is used together with the direction given by the tracking detectors to define $\vec{p}^{\,e}$, the electron momentum.

Neutrino Identification

The presence of neutrinos in W $\to$ eν decays is deduced by measuring the electron energy and the energies of all other particles (generally hadrons) produced in the event. The missing transverse momentum ($\not{p}_T$) is attributed to the undetected neutrino:

$$\vec{p}_T^{\,\nu} \approx \vec{\not{p}}_T = -\vec{p}_T^{\,e} - \vec{p}_T^{\,rec} . \qquad (1)$$

Here, $\vec{p}_T^{\,e}$ is the reconstructed transverse momentum of the electron candidate and $\vec{p}_T^{\,rec}$ is the total transverse momentum of the other particles, calculated as

$$\vec{p}_T^{\,rec} = (\sum E_{cell} \hat{v}_{cell})_T, \qquad (2)$$

where $\hat{v}_{cell}$ is a unit vector from the interaction vertex to the center of a calorimeter cell, E_{cell} is the weighted sum of compartment energies in that cell, and the sum extends over all cells in the calorimeter ($|\eta| < 3$) excluding the cells assigned to the electron. The W momentum is taken to be the sum of the electron and neutrino momenta, so p_T^W is measured only from the recoiling hadrons

$$\vec{p}_T^{\,W} = \vec{p}_T^{\,e} + \vec{p}_T^{\,\nu} \approx -\vec{p}_T^{\,rec} \qquad (3)$$

The total transverse energy of the recoil system is defined by the scalar sum

$$\tilde{E}_T = \sum |(E_{cell}\hat{v}_{cell})_T|, \qquad (4)$$

where the sum again excludes the cells associated with the electron.

W Sample

The final kinematic selection for the W sample was $p_T^e > 20$ GeV/c, $p_T^\nu > 20$ GeV/c and $M_T > 40$ GeV/c² where M_T is the transverse mass defined as

$$M_T = \sqrt{2\, p_T^e\, p_T^\nu (1 - \cos\Delta\phi_{e\nu})} \qquad (5)$$

where $\Delta\phi_{e\nu}$ is the angle between $\vec{p}_T^{\,e}$ and $\vec{p}_T^{\,\nu}$. The resulting p_T distribution for electrons in the central calorimeter is shown in Figure 1. After this selection, the QCD background was found to be much less than 1% and was therefore neglected in the subsequent analysis[2]. The measured cross section is determined from the equation

$$\sigma_W = \frac{N_W - N_\tau}{\varepsilon\, \eta\, L} \qquad (6)$$

where N_W is the observed number of W events (2041), N_τ is the contribution from $W \to \tau\nu_\tau$ followed by the decay $\tau \to e\nu_e\nu_\tau$ (75.7 ± 1.8 from Monte Carlo (MC) studies), η is the acceptance of the geometrical and kinematical selections (62.0 ± 1.9% from MC), ε is the combined electron detection efficiency (64.8 ± 1.1%), and L is the integrated luminosity (7.4 ± 0.4) pb⁻¹. The uncertainties on the acceptance were estimated by varying the structure functions, the p_T^W distribution and the calorimeter energy scale. The largest uncertainty was due to the structure functions because these affect the rapidity distribution of the W.

After combining the data samples from the different calorimeter regions the W cross section is measured to be : $\sigma_W^e = 660 \pm 15\text{(stat)} \pm 37$ (syst) pb. An alternative analysis method utilising the $\vec{p}_T$ spectrum without using the electron identification criteria resulted in : $\sigma_W^e = 656 \pm 27\text{(stat)} ^{+66}_{-80}$ (syst) pb.

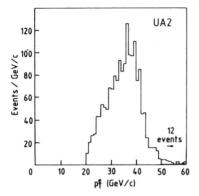 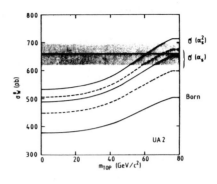

Figure 1. Transverse momentum distribution for electrons in the central calorimeter.

Figure 2. Comparison of the measured σ_W^e to the SM predictions as a function of M_T.

A comparison of the measured cross section to SM predictions is shown in Figure 2 as a function of the top quark mass where the shaded band represents the 1σ confidence interval combining statistical and systematic errors in quadrature. Calculations have been made of the partial widths for all decay channels and of the total cross section at three different perturbative levels; the Born level, including $O(\alpha_s)$ QCD corrections[3], and including a partial calculation of the $O(\alpha_s^2)$ corrections[4].

The theoretical predictions depend on a small number of basic parameters. The value for α_s was computed using Λ_{QCD} from the parton distribution functions and a scale $Q^2 = M_W^2$ (M_Z^2) for the case of the W(Z). The Z mass, computed from a weighted average of LEP[5] and SLC[6] results, was taken to be 91.15 GeV/c². The recent UA2 value of M_W/M_Z [7] was used to compute the weak mixing angle $\sin^2\theta_W = 0.220$ and to derive the W mass, $M_W = 80.5$ GeV/c². A serious uncertainty arises from the parton distribution functions[8]. Several recent sets were chosen to provide a plausible estimate of the uncertainty. The standard set was DFLM[9] with $\Lambda_{QCD} = 160$ MeV, which has been evaluated using next-to-leading order (N.L.O.) QCD calculations in the DIS regularisation scheme (solid curves in figure 2). Two alternative sets MRSE' (lower dashed curve) and MRSB' (upper dashed curve)[10] obtained from N.L.O. QCD calculations performed in the $\overline{MS}$ scheme were also used. The cross section calculation correctly accounts for the regularisation scheme dependence. It is apparent that the parton distribution function uncertainties are large but there are additional uncertainties due to the QCD corrections. The measurements clearly favour the inclusion of radiative corrections but are not sufficiently precise to distinguish between the $O(\alpha_s)$ and the $O(\alpha_s^2)$ QCD corrections.

3. σ_Z^e MEASUREMENT

The sample of events containing at least two electromagnetic clusters satisfying the third level Z-trigger requirements with invariant mass m_{ee} greater than 40 GeV/c^2 has been used to evaluate the cross section σ_Z^e for the process $Z \to e^+e^-$ (γ). When there was a third electromagnetic cluster in the event with a transverse energy greater than 5 GeV, it was included in the invariant mass calculation, thereby retaining candidates for the decay $Z \to e^+e^-\gamma$.

The final Z sample was obtained by requiring events in which at least one cluster satisfied the standard electron cuts and a second cluster which satisfied either the standard or a looser selection criteria. The looser selection criteria recovers events lost because of tracking inefficiency or inefficient track and preshower matching. The invariant mass distribution of the final Z sample is shown in Figure 3. The QCD background in the region $76 < M_{ee} < 110$ GeV/c^2 has been estimated to be 2.39 ± 0.30 events[2]. The Z cross section is then determined from

$$\sigma_Z^e = (N_Z - N_{QCD}) \cdot \frac{(1-f_{\gamma^*})}{\varepsilon \, \eta \, L} \qquad (6)$$

where N_Z is the number of Z candidates (169), N_{QCD} is the number of QCD background events, f_{γ^*} is the relative contribution from single photon exchange and γ^*Z interference terms (1.65%), ε is the combined electron efficiency ($63.3 \pm 1.5\%$), η is the acceptance ($49.6 \pm 1.0\%$), and L is the integrated luminosity (7.4 ± 0.4 pb^{-1}).

After combining the data samples from the different calorimeter regions the Z cross section is: $\sigma_Z^e = 70.4 \pm 5.5(\text{stat}) \pm 4.0(\text{syst})$ pb. In Table 1 the result is compared to SM predictions, at the three levels of perturbation theory as described earlier and using the different structure function sets. These predictions use $Q^2 = M_Z^2$ and, $M_{top} > M_Z/2$. The result is in good agreement using each structure function set. The statistical error will decrease with the inclusion of new data to be collected at the end of 1990, perhaps providing a more discriminating test between structure function sets.

Table 1. Comparison of the values of σ_W^e, σ_Z^e and their ratio R to the SM predictions.

UA2 ('88 & '89)	Perturbation level Structure Function Set Λ_{QCD} (GeV)	Born DFLM 0.16	$O(\alpha_s)$ MRSE' 0.10	$O(\alpha_s)$ DFLM 0.16	$O(\alpha_s)$ MRSB' 0.19	$O(\alpha_s^2)$ DFLM 0.16
$660 \pm 15 \pm 37$	σ_W^e	502	604	651	684	711
$70.4 \pm 5.5 \pm 4.0$	σ_Z^e	49.7	58.9	64.6	68.7	70.3
$9.38^{+0.82}_{-0.72} \pm 0.25$	$R = \sigma_W^e/\sigma_Z^e$	10.08	10.23	10.07	9.94	10.10

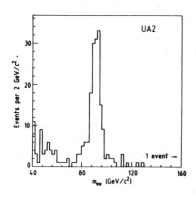
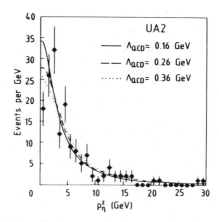

Figure 3. The invariant mass spectrum of the final Z sample.

Figure 4. The p_η^Z distribution compared to QCD calculations for three values of Λ_{QCD}.

The results for σ_W^e and σ_Z^e were used to calculate a value for the cross section ratio. In order to propagate the errors correctly, a simple Monte Carlo technique was used. The procedure correctly took into account the parts of the systematic error such as the luminosity that are fully correlated between σ_W^e and σ_Z^e and the parts that are partially correlated such as the electron efficiencies. The result is

$$R = \sigma_W^e / \sigma_Z^e = 9.38 \, ^{+\, 0.82}_{-\, 0.72} \text{ (stat)} \pm 0.25 \text{ (syst)}$$

and is compared to the SM predictions in Table 1 for $M_{top} > M_W$ and $N_\nu = 3$. Most of the theoretical uncertainties, including the effect of the higher order QCD corrections, are expected to almost completely cancel[4,11] in this ratio. The uncertainties on the absolute values of the structure functions cancel and the residual uncertainty is due to the lack of knowledge of the ratio of the valence structure functions $u_V(x)/d_V(x)$.

4. p_T^Z DISTRIBUTION

The $Z \to e^+e^-$ events are selected as for the Z production cross section analysis, except for the exclusion of the $Z \to e^+e^-\gamma$ events. The transverse momentum of Z bosons (p_T^Z) is evaluated from p_T^{ee}, the measured total transverse momentum of the two decay electrons. The measurement errors on p_T^{ee} are dominated by the energy resolution of the calorimeter, and are estimated to be about 2 GeV.

A more precise measurement can be made for the η component of p_T^{ee}, where the η direction is defined as the inner bisector of the angle between the transverse directions of the two electrons. This component is relatively insensitive to fluctuations in the electron energy measurement, relying mainly on the angles of the electrons which are well measured. A resolution of about 0.3 GeV is estimated. This distribution of p_η^Z is shown in Figure 4, with the predictions of Ref. 12 superimposed on the data. The curves have been modified to account for detector acceptance and resolution, and the predictions are normalized to the observed number of events. The principal theoretical uncertainties are due to the lack of precise knowledge of certain input parameters, namely Λ_{QCD}, the parton distribution functions, and the scale of the running coupling constant. The plausible range of variations is represented by changing the value of Λ_{QCD} used in the calculation; curves are shown for $\Lambda_{QCD} = 0.16, 0.26, 0.36$ GeV (four-flavor values) where the appropriate DFLM structure functions are used in each case and the scale $Q^2 = M_Z^2$ is chosen. The best agreement with the data, is obtained for $\Lambda_{QCD} \approx 0.26$ GeV, as determined from a maximum likelihood fit, but there is acceptable agreement over the range $\Lambda_{QCD} = 0.15$-0.4 GeV (90% CL, statistical errors only).

It must be emphasized that this should not be considered as a measurement of Λ_{QCD}. Instead, Λ_{QCD} has been treated as a parameter which reflects the theoretical uncertainties in the p_T^{IVB} calculation. In order to draw more quantitative conclusions about QCD, a resummed calculation matched to the second order perturbative expression is needed, along with additional statistics in the data sample.

5. p_T^W DISTRIBUTION

The $W \to e\nu$ events are selected as for the W production cross section analysis but using only events with the decay electron hitting the central calorimeter. The residual QCD background of less than 1% was studied using a special data sample. This study concluded that the fraction of background events does not accumulate at high values of p_T^W and therefore would not distort the p_T^W spectrum.

The p_T^W measurement is made only from the recoiling hadrons in the event, as given by Eq. 3. The p_T^{rec} measurement requires a careful study of the detector resolution and systematic error. The resolution of the p_T^{rec} measurement depends strongly on $\tilde{E}_T$ (defined in Eq. 4) and was estimated for W events from a study of the p_T resolution in minimum bias events and two jet events, where the total transverse momentum of the event is expected to balance. The relationship between $\tilde{E}_T$ and p_T^W in W events was determined from the data. The Monte Carlo simulation then generates a p_T^W from one of the theoretical distributions, enabling an associated $\tilde{E}_T$ value to be determined, which in turn enables the p_T^{rec} resolution to be determined[13].

The systematic effects in the p_T^{rec} measurement are examined with detailed Monte Carlo studies which include complete event generation and a detailed simulation of the calorimeter. These studies provide relations between $\langle p_T^{rec} \rangle$ and p_T^W which are parameterized and used in the detector response model in the simple Monte Carlo which is used to generate the curves in Figures 5 and 6. However these effects are difficult to model due to their sensitivity to the detailed energy and rapidity distributions of the particles in the underlying event. Therefore the momentum balance in $Z \to e^+e^-$ events is used to help constrain these relations.

Figure 5 shows the low momentum range of the p_T^W distribution. The mean of the intrinsic distribution corresponding to this range ($0 < p_T^W < 30$ GeV) is measured to be $\langle p_T^W \rangle = 6.4 \pm 0.1 \pm 0.8$ GeV, after unfolding detector acceptance and resolution. The first error is statistical and the second corresponds to variations in the corrections when the two extreme models of the detector response are adopted. The solid curve represents the calculation of Ref. 12 using the median detector response model while the dotted curves show the results using the extreme models.

For the high p_T tail of the distribution, the uncertainties in the detector response are less important. In addition, in this region perturbative calculations are expected to be reliable and the data can be compared with the $O(\alpha_s^2)$ calculation of Ref. 14. This comparison is shown in Figure 6, where the fraction of events is shown as a function of p_T^W for $p_T^W > 20$ GeV. Also shown are the possible variations in the theoretical predictions by varying Λ_{QCD}, structure functions, and the Q^2 scale. The solid and dashed curves are derived from the calculations of Ref. 14 with $Q^2 = M^2$, using DFLM structure functions with values of Λ_{QCD} of 0.160 and 0.360 GeV respectively. The dotted curve is the prediction of Ref. 15 which employs the calculation of Ref. 14 with an optimized renormalization scale of $Q^2 = (0.5 \ p_T^W)^2$, using MRSB' structure functions with $\Lambda_{QCD} = 0.2$ GeV. The results show good agreement with the theoretical predictions with no evidence for a significant excess of events at high p_T.

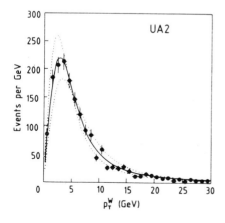

Figure 5. The p_T^W distribution for $p_T^W < 30$ GeV/c compared to QCD calculations (refer text).

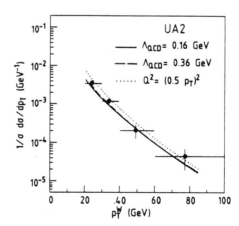

Figure 6. The fraction of high p_T^W events as a function of p_T^W (refer text).

6. SINGLE JET INCLUSIVE CROSS SECTION

An analysis of the inclusive jet cross section is presently being undertaken using a data set corresponding to an integrated luminosity of 7.8 pb^{-1}. The jets are defined according to a cluster merging algorithm. After a primary seed cluster is found, other clusters within a cone of $\sqrt{\Delta\eta^2 + \Delta\phi^2} < 1.3$ around the seed cluster axis are merged to it. This algorithm provides a sharply defined cone for the inclusion of hard, final state radiation which is necessary to compare the jet cross sections with the N.L.O. QCD calculations.

The preliminary acceptance corrected inclusive cross section is shown in Figure 7 for different pseudorapidity intervals, where the errors are statistical only. The steeply falling jet cross section has an overall p_T independent systematic error which consists of an uncertainty in the jet energy scale of 5% together with uncertainties on the luminosity measurement and the acceptance corrections. These result in a preliminary systematic error conservatively estimated at 50%. A comparison to leading order (L.O.) (solid curve in Figure 7) and N.L.O. (dashed curve in Figure 7) QCD calculations[16] using structure functions of Ref. 17 show good agreement with the data. However, at this stage of the analysis the overall systematic error precludes identification of the N.L.O. contributions to the cross section.

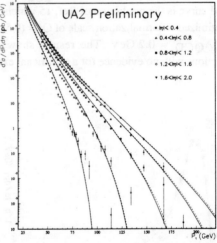

Figure 7. The inclusive jet cross section compared to QCD calculations at the L.O and N.L.O.

7. CONCLUSIONS

The W and Z production cross sections measured by UA2 show good agreement with N.L.O. SM predictions using a range of structure function sets currently available. The measurement p_T distributions for the W and Z also show good agreement with N.L.O. QCD calculations but are limited by systematic errors for the p_T^W distribution and statistical errors for the p_T^Z distribution. The inclusive jet cross section shows good agreement with a recent N.L.O. QCD calculation.

REFERENCES

1. G. Egan (UA2 Collaboration), "Direct Photon Production in UA2", these proceedings.
2. J. Alitti et al., (UA2 Collaboration), CERN-EP/90-20, to appear in *Z. Phys. C*.
3. G. Altarelli et al., *Z. Phys.* **C27** (1985) 617;
   W. J. Stirling, *private communication*.
4. T. Matsuura, "Higher Order Corrections to the Drell-Yan Process", *Ph. D. Thesis, University of Leiden,* 1989;
   T. Matsuura and W. L. van Neerven, *Z. Phys.* **C38** (1988) 623;
   T. Matsuura, S. C. van der Marck, and W. L. van Neerven, *Phys. Lett.* **B211** (1988) 171 and *Nucl. Phys.* **B319** (1989) 570.
5. ALEPH Collaboration, D. Decamp et al., *Phys. Lett.* **B231** (1989) 519 and CERN-EP/89-168;
   DELPHI Collaboration, P. Aarnio et al., *Phys. Lett.* **B231** (1989) 539;
   L3 Collaboration, B. Adeva et al., *Phys. Lett.* **B231** (1989) 509 and L3 Preprint #004;
   OPAL Collaboration, M. Z. Akrawy et al., *Phys. Lett.* **B231** (1989) 530.
6. MARK II Collaboration, G. Abrams et al., *Phys. Rev. Lett.* **63** (1989) 2173.
7. J. Alitti et al., (UA2 Collaboration), *Phys. Lett.* **B241** (1990) 150
8. A. D. Martin, R. G. Roberts and W. J. Stirling, *Phys. Lett.* **B207** (1988) 205.
9. M. Diermoz, F. Ferroni, E. Longo and G. Martinelli, *Z. Phys.* **C39** (1988) 21.
10. A. D. Martin, R.G. Roberts and W. J. Stirling, *Phys. Lett.* **B206** (1988) 327 and *Mod. Phys. Lett.* **A4** (1989) 1135.
11. D. A. Dicus and S. Willenbrock, *Phys. Rev.* **D34** (1986) 148.
12. G. Altarelli, R. K. Ellis, M. Greco and G. Martinelli, *Nucl. Phys.* **B246** (1984) 12;
    G. Altarelli, R. K. Ellis and G. Martinelli, *Z. Phys.* **C27** (1985) 617.
13. J. Alitti et al., (UA2 Collaboration), CERN-EP/90-52, submitted to *Z. Phys. C*.
14. P. B. Arnold and M. H. Reno, *Nucl. Phys.* **B319** (1989) 37.
15. A. C. Bawa and W. J. Stirling, *Phys. Lett.* **B203** (1988) 172 and private communication.
16. S. Ellis, Z. Kurszt and D. Soper, ETH-TH/90-3.
17. A. Martin, R. Roberts and W. Stirling, *Phys. Rev.* **D37** (1988) 1161.

WHAT CAN WE LEARN ABOUT PARTON DISTRIBUTION FUNCTIONS FROM THE DRELL-YAN PROCESS?

JOHN P. RUTHERFOORD

Department of Physics, University of Arizona
Tucson, AZ 85721, USA

ABSTRACT

The present status of the theoretical understanding of the Drell-Yan cross section, integrated over the final state lepton angles, is reviewed to highlight the remaining uncertainties in the numerical calculations. The kinematic range of the data is delimited where useful information on parton distribution functions can be extracted. Those data which are incisive for their information on distribution functions, both old and new, are presented.

1. Introduction

While at first sight virtual photon production in hadron-hadron collisions is an excellent probe of parton distribution functions, closer scrutiny reveals that there are theoretical uncertainties which are much larger than for the space-like regime. Also the kinematic range over which Drell-Yan data can be used is limited, and the precision of the data is not nearly as good as deep inelastic scattering data. Nevertheless the only information on the parton distribution functions in the pion come from this Drell-Yan process and information on the sea quark distributions in the nucleon are highly constrained.[1]

2. Status of the Theory

In recent years the meaning of the Drell-Yan process has been broadened to include, for instance, vector boson production in hadron-hadron collisions. Since such reactions are amply discussed in other sessions at this workshop, I will restrict my considerations to virtual photon production. The Drell-Yan process is $h_A h_B \to \gamma_v X$ followed by $\gamma_v \to \ell\bar{\ell}$ where ℓ is a charged lepton, in practice an electron or muon. The lowest order diagram is as in Fig. 1a and the cross section is

$$\frac{d\sigma}{dx_A dx_B} = \frac{1}{Q^2} \frac{4\pi\alpha^2}{9\tau} G(x_A, x_B)$$

or

$$\frac{d\sigma}{d\tau dx_F} = \frac{1}{Q^2} \frac{4\pi\alpha^2}{9\tau(x_A + x_B)} G(x_A, x_B) = \frac{1}{(x_A + x_B)} \frac{d\sigma}{dx_A dx_B}$$

where
$$G(x_A, x_B) = \sum_i e_i^2 x_A f_i^A(x_A) x_B f_i^B(x_B)$$

$$i = u, \bar{u}, d, \bar{d}, s, \bar{s}, \ldots$$

and $f_i^h(x)dx$ is the probability that quark i in hadron h has momentum fraction in the range $[x, x+dx]$. Also $\tau \equiv Q^2/s = x_A x_B$ and $x_F \equiv 2p_\gamma^*/\sqrt{s} = x_A - x_B$. Note that $Q^2 > 0$, i.e. the virtual photon is time-like as opposed to the case of deep inelastic scattering where it is space-like. This leads to problems at higher order.

Perturbative expansions in QCD are performed in two parameters, α_s and $\ln Q^2/\mu^2$. For the Drell-Yan cross section integrated over Q_T, the expansion can be represented as follows:

$$Q^2 \frac{d\sigma}{dx_A dx_B} =$$
$$+ \alpha_s^0 F_{00}(x_A, x_B)$$
$$+ \alpha_s^1 [F_{11}(x_A, x_B) \ln^1 \frac{Q^2}{\mu^2} + F_{10}(x_A, x_B) \ln^0 \frac{Q^2}{\mu^2}]$$
$$+ \alpha_s^2 [F_{22}(x_A, x_B) \ln^2 \frac{Q^2}{\mu^2} + F_{21}(x_A, x_B) \ln^1 \frac{Q^2}{\mu^2} + F_{20}(x_A, x_B) \ln^0 \frac{Q^2}{\mu^2}]$$
$$+ \alpha_s^3 [F_{33}(x_A, x_B) \ln^3 \frac{Q^2}{\mu^2} + \ldots$$
$$+ \vdots$$

Taking only the highest power of the log at each order in α_s gives the "leading log" approximation. Including the next highest power of the log gives the "next-to-leading log" approximation. At lowest order in α_s the next-to-leading log term is the so-called "constant term", i.e. $\ln^0$. The Drell-Yan cross section without QCD corrections, i.e. at order α_s^0, is the naive formula given on the previous page and represented by the diagram in Fig. 1a. This is the first line above. The leading log approximation gives the same formula as on the previous page but the parton distribution functions are now understood to have Q^2 dependence as in deep inelastic scattering. The full α_s calculation[2] includes all diagrams in Fig. 1 and corresponds to the first two lines above. The formula is not given here but the difference between the full α_s calculation and the leading log approximation can be summarized by a "K-factor" where

$$\frac{d\sigma}{dx_A dx_B}(\text{order } \alpha_s) = K(x_A, x_B, Q^2) \frac{d\sigma}{dx_A dx_B}(\text{leading log})$$

The K-factor at first order in α_s is dominated by the continuation π^2 part which comes from analytic continuation from space-like values of Q^2, where the parton

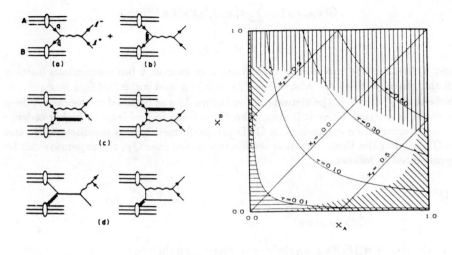

Fig. 1

Fig. 3

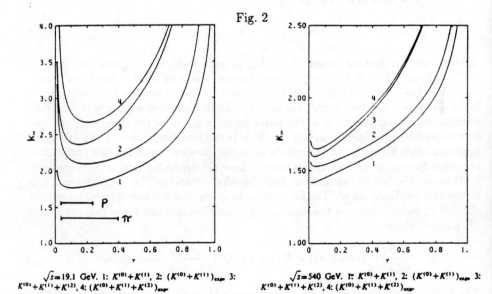

Fig. 2

$\sqrt{s} = 19.1$ GeV. 1: $K^{(0)}+K^{(1)}$, 2: $(K^{(0)}+K^{(1)})_{exp}$, 3: $K^{(0)}+K^{(1)}+K^{(2)}$, 4: $(K^{(0)}+K^{(1)}+K^{(2)})_{exp}$

$\sqrt{s} = 540$ GeV. 1: $K^{(0)}+K^{(1)}$, 2: $(K^{(0)}+K^{(1)})_{exp}$, 3: $K^{(0)}+K^{(1)}+K^{(2)}$, 4: $(K^{(0)}+K^{(1)}+K^{(2)})_{exp}$

distribution functions are defined, to time-like values of Q^2, the domain of the Drell-Yan process. Values of the K-factor near 2 led to worries that the perturbative series wouldn't converge. But it was noticed in these calculations that the so-called Abelian parts of the higher order terms seemed to exponentiate giving satisfactory convergence and that the left over pieces weren't alarmingly large.

The Drell-Yan cross section to second order in α_s has been calculated, in principle[3] (pieces of practical interest have been left out such as the gluon contribution). Fig. 2 shows the K-factor at two different energies without and with exponentiation. Over the kinematic range measured in experiments, these K-factors are roughly constant.

It is conventional to assign Q^2 for both the renormalization scale (argument to α_s) and for the factorization scale (scale argument to the parton distribution function). But this choice is not required and other choices have been investigated.[4] Recently Aurenche et al.[5], using the Principle of Minimal Sensitivity (PMS) found that they could reproduce the second order result of van Neerven et al.[3] (who use the conventional Q^2 for both scales), by optimizing the first order calculation.

Such considerations contribute to uncertainties in the calculated Drell-Yan cross section. If one starts from the best theoretical calculation, i.e. highest order in α_s and uses the best determined parton distribution functions and α_s, one is still left with several choices which affect the numerical answer. One can freely choose the fragmentation scale and the renormalization scale. And further there are some "scheme dependencies" even after a scheme is adopted for the distribution functions taken from deep inelastic scattering. And, of course, there are corrections from higher order terms which aren't calculated. These uncertainties are large compared to those in the case of deep inelastic scattering. Phenomenologists sometimes represent these additional corrections by a K'-factor which multiplies the K-factor.

Much of these uncertainties are introduced because we wish to use parton distribution functions obtained from deep inelastic scattering. Recent motivation to better determine the distribution functions comes from hadron collisions. If one were to define the parton distributions in the time-like region, e.g. in the Drell-Yan process, then the theoretical uncertainties would be reduced. Unfortunately the data one could use to extract distribution functions in the time-like region aren't extensive enough yet.

3. Useful Kinematic Regions

Fig. 3 shows the kinematic region open to the Drell-Yan process. For fixed target experiments it is conventional for hadron A to represent the beam particle and hadron B the target. At fixed center-of-mass energy, the curves of constant τ are also curves of constant Q^2. Fig. 4 shows an example of data superimposed on such a kinematic plot. The shaded region in the lower left of Fig. 3 (below the curve $\tau = 0.01$) represents schematically a region that can't be used to extract parton distribution function data because there are sizeable backgrounds. For most of the fixed target data the ψ and ψ' cut a band out of this region (and the Υ's cut a band out of the kinematic region at a higher value of τ which is not shown). Furthermore there are backgrounds coming from heavy quark decays. Fig. 5 shows an estimate of the contributions to the dimuon yield from various sources[6]. At fixed target energies less is known about the sources of the dimuon yield below the ψ but the sharp departure of the yield from the expected A-dependence as shown in Fig. 6 provides a clear warning.

Just as in deep inelastic scattering, there are higher twist contributions to the yield that here dominate in the kinematic regions near the axes. The E615 group at Fermilab[7] has explored such effects extensively for pion beams. Fig. 7 shows their measurements of the dimuon angular distribution. Large deviations from the expected $1 + \cos^2 \theta$ distribution indicate large higher twist contributions.[8] The kinematic region where this effect significantly contaminates the lowest twist data is near the x_A axis. There are presumably similar contributions near the x_B axis as well but these don't show up in the angle distribution. Fig. 8 shows the quark distribution function in the pion, including the higher twist contribution, at three different values of Q. At $Q = 3.5$ GeV the higher twist contribution dominates a large region near $x_\pi = 1$. At $Q = 5.16$ GeV the higher twist contribution is smaller and at $Q = 8.5$ GeV it is quite small. Fixed target Drell-Yan data are the *only* source of parton distribution functions in the pion and, in principle, in any other unstable hadron as well.

The shaded region in the upper right of Fig. 3 shows schematically where the data is statistics limited.

4. Parton Distribution Functions in the Nucleon

Experiments with good parton distribution function data for the nucleons are fewer than one might expect because these experiments tend to concentrate on resonance searches. Fig. 9 shows the high resolution data, not yet corrected for acceptance, from Fermilab experiment E605 with the three Υ states clearly resolved. The quest for good resolution dictates smaller acceptance which restricts the kinematic coverage. So most of the data is concentrated near $x_F = 0$, i.e. where $x_A = x_B$. Fig. 10 shows a scaling form of the cross section near $x_F = 0$ for protons on a nuclear target. The nuclear cross section has been divided by the atomic number A of the target, i.e. it is assumed that the cross section has an $A^{1.0}$ dependence. Also shown is older data from Fermilab experiment E288, the experiment that discovered the Υ resonances.[9] In both cases the Υ region is excluded. The small discrepancies between the two data sets are consistent with the expected scale violations as can be seen more clearly in Fig. 11. The two curves correspond to the lowest and highest energies respectively.

The HMRS[10] collaboration has used this data in their global fits. They constrain the valence quark distributions from deep inelastic data and fit the Drell-Yan data with a parameterization of the sea quark distributions and a free K' factor. The power of the $(1-x)$ factor in the sea quark distribution is η_S and the magnitude is A_S. Fig. 12c shows χ^2 contours for A_S and η_S from deep inelastic data alone.

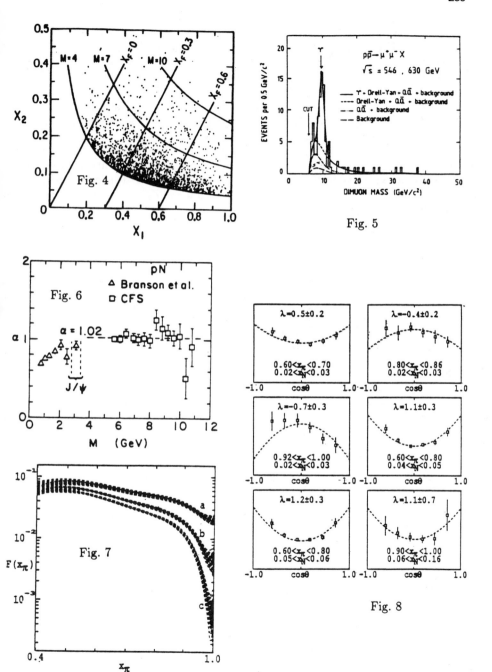

Fig. 4

Fig. 5

Fig. 6

Fig. 7

Fig. 8

Fig. 12b shows contours using deep inelastic muon data (which constrain the valence quarks) and E605 Drell-Yan data (which constrain the sea). This shows the strong correlation between A_S and K'. This data does almost nothing to pin down A_S but it does help (about as well as charged current neutrino data) to determine η_S. The optimum values of K' are shown in Fig. 12a. Here the strong correlation between A_S and K' is obscured by a very strong correlation between A_S and η_S. Combining all the data gives the contours in Fig. 12d with the best fit values shown by a dot.

Global distribution function fits, such as the example in the previous paragraph, might glean a bit more information from the data shown in Fig. 13. This data shows the dependence of the Drell-Yan cross section on rapidity y (or on x_F). Such data are sensitive to the $d(x)/u(x)$ ratio in the nucleon. It is also sensitive to any asymmetries in the sea distributions between $\bar{u}$ and $\bar{d}$, i.e. isospin breaking sea distributions as suggested by Feynman.[11]

5. A-dependence

The EMC collaboration generated much excitement several years ago when they found interesting A-dependence to the nucleon structure functions.[12] Models constructed to explain the effect differed somewhat in their predictions for the A-dependence of the Drell-Yan yield.[13] New data from Fermilab experiment E772[14] cover a wide kinematic range as seen in Fig. 14. This group has succeeded in extracting a precise measurement of the A-dependence of the target as shown in Fig. 15, ruling out some models of the EMC effect.

6. Conclusions

The present theoretical uncertainties are larger for the Drell-Yan process than for deep inelastic scattering but the precision of the data is also not as good. Still the Drell-Yan process is the only handle on quark distribution functions in unstable hadrons such as the charged π. The linear dependence of the cross section on the anti-quark distribution function allows competitive precision in determinations of the nucleon sea. And it gives complementary information on the EMC A-dependence effect.

7. References

1. For reviews of the Drell-Yan process, see J.Huston, Proceedings of the 1989 International Symposium on Lepton and Photon Interactions at High Energies, Stanford Univ., Aug. 7-12 (1989); C.Grosso-Pilcher and M.L.Shochet, Ann. Rev. Nucl. Part. Sci. **36**, 1 (1986); J.P.Rutherfoord, Proceedings of the 1985 International Symposium on Lepton and Photon Interactions at High Energies, Kyoto (1985); Proceedings of the Drell-Yan Workshop, Fermilab, Oct. 7 and 8, (1982); I.R.Kenyon, Rep. Prog. Phys. **45**, 1261 (1982); G.Matthiae, Riv. Nuovo Cim. **4**, No. 3, 1 (1981); I.Mannelli, Proceedings of the 1981 International Symposium on Lepton and Photon Interactions at High Energies, Bonn (1981); L.Lyons, Prog. Part. Nucl. Phys. **7**, 169 (1981); R.Stroynowski,

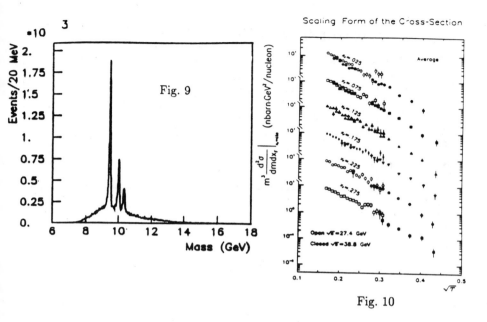

Fig. 9

Fig. 10

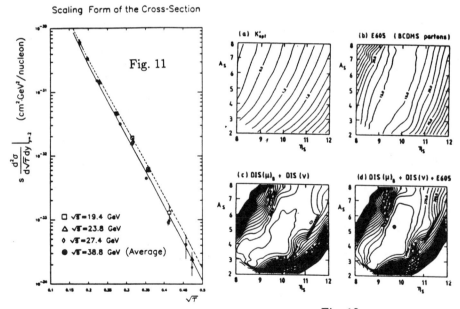

Fig. 11

Fig. 12

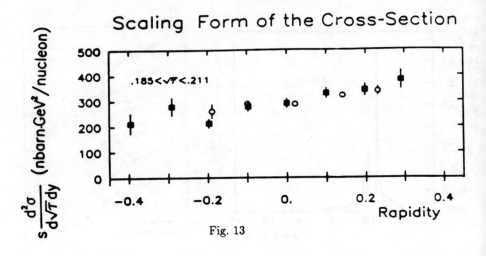

Fig. 13

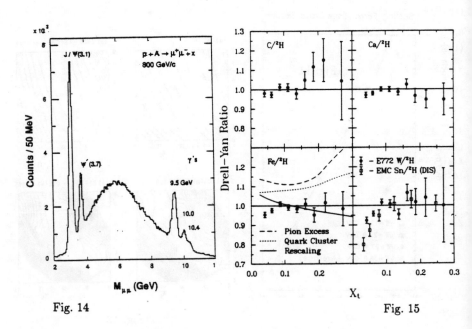

Fig. 14

Fig. 15

Phys. Rep. **71**, 1 (1981); Lepton Pair Production, Moriond Workshop, Les Arcs, France (1981); J.Rutherfoord, Proceedings of the X International Symposium on Multiparticle Dynamics, Goa, India, 25-29 Sept. (1979); J.E.Pilcher, Proceedings of the 1979 International Symposium on Lepton and Photon Interactions at High Energy, Fermilab (1979); L.M.Lederman, Proceedings of the 19th International Conference on High Energy Physics, Tokyo (1978).

2. G.Altarelli, R.K.Ellis, and G.Martinelli, Nucl. Phys. **B143**, 521 (1978); erratum **B146**, 544 (1978); K.Harada, T.Kaneko, and N.Sakai, Nucl. Phys. **B155**, 169 (1979); J.Kubar-Andre and F.E.Paige, Phys. Rev. **D19**, 221 (1979); G.Altarelli, R.K.Ellis, and G.Martinelli, Nucl. Phys. **B157**, 461 (1979); J.Kubar, M.LeBallac, J.L.Meunier, and G.Plant, Nucl. Phys. **B175**, 251 (1980); B.Humpert and W.L.van Neerven, Nucl. Phys. **B184**, 225 (1981).

3. W.L.van Neerven, Phys. Lett. **B147**, 175 (1984); T.Matsuura and W.L.van Neerven, Z. Phys. **C38**, 623 (1988); T.Matsuura, S.C.van der Marck, and W.L.van Neerven, Phys. Lett. **B211**, 171 (1988); Nucl. Phys. **B319**, 570 (1989).

4. P.M.Stevenson, Phys. Rev. **D23**, 2916 (1981); P.M.Stevenson and H.D.Politzer, Nucl. Phys. **B277**, 758 (1986).

5. P.Aurenche, R.Baier, M.Fontannaz, FERMILAB-PUB-90/27-T (Jan. 1990).

6. C.Albajar *et al.*, Phys. Lett. **B186**, 237 (1987).

7. J.G.Heinrich *et al.*, Fermilab experiment E615, Private communication, to be published; J.S.Conway *et al.*, Phys. Rev. **D39**, 92 (1989); S.Palestini *et al.*, Phys. Rev. Lett. **55**, 2649 (1985); K.J.Anderson *et al.*, Phys. Rev. Lett. **43**, 1219 (1979).

8. E.L.Berger and S.J.Brodsky, Phys. Rev. Lett. **42**, 940 (1979); E.L.Berger, Z. Phys. **C4**,289 (1980).

9. A.S.Ito *et al.*, Phys. Rev. **D23**, 604 (1981).

10. P.N.Harriman, A.D.Martin, R.G.Roberts, and W.J.Stirling, DTP/90/04 (Jan. 1990).

11. R.D.Field and R.P.Feynman, Phys. Rev. **D15**, 2590 (1977).

12. J.J.Aubert *et al.*, Phys. Lett. **B163**, 275, (1983); A.Bodek *et al.*, Phys. Rev. Lett. **50**, 1431 (1983); **51**, 534 (1982); R.G.Arnold *et al.*, Phys. Rev. Lett. **52**, 727 (1984); G.Bari *et al.*, Phys. Lett. **B163**, 282 (1985); J.Ashman *et al.*, Phys. Lett. **B202**, 603 (1988); M.Arneodo *et al.*, Phys. Lett. **B211**, 493 (1988).

13. T.Sloan, G.Smadja, and R.Voss, Phys. Rep. **162**, 47 (1988); L.Frankfurt and M.Strickman, Nucl. Phys. **B316**, 340 (1989); L.V.Gribov, E.M.Levin, and M.G.Ryskin, Nucl. Phys. **B188**, 555 (1981); A.H.Mueller and J.Qiu, Nucl. Phys. **B268**, 427 (1986); E.L.Berger and J.Qiu, Phys. Lett. **B206**, 141 (1988); S.J.Brodsky and H.J.Lu, Phys. Rev. Lett. **64**, 1342 (1990); R.P.Bickerstaff *et al.*, Phys. Rev. Lett. **53**, 2531 (1984); E.L.Berger, Nucl. Phys. **B267**, 231 (1986); M.Ericson and A.W.Thomas, Phys. Lett. **B148**, 191 (1984).

14. D.M.Alde *et al.*, Phys. Rev. Lett. **64**, 2479 (1990).

DIRECT PHOTON AND HEAVY QUARK PRODUCTION

THE GLUON STRUCTURE FUNCTION
FROM THE TAGGED PHOTON LAB

M. V. Purohit
Physics Dept.
Princeton University
Princeton, NJ 08544

ABSTRACT

Results on the photoproduction of 10000 charmed particles from the 10^8 recorded triggers of Fermilab experiment E691 have been analyzed in the photon-gluon fusion model. We find that the total cross-section, its rise with energy, and the p_T^2 and x_F distributions can be explained by a high mass for the charm quark ($m_c = 1.74^{+0.13}_{-0.18}$ GeV/c^2) and a soft gluon distribution ($G(x) \sim (1-x)^n$ where n=7.1±2.2).

INTRODUCTION

This is a report of an analysis of charm photoproduced by experiment E691 at Fermilab. The analysis is of four quantites: the total charm cross-section, its rise with energy, the p_t^2 distribution and the x_F distribution. These quantities are analyzed in next-to-leading order QCD.

Photoproduction of heavy quarks[1] as described by PGF[2] is uniquely suited for the measurement of the gluon structure function because only one structure function, $G(x)$, is involved, because in leading order α_S enters in its first power, and mainly because measurement of a *single* outgoing parton is sufficient to determine the entire kinematics. The uncertainties that remain include the intrinsic k_T^2 of the initial state partons and the fragmentation of the outgoing heavy quark. The uncertainty in intrinsic k_T^2 is minimal because only one initial state particle has this k_T^2, which in turn is shared by the two outgoing heavy quarks (reducing the effect on each) and can be further minimized by restricting analysis to the region of high p_T^2 where perturbative QCD predictions are, in any case, on firmer ground. Thus the only remaining uncertainty is the heavy quark fragmentation, which is reasonably well understood from e^+e^- production. In a recent publication[3] it was shown that the next-to-leading order corrections for heavy quark photoproduction are indeed small as compared to those in heavy quark hadroproduction, Drell-Yan and prompt-photon production. Specifically, the total cross-section estimate is increased by only 32% relative to the leading-order calculation and the differential distributions are essentially unaffected.

EXPERIMENT E691

Our charm production experiment, E691, collected data in 1985 using the Tagged Photon Spectrometer at Fermilab and has already been described elsewhere. In summary, the photons ranged in energy from 80 to 230 GeV ($< E_\gamma >=145$ GeV), and were incident on a beryllium target followed by a high resolution silicon microstrip vertex detector and a spectrometer.[4] This analysis is of the total cross-section and differential distributions in p_T^2 and x_F described in a previous publication[1] for the high statistics modes $D^0 \to K^-\pi^+$, $D^0 \to K^-\pi^+\pi^+\pi^-$ and $D^+ \to K^-\pi^+\pi^+$. (Charge-conjugate states are included throughout this paper).

PGF PREDICTIONS

As can be seen from fig. 1, the kinematics in the parton frame is given entirely by the photon energy, E_γ and by x, the fraction of the nucleon's momentum that is carried by the gluon. The parton-level cross-section, $\hat{\sigma}(\hat{s})$, rises roughly logarithmically with the square of the energy in the photon-gluon collision, $\hat{s}$. The total cross-section is most sensitive to m_c, the charm quark mass. The minimum x that is probed by a given photon energy is given by

$$x_{min} \approx \frac{m_{thresh}^2}{2 m_N E_\gamma} \qquad (2)$$

where m_N is the nucleon mass. As the photon energy is increased, more and more of the gluon distribution is probed by going to lower x and thus the rise in cross-section is a measure of the shape of the gluon structure function and is sensitive to the power n_g in the assumed shape:

$$G(x) = (1-x)^{n_g} \qquad (3)$$

where G(x) is the momentum density of gluons in a nucleon. Although this shape (motivated by counting rules) is expected to be valid only at large x, we simply follow the convention of using this form for all x. In fact what our data are most sensitive to is the slope of G(x) in the range x=0.04 to x=0.08. A scaling distribution is assumed at least in part because the range of Q^2 available is narrow ($\approx$ 16 – 25 GeV^2) since the mass of the produced $c\bar{c}$ pair peaks sharply near threshold.

A lighter charm quark leads to more forward-backward peaking of the quarks and hence a softer p_t^2 distribution. The x_F distribution in the range measured is not very sensitive to either m_c or n_g, but is sensitive to the fragmentation scheme used. Because of the limited range of Q^2, α_S is effectively a constant. Thus the cross-section is insensitive to a choice of the argument of α_S which we assume to be $\hat{s}$ (the only scale in the process).

Recently Ellis et al.[3] calculated the next-to-leading order (NLO) corrections to the total and differential cross-sections. It is clear from that work that the differential cross-sections are virtually unchanged in shape from the leading order (LO) case, so we use the simpler leading order results to fit our differential cross-sections. The NLO total cross-section however is 32% larger than the LO result at our energies and hence we use the NLO result to fit our total cross-section. Our $\sigma_{c\bar{c}} = 0.58 \pm 0.01 \pm 0.06 \mu$b when fit to the NLO results yields a charm quark mass of $1.6^{+0.3}_{-0.1}$ GeV/c^2. The errors include the theoretical errors due to the renormalization scale and the gluon distribution as mentioned in ref. 2. The gluon distribution used in ref. 2 is not strictly a scaling distribution of the form (3), but is well approximated by that form with $n_g=7.4$. As described earlier[1] the total cross-section is derived assuming an A-dependence of the form $A^{0.93}$, but is consistent with A^1 within systematic errors.

DATA ANALYSIS

Fixing m_c to 1.74 GeV/c^2, (our final combined fit value, see below) we next fit the slope of the cross-section to n_g and obtain $n_g = 8.8^{+2.3}_{-2.3}$. A nice feature of this result is that it arises from the slope of the *total* cross-section and hence is independent of uncertainties arising from fragmentation and the intrinsic k_T^2 of partons.

We view fragmentation as having three kinds of uncertainties:
(1) The fragmentation scheme could be independent fragmentation of the two charm quarks or string fragmentation with one string between the charm quark and the target diquark and another between the anticharm quark and the target quark. Our charged particle multiplicity distribution favors the latter.
(2) There are errors on the parameters[5] of the charm fragmentation function as measured by e^+e^- experiments. These are propagated through the analysis done with the help of the Lund Monte Carlo using string fragmentation[6,7].
(3) The exact nature of the splitting of the remnant nucleon into a quark and diquark is not known. This splitting was varied within the entire allowed range to determine a systematic error from this source.

We have also explored the effect of a spin-zero gluon on the differential cross-section. The shape is not very sensitive to the spin. Fitting the p_T^2 distribution indicates a weak preference for a spin-1 gluon.

The four quantities mentioned above are fit together in a joint fit. The combined fit yields $m_c = 1.74^{+0.13}_{-0.18}$ GeV/c^2 and $n_g = 7.1 \pm 2.2$. These final values are within the 1σ ranges of the four individual fits. The errors are mainly systematic, with the intrinsic k_T^2 of the gluons being the dominant effect. The systematic error includes the change in the parameters when the entire p_T^2 distribution is included in the fit.

CONCLUSION

In conclusion, fits are made to $\sigma_{c\bar{c}}$, its rise with energy, the x_F distribution and the p_T^2 distribution. The most reliable result is the combined fit including data only for the region $p_T^2 > 2$ GeV^2/c^2 which yields $m_c = 1.74^{+0.13}_{-0.18}$ GeV/c^2 and $n_g = 7.1\pm 2.2$ at $Q^2 = 20$ GeV^2. It would be more accurate to say that we have measured the ratio $G'(x)/G(x)$ at x=0.06 to be 7.6±2.3; we quote the result for n_g because it is conventional to do so. Thus we have determined the mass of the charm quark to fair accuracy and contributed valuable additional information on the shape of the gluon distribution in light of present-day knowledge. The value of m_c should be useful in describing other processes involving similar perturbative QCD calculations, e.g., in neutrino-induced charm production provided next-to-leading order calculations are used.

I would like to acknowledge useful discussions with E. L. Berger and R. K. Ellis.

References

1. J. C. Anjos et al., Phys. Rev. Lett. **62**, 513 (1989).
2. L. M. Jones, and H. W. Wyld, Phys. Rev. **D17**, 759 (1978).
   M. A. Shifman, A. I. Vainstein and V. I. Zakharov, Phys. Lett. **65B**, 255 (1976).
3. R. K. Ellis and P. Nason, Nucl. Phys. **B312**, 551 (1989).
4. J. R. Raab et al., Phys. Rev. **D37**, 2391 (1988) and additional references therein.
5. S. Bethke, Z. Phys. **C29**, 175 (1985).
6. T. Sjöstrand, LU TP 85-10, University of Lund, Sweden.
7. T. Sjöstrand, Int. J. of Mod. Phys. **A3**, 751 (1988) and references therein.

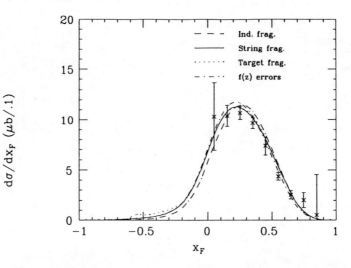

Fig 1. The x_F dependence of the cross-section on Be for production of charm events ($d\sigma_{c\bar{c}}/dx_F$). The curves are for string fragmentation (solid), independent fragmentation (dashed), for the case where the diquark carries a fixed fraction of the remnant nucleon momentum (dotted) and for the fragmentation function parameters changed by one standard deviation (dotdashed).

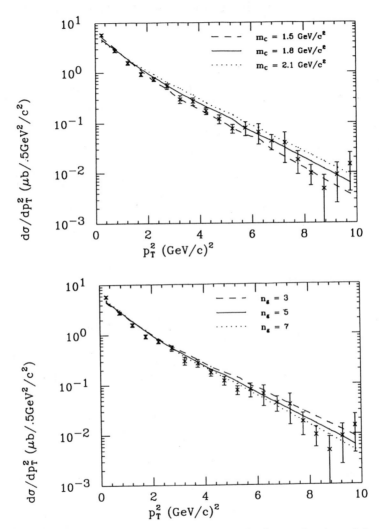

Fig 2. The p_T^2 dependence of the cross-section on Be for production of charm events ($d\sigma_{c\bar{c}}/dp_T^2$). The curves are for different values of m_c, the charm quark mass and of n_g, the power of (1-x) in the gluon structure function.

Photoproduction of Heavy Quarks, Preliminary Results from E687*

P. Lebrun

Fermi National Accelerator Laboratory
P.O. Box 500, Batavia, Illinois 60510, USA

*for the E687** Collaboration.*

Abstract : Photoproduction of Charm and beauty at high energy is discused within the contex of the photon-gluon fusion model. Early results on J/Ψ, D and D* photoproduction cross section from the 1987 fixed target run are presented. Fermilab E687 used the highest energy photon beam thus far.

I. Introduction.

In this talk, we consider the production of heavy quarks by a beam of real photons, at high energy. The fixed target program at Fermilab is a unique opportunity to study photoproduction at the highest energy so far, prior to HERA . In the context of the parton model, perturbative QCD can be used to estimate cross sections for open charm and beauty. A full calculation of the QCD $O(\alpha^2_s \alpha_{em})$ has been recently published [1], showing that, within large theoretical uncertainties, the calculated charm cross section is in reasonable agreement with existing data (up to a photon energy of $\omega \approx 200$ GeV/c). It is also shown that the energy dependance of this cross section is sensitive to the mass of the Charm quark M_C. In addition, in photoproduction, one probes directly the gluon structure function at low x : the invariant mass of the c - $\bar{c}$ pair is simply of the order of 2 * M_C, leading to the fraction of momentum carried by the gluon ρ to be equal to 4 * M_C^2/S; for $M_C \approx 1.5$ GeV, $\rho \approx 0.02$.

Several factors allows us to measure accurately differential cross section for Charm and, hopefully, to give an estimate of the total cross section

---

* Supported by US DOE and NSF and by INFN.
**E687 Collaboration : INFN Bologna; University of Colorado; Fermilab; INFN Frascati; University of Illinois; University of Notre Dame; University of Milan; Northwestern University; INFN Pavia; University of Pavia, University of Puerto Rico - Mayaguez; Cinestuo, Univ. of Mexico.

for beauty at $\omega \approx 250$ to 300 GeV/c : (i) a new high energy, high intensity wide band photon beam (ii) a multi-particle spectrometer equipped with high resolution Silicon Microstrip Detector (SSD) with a point-back accuracy at the vertex of the order of 10 µm transverse to the beam direction , (iii) a high bandwidth data acquisition system capable of recording 60 million triggers in a few weeks of running time (iv) a large amount of compute power to reconstruct these events in a reasonable amount of time (of the order of 150 Mips/year).

We will first briefly describe the E687 spectrometer. Emphasis is placed on the spectrometer elements used in the cross section measurements of the following processes :

γ + Be $\rightarrow$ J/Ψ + X, J/Ψ $\rightarrow$ μ^+ μ^-

γ + Be $\rightarrow$ D^{*+-} + X, D^{*+-} $\rightarrow$ D^0 π^{+-}, D^0 $\rightarrow$ K^{-+} π^{+-}

γ + Be $\rightarrow$ D^{+-} + X, D^{+-} $\rightarrow$ K^{-+} π^{+-} π^{+-}

As systematic error calculations are in progress, our current results are preliminary. In addition, many decay modes have not yet been fully exploited and the B search is only starting. Finally, as we are currently taking our 1990 data, only results from the 1987 run are presented.

II. The Wide Band Photon Beam and the E687 spectrometer.

The Fermilab wide band photon beam is a bremstrahlung beam with a large momentum bite ($\sigma \approx 13\%$). Primary protons of 800 GeV/c were used to generate an intense beam of photon from π^0 decay. These photons passed through a lead converter and the electrons produced were steered around a neutral dump and directed onto a 20% radiation length lead radiator. The scattered electrons passed through a sweeping magnet and were detected in a recoil shower counter (RESH). The bremstrahlung photons produced in this radiator continued onto the E687 experimental target. Photons which did not interact in this 10.% radiation length target deposited their energy in a beam dump calorimeter (BGM) located at the back of the E687 spectrometer. Fig 1 shows the inferred energy distribution of these photons. The central energy of the electron beam was set to 350 GeV/c. The detection of the recoil electron in the RESH array counter imposes a mean photon energy around 220 GeV. Because of the large momentum bite of the electron beam, a sizeable number of photons over 350 GeV were produced.

Although the 1990 wide band beam upgrades are irrelevant for results presented in this talk, they are certainly worth mentioning : (i) in conjunction with the electron beam, a positron beam enhances the photon flux (ii) the momentum bite of the electron beam increased, also enhancing the photon flux (iii) the primary target is now made of liquid deuterium, in order to maximise the secondary photon yield (iv) the electron/positron momentum is now measured on a event by event basis by recording particle

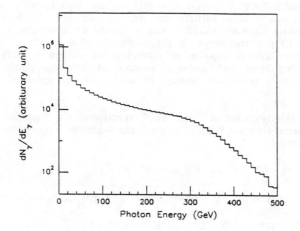

Figure 1. The photon energy spectrum.

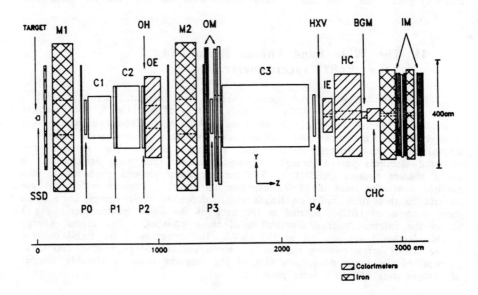

Figure 2. The layout of the E687 Spectrometer.

positions along the beam transport system with microstrip detectors. In conjunction with the RESH and BGM counters, this tracking system allows us to measure the bremstrahlung photon energy accurately.

The E687 spectrometer is shown on Fig 2 [2]. The main tracking components consist of two vertical bending dipole magnets (M1 and M2) and a system of 5 multiwire proportional chambers (PWC), in addition to the SSD system. The primary use of the PWC system was to determine the momentum of the tracks which linked to tracks found in the SSD system. For tracks which traversed the entire spectrometer the momentum (p) resolution can be described as follow :

$\sigma_p/p = 1.4\%$ (p/100 GeV) $\sqrt{1. + (23. \text{ GeV}/p)^2}$ for the P0-P4 tracks.

$\sigma_p/p = 3.4\%$ (p/100 GeV) $\sqrt{1. + (17. \text{ GeV}/p)^2}$ for the P0-P2 tracks.

The $1/p^2$ term is due to the contribution of multiple scattering to the resolution function. This tracking system was used in conjunction with the muon detector system, consisting of 3 arrays of scintillation counters and 4 arrays of proportional tubes covering approximately 30 mr polar angle.

For open charm, we also used a particle identification system consisting of 3 gas filled Cerenkov counters, running at atmospheric pressure. The gas compositions were He N_2, N_2 O, He, and the corresponding π thresholds were 6.7, 4.4 and 17.0 GeV respectively. This gave us good Kaon identification up to a momentum of 60 GeV. In addition, a hadrometer consisting of proportional tubes with pad readouts was used in triggering on hadronic interactions.

The first level trigger required a three-fold coincidence between (i) a scintillating counter placed downstream of the target and upstream of the SSD (ii) a counter located downstream the SSD system (iii) 2 hits in the scintillating counter hodoscope HxV located just downstream of the last proportional chamber. A gap in this HxV hodoscope discriminates against $e^+ e^-$ pairs created in the target or in the spectrometer itself. In addition, charged beam halo particles were vetoed using scintillating counters located upstream of the target. The second level trigger used in collecting the dimuon sample required signals from the muon hodoscope scintillating counters consistent with 2 muons. For open charm, the second level trigger required (i) a signal from the hadrometer corresponding to at least 35 GeV energy deposition (ii) at least one hit in the first proportional chamber, outside the pair region. In addition, in order to optimize the yield for high energy photons, this trigger required a RESH signal corresponding to a photon energy greater than 135 GeV.

In 1987, the data acquisition was capable of recording up to 2,500 events per Tevatron spill. Using modern technologies based on dedicated VME based I/O servers and high density helicoidal magnetic tape (8mm Exabytes), we are now capable of writing more than 5,000 events a spill. This data acquisition system, although well matched to the beam luminosity and the rate capability of the spectrometer, is now limited by the front end electronic (namely, digitization time of the Fastbus ADC's) .

During the 1987 run, about 60 millions good trigger were recorded. We are now capable of recording equivalent samples in less than 2 weeks running time. Therefore, while the cross sections deduced from the 1987-1998 have sizeable statistical errors, we are confident that these errors will be entirely be dominated by systematic uncertainties, mostly on luminosity estimations and branching ratios once the 1990 run data is reconstructed.

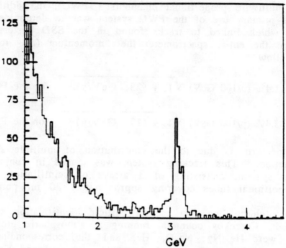

Figure 3. The dimuon mass spectrum, 1987 data.

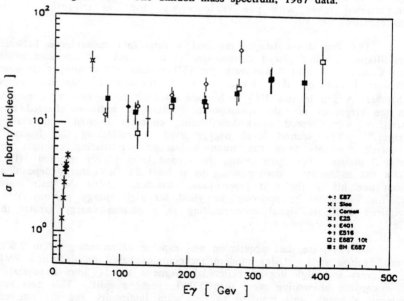

Figure 4. Energy dependance of the photoproduction of J/Ψ mesons.

III. Preliminary Results on J/Ψ and D Total Cross Section Measurements.

a. $\gamma + Be \rightarrow J/\Psi + X,$
$$J/\Psi \rightarrow \mu^+ \mu^- \quad [3]$$

Events corresponding to a muon trigger were stripped off the raw data tapes and reconstructed separately. Further event selections took place, requiring (i) there must be two tracks of opposite sign coming from a vertex located in the target (A third track was allowed if it did not come from this vertex) (ii) the muon counter hits must be associated with the tracks measured in the PWC system. The invariant mass spectrum for these muon pairs is shown on Fig. 3. We calculate, for events above an invariant mass of 2.5 GeV the error on the mass measurement, $\sigma(M)$, event by event, and cut on the three sigma of the Gaussian fit to the quantity $(M-3.097)/\sigma(M)$. Further, a cut on extremely low momentum transfer is applied to remove the remaining Bethe-Heitler dimuons and the coherent part of the cross section. This cut removes about 15 % of the events. The final sample consist of 257 events called ' elastic incoherent' J/Ψ 's.

Two independent normalization methods were used to deduce the cross section : (i) Determine the photon flux from the BGM counter, the measured electron spectrum shape and the measured energy loss spectrum recorded by the RESH counters; (ii) Normalize to the calculated Bethe-Heitler dimuon background for dimuon masses above 1.0 GeV. A special Monte-Carlo was written to simulate this process in the spectrometer and to compute the ratio of the J/Ψ's acceptance to the Bethe-Heitler acceptance. The photon flux was calculated using both methods and shows very good agreement. Assuming an A dependence of $\sigma \approx A^{0.94}$, the deduced cross section/nucleon is shown on Fig. 4. From a simple linear fit to this data, we deduced that $\partial \sigma(\omega) / \partial \omega \approx 0.05 \pm 0.02$ nb/GeV at $\omega \approx 250$ GeV.

b. $\gamma + Be \rightarrow D^{*+-} + X,$
$$D^{*+-} \rightarrow D^0 \pi^{+-}, D^0 \rightarrow K^{-+} \pi^{+-} [4]$$

This open charm decay chain yields very clean signals, and has been used to study Charm in the past, in hadroproduction [5] as well as in photoproduction [6]. Contrary to hadroproduction experiments, this signals can easily be seen without a vertex detector in photoproduction experiments. In this analysis, we used only the Cerenkov Counters to identify the charged Kaon and reduce the combinatoric background. This signal is shown on fig. 5. Three independent factors explain such a cleanliness with respect to hadron experiments (i) the ratio between the total inelastic cross section and the open charm cross section is 5 to 10 times more favorable in photoproduction (ii) the average charge multiplicity is lower by a few units in photoproduction compared to hadroproduction, reducing the combinatoric background (iii) in photoproduction, D mesons tends to be produced at higher x_f than in hadroproduction, i. e., in a kinematical region where fixed target spectrometer acceptance can easily be maximized. In order to deduce the cross section for c - $\overline{c}$ pairs, from this signal, we made the following assumptions :

- The branching ratio for the $D^{*+-} \rightarrow D^0 \pi^{+-}$ and the $D^0 \rightarrow K^{-+} \pi^{+-}$ transitions are 0.49 and 0.038 respectively [7]

- The A dependance of the cross section : $\sigma \approx A^{0.93}$

- Clearly, not all c - $\overline{c}$ pairs end up making D*'s. We used the LUND/LUCIFER Monte-Carlo program [8] to estimate the probability of generating a D* (either a D^{*+} or a D^{*-}) per produced c - $\overline{c}$ pair. This probability was found to be ≈ 0.6.

The result of this calculation is shown on fig. 6. The statistical errors are still not negligible with respect to the systematic errors, which are largely driven by luminosity uncertainties. This analysis is based on only $\approx$ 300 D*. We obviously could take advantage of other decay modes of the D^0, such as $D^0 \rightarrow K^{-+} \pi^{+-} \pi^{-+} \pi^{+-}$. Unfortunately, this decay mode is more susceptible to combinatoric background and the SSD information is required. Such an analysis is considered in the next topic.

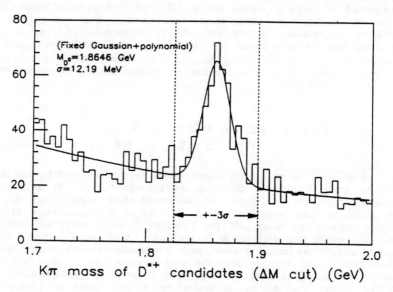

Figure 5. K- π invariant mass distribution after the $|(\Delta M - 145.6 \text{ MeV})| < 2.3$ MeV cut.

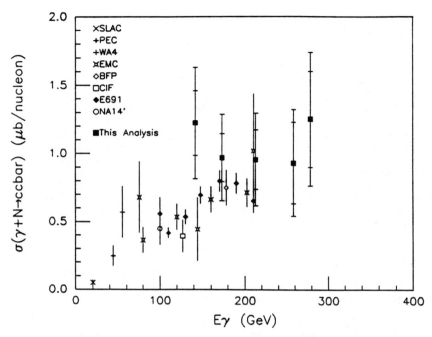

Figure 6. The photon energy dependence of the total cross section per nucleon for open charm events. References for other experiments can be found in Ref. 1.

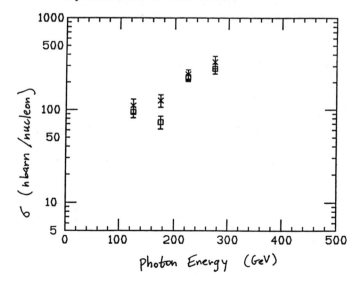

Figure 7. The photon energy dependence of the cross section $\gamma + N \rightarrow D^{+-} + X$.

c. $\gamma + Be \rightarrow D^{+-} + X$,
$D^{+-} \rightarrow K^{-+} \pi^{+-} \pi^{+-}$

This decay mode has several advantages : (i) the branching ratio is large (7.8 %) [7] ; (ii) the signal can be seen without Kaon identification cuts, when stringent vertex cuts are applied. This is due to the relatively long lifetime of the D^{+-} compare to other charm states. Of course, these vertex cuts can be relaxed if the Kaon identification cuts are applied. The cross section presented on Fig. 7 was derived from a D sample obtained using 2 distinct vertex algorithms : a candidate driven vertex algorithm (CDV) and a global stand-alone vertex filter. In the former case, we considered the Charm candidate tracks satisfying the particle identification criteria, fit them to form a secondary vertex and them formed a track vector for the D candidate. This track vector was then used as a seed for finding a primary vertex. Since full covariance matrices with multiple scattering contributions were used in performing these vertex fits, the best way to assess if the these two vertices are distinct is to cut on the quantity $L/\sigma(L)$, where L is the Cartesian distance between these 2 vertices ($L/\sigma(L) \geq 8$.). In the latter case, the SSD information is used in a stand-alone mode, looking for multiple vertices characterized by good χ^2. We required significant separation between at least 2 of these vertices, in addition, the D vertex must point back the upstream vertex.

Clearly, we now have to compare this cross section with the cross section obtained through the D* channel and understand systematic errors in acceptance estimations. In addition, we have to perform similar analysis for other decay modes of the D^0 and the D^{+-}. The final cross section measurement will be based on approximately 3,000 to 4,000 D decays. Finally, the 1990 data sample will considerably reduce the statistical errors and overall normalisation uncertainties.

IV. Search for beauty.

Perturbative QCD calculations for the photoproduction of bottom are expected to be reliable, and show that roughly 1 out of 500 charm pairs comes from B decay at $\omega \approx 250$ GeV. (The ratio of the cross section is about :1 nbarn/ 0.5 μbarn and to first order all B's decay to Charm) . This ratio is actually smaller than the ratio of Charm production to Kaon production. In addition, the branching ratios to exclusive decay modes are always smaller in B decay than in Charm decay. Therefore, substantial combinatoric background and very small statistics   are expected.

We are currently investigating three classes of decay modes of the B mesons. The first one is simply the inclusive decays B $\rightarrow$ D (or D*) + n π , where n = 1,2...4, with allowance for missing π^0(s).   The CDV algorithm described above has been upgraded to handle a sequence of 3 vertices, the $L/\sigma(L)$ referring to the primary and the B vertex. While further cuts severely constraining the vertex topology will be included in the analysis, it is already clear that the combinatoric background is severe, and very good resolution is required. This is true, despite of the fact that decay products of the B mesons tends to have higher transverse momentum than particle produced in Charm decay. The only hope is that the rate might be sufficient (sample are of ≈ 100 is not hopeless using the anticipated 1990 data), allowing us to tighten the cuts and reduce this background.

The second class of decay modes we are considering is based on semileptonic decays, either one possibly, two) lepton(s) from a B meson and one (possibly 2) leptons from the Charm mesons produced in the B decay. As in the first case, we will severely cut on transverse momentum and vertex. Unfortunately, the price to pay for each required lepton is the semileptonic branching ratio, 20% at best. Finally, the cleanest decay modes are undoubtedly the inclusive $B \rightarrow J/\Psi + X$ decays, simply because the charm background is naturally suppressed. The rate is clearly a problem : the projected number for such 'gold-plated' events expected in the 1990 sample is only ≈ 5, assuming our sensitivity is such that we can collect and reconstructe 100,000 open charm decays.

References :

1. Ellis, R. K. and P. Nason, *Nucl. Phys.* **B312** (1989) 551.

2. J. R. Wilson *et al*, Proceeding of the *DPF'90 conference, Houston, Jan 3-7 1990,* to be published.

3. R. Yoshida *et al*, Proceeding of the *DPF'90 conference, Houston, Jan 3-7 1990,* to be published.

4. S. Park, *Ph D. Thesis*, NorthWestern University, june 1990.

5. Thomas Knight Kroc, *Ph D. thesis, PACS 13.25 +M, 14.40.JZ, univeristy of Ill., 1989, page 82.*; P. Mooney *et al, Phys. Rev.* **D 39**, 2494 (1989) ; F. Abe *et al, FERMILAB -pub- 89/171-E* (1989), G. Arnison *et al, Phys. Lett.* **147B,** 222 (1984);,

6. C. J. Anjos *et al, Phys Rev. Lett.* **60** (1988) 1239;, P. Avery *et al, Phys. Rev. Lett.* **44** (1980) 1309.

7. Particle Data Group. "Review of particle Properties" *Phys. Lett.* **B204** (1988) 1.

8. G. Ingelman and A. Weigend. "LUCIFER - A Monte-Carlo for High-Pt Photoproduction " *Compt. Phys. Commun.* **46** (1987) 247.

Next-to-leading-logarithm Calculations of Direct Photon Production [*]

J. F. Owens
Department of Physics, B-159, Florida State University
Tallahassee, Florida 32306, United States

ABSTRACT

A method for calculating direct photon production in the next-to-leading-logarithm approximation using a mix of analytic and Monte Carlo integration techniques is described. The flexibility afforded by the use of Monte Carlo techniques allows a variety of observables to be calculated relatively easily. Some examples are discussed, including the photon-jet cross section. The question of photon isolation cuts is also addressed.

1. Introduction

Observations of direct photon production in hadronic collisions have the potential to yield valuable information concerning the gluon distribution in the proton.[1] When combined with measurements of other processes, such as deep-inelastic lepton-nucleon scattering or the production of high-mass lepton pairs, it is also possible to place constraints on the various quark distributions.[5] However, most of the existing direct photon measurements have been for the invariant cross section, the calculation of which involves a convolution of parton distributions. It has long been realized that more precise information concerning the gluon distribution could be obtained from data for the double inclusive photon-jet cross section[6] since, in the leading-logarithm approximation, the arguments for both parton distributions would be fixed by the kinematics of the observed jet and photon. On the theoretical side, existing next-to-leading-logarithm calculations for large momentum transfer processes have also focussed on the single particle, jet, or photon inclusive cross section. The information contained in the correlation between the various partons in the process is lost when the unobserved partons are integrated over the available phase space. On the other hand, it is difficult to know what observable should be calculated ahead of time, since there are many variations in jet definitions, acceptance corrections, etc. Ideally, one would like to have a method of performing next-to-leading-logarithm calculations which can be easily adapted to the particular experimental situation being studied. In this report a new calculation of direct photon production is described which explicitly includes the point-like photon subprocesses through $O(\alpha\alpha_s^2)$. The various two-body and three-body phase space integrations are performed using a mix of analytic and Monte Carlo methods, thereby allowing one to calculate observables which involve correlations between the photon and another parton. A variety of observables can be calculated simultaneously by forming suitable histograms within the same program.

A number of processes have been treated using this Monte Carlo technique,

---

[*] This work was supported in part by the U. S. Department of Energy.

including symmetric dihadron production with non-singlet fragmentation[7], the photoproduction of jets[8], and direct photon production[9,10]. The techniques utilized in the calculation are discussed extensively in Refs. 7-10, so only a brief description will be given here. Following that, the treatment of the bremsstrahlung component will be reviewed as this topic is relevant to the data being obtained by current collider experiments. Additional comments concerning various observables involving both photons and jets will also be given.

2. Next-to-leading-logarithm Calculation

In a typical leading-logarithm calculation for high-p_T photon production, there are two different types of production mechanisms to be considered. The first consists of the $O(\alpha\alpha_s)$ subprocesses $gq \to \gamma q$ and $q\bar{q} \to \gamma g$ and is referred to as the direct or pointlike component. This component is characterized by an isolated high-p_T photon recoiling against a hadronic jet. The emission of additional partons from the incoming legs is treated in the collinear approximation to all orders, giving rise to the usual scaling violations in the parton distributions. The second mechanism arises because quarks participating in a hard scattering process can radiate photons. This is taken into account by introducing photon fragmentation functions for both quarks and gluons. These can be obtained by solving appropriately modified Altarelli-Parisi equations.[11] For this photon bremsstrahlung component the hard scattering subprocesses are $O(\alpha_s^2)$ while the photon fragmentation function is $O(\alpha/\alpha_s)$ so that the results are of the same order as for the direct photon case. Both of these constitute all orders calculations, but only the leading-logarithm from each of the terms beyond $O(\alpha\alpha_s)$ is retained.

In order to improve the precision of the predictions, the next step is to include completely the effects of the various $O(\alpha\alpha_s^2)$ subprocesses involving a single photon. Specifically, the one-loop contributions to $gq \to \gamma q$ and $q\bar{q} \to \gamma g$ and the three-body subprocesses $qq' \to \gamma qq'$, $q\bar{q}' \to \gamma q\bar{q}'$, $q\bar{q} \to \gamma q'\bar{q}'$, $qq \to \gamma qq$, $q\bar{q} \to \gamma q\bar{q}$, $gq \to \gamma qg$, $gg \to \gamma q\bar{q}$, and $q\bar{q} \to \gamma gg$ have been included in this calculation using expressions obtained from Refs. 12 and 13. The calculation proceeds along much the same lines as the usual inclusive single photon calculation, in which one integrates over the unobserved final state partons. Such integrations can be performed analytically at the parton level. For a suitably defined inclusive observable, the infrared singularities associated with the one-loop contributions will cancel against the soft singularities associated with the three-body tree graphs. Furthermore, the hard collinear singularities associated with the initial partons or the produced photon can be factorized and included in the relevant parton distribution or fragmentation functions, respectively. The remaining hard collinear singularities will cancel against corresponding singularities from the one-loop graphs. If, however, one wants to constrain one or more of the final state partons to be in a singularity-free region of phase space, then the integration regions are restricted and Monte Carlo methods offer an alternative to the purely analytic approach. The basic challenge, then, is to find a way of ensuring that all of the required cancellations can take place within the context of a Monte Carlo calculation. In order to discuss the technique for isolating the various singularities, let the four-vectors of the two-body and three-body subprocesses be labelled by $p_1 + p_2 \to p_3 + p_4$ and $p_1 + p_2 \to p_3 + p_4 + p_5$, respectively, and define the following Lorentz scalars $s_{ij} = (p_i + p_j)^2$ and $t_{ij} = (p_i - p_j)^2$. Briefly, the calculation proceeds via the following steps. The ultraviolet singularities associated with the one-loop contributions are regulated using the method of dimensional regularization[14] and subtracted using the $\overline{\text{MS}}$ scheme.[15] Similarly, dimensional regularization is used in treating the infrared, soft, and collinear divergences. Next, two cut-off parameters, δ_s and δ_c, are introduced whose purpose is to allow the separation of the regions of phase space which contain the singularities. For the three-body subprocesses, the soft singularities are associated with the phase space region where one final state gluon becomes soft. The soft region is defined to be that where the relevant parton energy in the subprocess rest frame becomes less than $\delta_s \sqrt{s_{12}}/2$. If δ_s is chosen to be sufficiently small, then the relevant three-body subprocesses can be evaluated using the soft-gluon approximation wherein the the gluon energy is set to zero in the numerator of the expression. The resulting expression is then easily integrated over the soft region of phase space. At this stage, this integrated soft piece contributes to the two-body part which contains the one-loop terms. The soft and infrared singularities can then be cancelled explicitly. Next, the collinear regions of phase space are defined to be those where any invariant (s_{ij} or t_{ij}) becomes smaller in magnitude than $\delta_c s_{12}$. If δ_c is chosen to be sufficiently small, then in each collinear region the relevant subprocess can be evaluated using the leading-pole approximation. The result is easily integrated in n-dimensions, thereby explicitly displaying the collinear singularities. These are then factorized and included in the relevant structure functions or cancelled with corresponding singularities in the two-body expressions. At this point, the remainder of the three-body phase space contains no singularities and the subprocesses can be evaluated in four dimensions.

The calculation now consists of two pieces – a set of two-body contributions and a set of three-body contributions. Each set consists of finite parts, and all singularities having been cancelled, subtracted, or factorized. However, each part depends separately on the two theoretical cut-offs δ_s and δ_c. Each by itself has no intrinsic meaning. However, when the two- and three-body contributions are combined to form a suitably inclusive observable, $e.g.$, an inclusive single photon invariant cross section, all dependence on the cut-offs cancels. It turns out that the answers are stable against variations of these cut-offs over quite a wide range, which is as it should be. The cut-offs merely serve to distinguish the regions where the phase space integrations are done by hand from those where they are done by Monte Carlo. When the results are added together, the precise location of the boundary between the two regions is not relevant. The results obtained with this technique are stable to reasonable variations in the cut-offs, thus providing a check on the calculation.

One question which repeatedly comes up when discussing next-to-leading-logarithm calculations is the choice of scales. In each of the cases calculated to date we have found that the next-to-leading-logarithm calculations are, in general, less sensitive to the choice of scale. Detailed studies have been made taking the factorization and renormalization scales to be the same. The reduced scale dependence is most noticeable in the intermediate x_T region; for very small values of x_T the sensitivity remains comparable to that of the leading-logarithm calculation. From a practical standpoint this means that the theoretical uncertainty due to the scale dependence is often reduced by including the next-to-leading-logarithm corrections which, after all, was one of the reasons for performing the calculation in the first place. Moreover, the variations in the scale dependence means that there is no unique value for the ratio of the leading-log and next-to-leading-log results, *i.e.*, there is no unique "K-factor". Furthermore, this ratio is different for different observables. As a case in point, consider the photon-jet cross section. The amount which the three-body matrix elements contribute to the final result depends in part on the fiducial cuts used to define the recoiling jet, the jet clustering or coalescence algorithm, and the photon isolation criteria, for example. Since the three-body contribution is positive definite, changing these factors alters the ratio of the $O(\alpha\alpha_s)$ and $O(\alpha\alpha_s^2)$ terms. This variation would be missed in any approximate approach which simply rescaled the two-body leading-log contributions.

3. Photon Isolation Cuts and the Bremsstrahlung Contribution

In principle, the $O(\alpha\alpha_s^2)$ calculation described previously is not complete, as there are additional terms which come from $O(\alpha_s^3)$ $2 \to 3$ subprocesses convoluted with photon fragmentation functions. These terms provide a correction to the bremsstrahlung component. However, the bremsstrahlung component does not provide a significant contribution except in the small x_T region. Therefore, the neglect of these additional terms is justified for those experiments which do not go to very small values of x_T. Current fixed target experiments generally cover a kinematic region which starts above approximately $x_T = 0.2$, so that the bremsstrahlung corrections are not an important issue. On the other hand, collider experiments do cover the small x_T region. Indeed, the region of very small x_T values is important for distinguishing between extrapolations of different types of gluon distributions. In order to use direct photon production as a reliable means for determining the gluon distribution, the bremsstrahlung contribution must be well understood.

In some experiments an isolation cut is placed on electromagnetic triggers as part of the definition of single photon events. Events are rejected which have more than a certain amount of hadronic energy in a cone about the electromagnetic trigger direction. This has the effect of reducing the bremsstrahlung contribution, which would seem to be desirable. However, in regions where the bremsstrahlung and pointlike contributions are comparable this amounts to making a large correction to a major component of the cross section. It will be necessary to have a good understanding of the effects of such cuts in order to obtain reliable results which can be used in determining the gluon distribution.

In a leading-logarithm calculation it is easy to distinguish between the pointlike and bremsstrahlung contributions; the latter involves a photon fragmentation function while the former does not. When the contributions from higher order subprocesses are included beyond the leading-logarithm approximation, the distinction becomes less clear. The reason is that different regions of the allowed phase space for a specific $2 \to 3$ subprocess contribute to the pointlike and bremsstrahlung leading-logarithm expressions. Consider, for example, the subprocess $gq \to \gamma qg$. The logarithmic terms arising from the integration over the region where the final state gluon is nearly collinear with the initial quark or gluon are included in the Q^2 dependence of the quark and gluon distribution functions and, therefore, contribute to the pointlike leading-logarithm contribution. For this contribution the photon recoils against the final state quark and is therefore relatively isolated. There is another region of phase space where the photon and final quark are nearly collinear and they both recoil against the final gluon. The integration over this region builds up a logarithm which enters the photon fragmentation function. This particular term is included in the leading-logarithm calculation by convoluting the photon fragmentation function with the $gq \to gq$ subprocesses. Thus, this is considered as a bremsstrahlung contribution. However, both of these terms have originated from the *same* $2 \to 3$ subprocess. In a fully inclusive calculation one never distinguishes between the pointlike and bremsstrahlung contributions and so the question of labelling each contribution does not arise. However, when isolation cuts are included as part of the experimental trigger, one must estimate the fraction of the true inclusive cross section which is removed.

The effects of a photon isolation cut can be simulated using the Monte Carlo approach described previously. Since the calulation is being performed at the parton level, the first approximation is that the cuts must be made using parton rather than hadron four-vectors. Let $\Delta\eta$ and $\Delta\phi$ represent the differences in rapidity and azimuthal angle between the photon and one of the final state partons and define

$$R = \sqrt{(\Delta\eta)^2 + (\Delta\phi)^2}.$$

Consider, first, a simple cut which excludes contributions for which $R < R_\gamma$ where R_γ is taken to be of $O(1)$. Recall that the logarithm in the photon fragmentation function is obtained by integrating over the transverse momentum of the photon relative to the parton from which it came. This isolation cut limits that integration and, therefore, the argument of the logarithm is modified from being of $O(p_T^2)$ to one of $O(R_\gamma)$. This greatly reduces the size of the fragmentation contribution since the large logarithm has been removed. This point has been stressed in a recent study of photon isolation cuts.[16] Next, consider what happens for the various three-body contributions. Here, too, one would simply disregard any contribution for which $R < R_\gamma$. This is eaily done for the $2 \to 3$ tree graphs. There is no problem with collinear singularities involving the photon since one never includes the region of phase space where a quark and a photon are parallel. However, at this stage one encounters a problem with the simple isolation cut outlined above. Consider a subprocess such as $gq \to \gamma qg$. The region where the final state gluon energy goes to zero is singular and this singularity cancels a corresponding one coming from the one-loop contribution to $gq \to \gamma q$. In order for the cancellation to occur, the soft gluon must be integrated over the full solid angle in the three-body phase space. However, the isolation cut limits this integration and, as a result, the required cancellation would not be complete. In the terminology used to describe the Monte Carlo calculation, there would be a residual dependence on the soft cutoff δ_s which would not cancel between the two- and three-body contributions. Therefore, it is necessary to modify the isolation cut to allow for the possibility of some soft hadronic energy within the cone. Typically, what is done is to require that the hadronic energy in the region $R < R_\gamma$ be no more than a fraction ϵ_h of the photon's energy. Treating the hadronic energy within the cone as being collinear with the photon, this amounts to a restriction on the fraction, z, of the parent parton's momentum taken by the photon

$$z > z_{min} = \frac{1}{1 + \epsilon_h}.$$

The simulation of the isolation cut proceeds now as follows. For the two-body bremsstrahlung contribution the fragmentation functions are unchanged for $z > z_{min}$ while for $z < z_{min}$ they are set equal to zero. For the three-body subprocesses, all contributions with partons having $R < R_\gamma$ and having an energy greater than ϵ_h times the photon energy are discarded. In the region corresponding to $z > z_{min}$ the usual collinear singularity must be factorized and absorbed into the photon fragmentation function, as is done in the fully inclusive case. This completes the algorithm for simulating the isolation cut. There are no problems with remaining collinear singularities involving the photon, as they have been factorized into the photon fragmentation function.[17]

The effects of various photon isolation cuts have been studied in Ref. 16. There the fragmentation functions entering the bremsstrahlung contribution were modified as described above. They find, for example, that with $\epsilon_h = 0.15$ and $R_\gamma = 0.7$ that the cross section is reduced by 35% at $p_T = 10$ GeV and $\sqrt{s} = 1.8$ TeV for $\bar{p}p$ collisions. The effect decreases with increasing p_T to about 5% at $p_T = 200$ GeV. I have obtained similar results using the Monte Carlo approach just described.

From a theoretical standpoint, it is best to calulate fully inclusive observables without having to simulate the effects of isolation cuts. However, if such cuts are required from an experimental standpoint, then techniques such as those described herein can be used to simulate their effect.

4. Photon-Jet Cross Section

In order to determine the gluon distribution in the initial hadrons, it is desirable to have data for the photon-jet cross section. The inclusive photon cross section involves integrating over the four-vectors of all the unobserved partons. This, in turn, means that the gluon and quark distributions are integrated over a range of x values. However, by simultaneously constraining the four-vectors of both the photon and a jet, the number of integrations is reduced. The photon-jet cross section has recently been measured by the AFS Collaboration.[18] The results are in good agreement with the next-to-leading-logarithm calculation presented in Ref. 9 performed using the technique discussed above. There are, however, several points concerning this observable which are not as widely known as they should be.

If only the lowest order contributions to the photon-jet cross section are retained, then the photon and the recoiling jet have equal and opposite transverse momenta. When the effects of higher order contributions are included in the leading-logarithm approximation, this p_T balance is unchanged, since the various emissions by the incident partons are treated as being collinear. However, this balancing of transverse momenta is actually an artifact of the approximations used in the calculation and does not persist once higher order contributions are included beyond the leading-logarithm approximation. For the single photon inclusive cross section the unobserved partons are integrated over. They can have transverse momenta up to a limit which is of the order of the transverse momentum of the observed photon. That is why the scaling violations are governed by a scale of $O(p_T^2)$. Similarly, for the photon-jet cross section there will be additional accompanying emissions which will generate a net transverse momentum imbalance for the photon-jet system. This p_T imbalance sets the scale for the resulting scaling violations in the parton distributions.[19]

Consider the photon-jet cross section $d\sigma/dp_{T\gamma}d\eta_\gamma d\eta_{jet}$. Since the photon transverse momentum is specified, but not that of the jet, the p_T imbalance distribution is integrated over. This is easily done in the Monte Carlo calculation discussed above. The tail of the p_T imbalance distribution is given by the configuration where there is one unobserved hard quantum accompanying the away-side jet and this is described by the $O(\alpha\alpha_s^2)$ $2 \to 3$ contributions. However, if one wants a detailed description of the shape of this distribution in the region where the photon and jet nearly balance each other, then the effects of multiple soft-gluon emission must be included, as well.

The situation described above is very similar to the well-known case of massive lepton pair production where the roles played by the photon-jet and dilepton systems are similar. In lowest order the p_T distribution of the lepton pair is generated solely by intrinsic parton transverse momenta. Higher order effects generate a p_T distribution, the tail of which is given by a single hard parton recoiling against the lepton pair. However, in order to study the lepton p_T distribution for small values of p_T, multiple soft-gluon emission must be included.

It has sometimes been suggested that one could use photon-jet events to test jet reconstruction algorithms since the jet and photon transverse momenta would be equal and opposite. If the measurement of the photon transverse momentum had smaller errors, then the jet algorithm could be calibrated using a sample of such events. However, the above discussion shows the fallacy in such a strategy. There will be, in general, a net transverse momentum imbalance which is dynamically generated and which contains useful information concerning the underlying scattering process. This information would be lost if the jet and photon were forced to have balancing transverse momenta and the results would be misleading.

5. Event Structures

One limitation of the present calculation is that fragmentation is not included, in the sense of an event generator. The Monte Carlo procedure is used simply to perform integrations over limited regions of phase space. The weights generated do not correspond to individual weight one events and, indeed, the individual contributions to the histograms are not positive definite – only their sum should be. Finally, all different flavor contributions and subprocesses are summed over for each kinematic configuration, making it difficult to put in a specific model for fragmentation. Note, however, that it is relatively easy to put in the effects of collinear fragmentation functions. Therefore, this technique in its present form is not suited for studying global event structures. Rather, it should simply be viewed as a means for calculating inclusive observables in a flexible manner.

A second point, alluded to in the discussion of the photon-jet cross section, is that the effects of multiple soft-gluon emission are not included. Therefore, this technique is not suited for the calculation of those observables for which multiple soft-gluon effects are important. For example, the azimuthal distribution of jets with respect to the photon should have a peak at 180° with a tail going to smaller angles. The tail of the distribution away from 180° can be calculated using the $O(\alpha\alpha_s^2)$ matrix elements, but for the detailed shape of the distribution near 180° multiple soft-gluon effects would have to be added. Similarly, attempts to estimate the effects of parton intrinsic transverse momenta require the inclusion of multiple soft-gluon emission.

6. References

1. See Refs. 2-4 and references therein.
2. J. F. Owens, *Rev. Mod. Phys.* **59**, 465 (1987).
3. E. Berger, E. Braaten, and R. D. Field, *Nucl. Phys.* **B239**, 52 (1984).
4. T. Ferbel and W. R. Molzen, *Rev. Mod. Phys.* **56**, 181 (1984).
5. P. Aurenche, R. Baier, M. Fontannaz, J. F. Owens, and M. Werlen, *Phys. Rev.* **D39**, 3275 (1989).
6. L. Cormell and J. F. Owens, *Phys. Rev.* **D22**, 1609 (1980).
7. L. Bergmann, FSU-HEP-890215, Ph. D. Dissertation.
8. H. Baer, J. Ohnemus, and J. F. Owens, *Phys. Rev.* **D40**, 2844 (1989).
9. H. Baer, J. Ohnemus, and J. F. Owens, *Phys. Lett.* **B234**, 127 (1990).
10. H. Baer, J. Ohnemus, and J. F. Owens, FSU-HEP-900214; *Phys. Rev.* **D**, in press.
11. K. Koller, T.F. Walsh, and P.M. Zerwas, *Z. Phys.* **C2**, 197 (1979).
12. P. Aurenche, R. Baier, A. Douiri, M. Fontannaz, and D. Schiff, *Nucl. Phys.* **B286**, 553 (1987).
13. The expressions in Ref. 8 for the $qq \to qq\gamma$ and $qq' \to qq'\gamma$ subprocesses contain a number of typographical errors where the variable a_3 appears in place of a_5. The original preprint, LPTHE Orsay 86/24, contains the correct expressions.
14. G. 't Hooft and M. Veltman, *Nucl. Phys.* **B44**, 189 (1972).
15. W. A. Bardeen, A. J. Buras, D. W. Duke, and T. Muta, *Phys. Rev.* **D18**, 3998 (1978).
16. P. Aurenche, R. Baier, and M. Fontannaz, FERMILAB-PUB-89/226-T.
17. This point was the subject of some debate during the discussion session on direct photons.
18. T. Åkesson et. al., CERN-EP-89/98R, to be published in *Yad. Fiz.*
19. Yu.L. Dokshitzer, D.I. Dyakonov, and S.I. Troyan, *Phys. Rep.* **58**, 269 (1980).

RESULTS ON DIRECT-PHOTONS FROM FIXED-TARGET EXPERIMENTS

C. BROMBERG[*]
Michigan State University
E. Lansing, MI 48824

ABSTRACT

A review of the contributions of recent direct-photon measurements in fixed-target experiments to our knowledge of the gluon structure function of the nucleon is presented. The direct-photon data constrain the gluon in a way that deep-inelastic experiments can not match.

1. Introduction

Fitting deep-inelastic lepton scattering (DIS) and lepton-pair production data to the behavior predicted by QCD is the method by which the quark structure functions of the nucleon have been determined[1]. With this procedure a wide variety of data have been successfully reproduced with a common set of structure functions. However, only poorly constrained determinations of the gluon structure function are possible from this data. This is a consequence of the fact that the gluon structure function enters the behavior of deep-inelastic scattering only through a small contribution to the Q^2-evolution of the cross sections at fixed-x. Since the gluon structure function is needed to predict most processes at hadron colliders obtaining a better determination is quite important.

To constrain the gluon a number of groups have recently incorporated fixed-target measurements of the hadronic production of direct-photons into fits for the gluon structure function alone or in global fits to both the quark and gluon distributions[2-4]. These measurements have the desired constraining effect on the gluon because the production of direct-photons in proton-proton scattering is dominated in leading order by the contributions of the $qg \rightarrow \gamma q$ Compton subprocess. Also important for these most recent determinations is the development of procedures for full next-to-leading order QCD calculations of direct photon cross sections[5], however, some uncertainty still exists in the procedure for choice of the appropriate Q^2-scale.

A number of fixed-target experiments have either completed or are in the process of analyzing recent data on direct-photon production[6-10], most notably the CERN fixed-target experiment WA70. In these experiments the cross section for direct photon production is small compared to π^0 production which is the major source of photon background (in pp collisions $\gamma/\pi^0 \approx 0.1$ over most of the p_t-range). However, photon detectors with excellent energy and pair resolution are utilized in these experiments so that a majority of the background can be identified, studied, and removed. Also, the fixed-target data covers a kinematical range of x which is inaccessible elsewhere. This complimentary kinematical regime and the superior background rejection capabilities distinguish the fixed-target data from the higher-p_t data of the colliders.

2. Fixed-Target Direct-Photon Experiments

The heart of a fixed-target direct-photon experiment is its π^0 detector. Most of the experiments use a two-dimensional analog readout of the electromagnetic showers in either two or three depth segments, WA70, NA24, and UA6 use a Cartesian coordinate system

---

[*] Research Supported by NSF grant 8922164.

while E706 uses a polar coordinate system for readout. The strip size of the detectors is in most cases sufficiently small ($\approx$ 5mm in the E706 case) to resolve symmetric π^0-decays into two photon showers up to the highest p_t-values reached in the experiment. Therefore, it is not the coalescence of two photons which causes the most serious background to direct-photons but the loss or distortion of the lowest energy photon in highly asymmetric π^0-decays. This latter source of background is sensitive to the behavior of the detector at low energies and to the ability of the reconstruction programs to identify a low-energy photon in the tail of a much larger shower. This background source is most prevalent at the lowest p_t-values where the ratio of γ/π^0 is smallest. Therefore, uncertainties in the corrections for these effects result in large systematic errors on the direct-photon cross section at low p_t-values while at large p_t-values statistical errors dominate.

It has been recognized for some time that the measurement of direct-photons in hadronic collisions would constrain the the gluon structure function of hadrons[11]. Recent fixed-target experiments, some of which were proposed nearly a decade ago, study the production of direct-photons for transverse momenta above 3 GeV/c and up to about 10 GeV/c, which is in the range of momentum transfers where hard-scattering phenomenology using perturbative QCD is thought to be valid. A comparison of the x-range which the various

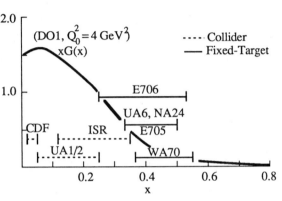

Fig. 1. Sensitivity range to the gluon structure function of various experiments measuring direct-photons.

experiments are sensitive to the gluon structure function is shown in Fig. 1. As is evident from the figure, kinematics and available luminosity locate the measurements of the gluon structure function made by fixed-target experiments to the range of x between 0.25 and 0.55 which is complimentary to the range of x accessible to modern collider experiments[12] (x<0.25).

Charged particles are also measured by these fixed-target experiments in order to verify that the direct-photon (and π^0) high-p_t triggers exhibit the jet-characteristics expected for the parton-parton scattering origin of the events. Representative plots from E706 are shown in Fig. 2. The data in Fig. 2a. shows that accompanying the trigger particle is a jet of hadrons clustered sharply about an azimuthal angle (ϕ) 180° away from the trigger direction. A small enhancement of associated particles in the trigger direction for π^0-triggers is evidence of an associated jet in those events which is absent in direct-photon events. The recoil-jets are also seen in rapidity. In Fig. 2b the rapidity difference (Δy) between the charged particles with the largest and next-largest p_t-values ($p_t > 0.5$ GeV/c) in the azimuthal hemisphere opposite the trigger exhibits a clear clustering of particles about the rapidity of the leading-p_t particle (a background shape taking particles from random events has been normalized to large Δy and subtracted from the distribution). The correlation between these large-p_t particles suggests that the recoil-jet covers about 1.2 units of rapidity.

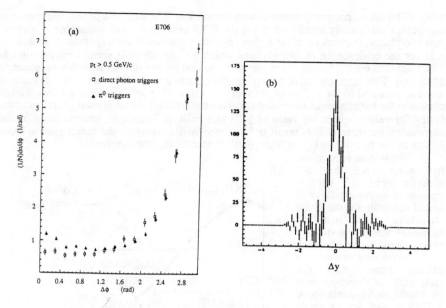

Fig. 2. (a) The azimuthal angle difference of charged particles (p_t>0.5 GeV/c) with respect to the trigger direction. (b) The rapidity difference between the leading-p_t and next-to-leading-p_t charged particles; a random event background has been subtracted.

3. Gluon Structure Function Determinations

Commonly used parameterizations of the parton structure functions have been available for some time, eg., sets 1 and 2 of Duke and Owens (DO1 and DO2)[13]. The direct photon data can be directly compared with QCD predictions based on these structure functions using a fixed-scale and the next-to-leading order terms in the cross section[4]. Alternatively, in global fits to both quark and gluon structure functions[2,3] also using next-to-leading order terms (with the PMS-procedure[14] used to set the Q^2-scale) the current measurements are sufficient to establish the power η_g in the parameterization of the gluon structure function $xg(x)=A_g(1-x)^{\eta_g}$ with the normalization, A_g, constrained by the momentum sum rule.

From the two recent papers (HMSR[3] and BO2[4]) comparisons of the results of both calculations with the WA70 data are shown in Fig. 3. One should note that the data in the HMSR paper (corrected to a y=0 cross section) have the systematic errors included in the error-bars (and the fits) while the comparison made in the BO2 paper shows only the statistical errors. The conclusions which can be reached from these fits are that the WA70 data is consistent with the DO1 parameterization of the gluon structure function if the Q^2-scale is set at $Q^2 = p_t^2/4$ and that the PMS-procedure yields a scale choice very similar to this in the global fits of HMSR.

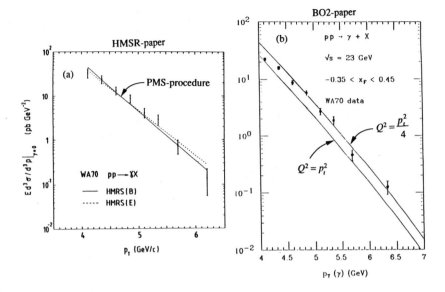

Fig. 3 A comparison of the direct-photon data of WA70 with the predictions of next-to-leading order calculations of the cross section using (a) a global fit to both the quark and gluon structure functions with the PMS-procedure used to set the Q^2-scale and (b) two fixed Q^2-scales and the DO1 structure functions.

In the global fits of HMSR values of η_g are only weakly correlated with the DIS data chosen with the fits to the BCDMS-collaboration[15] data yielding an η_g near 5 while the EMC[16] data favor η_g closer to 4.

In Fig. 4 the preliminary results of E706 are compared with the predictions of DO1(and DO2) and again the DO1 set is preferred.

The WA70 data and the data of two other experiments, NA24 and UA6, were used by ABFOW[2] in their determination of the power η_g. Also, they have, independently, determined η_g from the deep-inelastic scattering data of the BCDMS collaboration alone. The insensitivity of the deep-inelastic scattering data to the functional form of the gluon structure function and the consistency of the results from three fixed-target

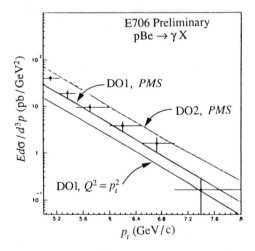

Fig. 4 Preliminary E706 data compared with DO1 and DO2 structure functions with various Q^2 - scale choices.

experiments in their determinations of η_g near 4 is clear in their results shown in Fig. 5. The value of Λ from the fits is also shown to be only weakly coupled to the value of η_g.

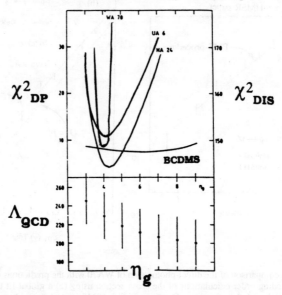

Fig. 5 The χ^2 of the fit to direct-photon (DP) or deep-inelastic (DIS) experimental data with respect to the gluon structure function parameter η_g. The correlation of η_g with the QCD scale parameter Λ is shown below.

An important test of the procedures used to determine the gluon structure function will be the study of a related process, the production of two direct-photons. The WA70-collaboration has published results[17] on the two-photon process as observed in π^-p collisions. The transverse momentum of the pair shown in Fig. 6 is consistent with a large intrinsic transverse momentum for the scattering partons, $k_t \sim 0.9$ GeV/c. This is considerably larger than the value k_t predicted by next-to-leading order calculations for the process even when smeared with an intrinsic $k_t \sim 0.35$ GeV/c. This discrepancy may indicate that a feature of hadronic scattering is still missing from the hard-scattering formalism used in direct-photon analyses.

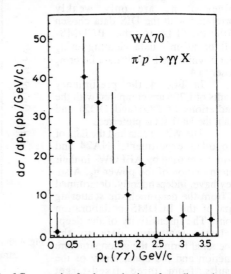

Fig. 6 Cross section for the production of two direct-photons as a function of the transverse momentum of the pair.

4. Conclusions

Recent inclusion of the direct-photon data from fixed-target experiments, in particular the data of the WA70 group, into global fits for the quark and gluon structure functions has significantly altered the character of the gluon structure function as previously determined by deep-inelastic scattering experiments alone. The new determinations of the gluon structure function power $\eta_g \approx 4$, yield a shape similar to the DO1-parameterization and are consistent with all recent fixed-target data.

Final data from the recent fixed-target experiments will broaden the range of x over which the gluon structure function is determined and, if systematic errors are carefully controlled, add significantly to the constraints on its shape.

5. Acknowledgements

I would like to thank the conference organizers for the opportunity to speak at the conference and in particular to D. F. Geesaman for his patience during the preparation of this manuscript. I also thank P. Aurenche, J. Owens, and W.J. Sterling for early access to and conversations regarding their work on the determination of structure function parameters.

6. References

1. A. D. Martin, R. G. Roberts and W. J. Sterling, Phys. Lett. **206B** (1988) 327.
2. P. Aurenche, R. Baier, M. Fontannaz, J. F. Owens and M. Werlen, Phys. Rev. **D39** (1989) 3275.
3. P. N. Harriman, A. D. Martin, J. W. Sterling, and R. G. Roberts, Durham preprint DTP/90/04, Rutherford preprint RAL/90/007, January 1990.
4. H. Baer, J. Ohnemus, and J. F. Owens, Florida preprint, FSU-HEP-900214.
5. P. Aurenche, R. Baier, M. Fontannaz and D. Schiff, Phys. Lett. **140B** (1984) 87, and Nucl. Phys. **B297** (1988) 661.
6. WA70 collaboration: M. Bonesini et al., Zeit. Phys. **C38** (1988) 371.
7. NA24 collaboration: C. De Marzo et al., Phys. Rev. **D36** (1987) 8.
8. UA6 collaboration: A. Bernasconi et al., Phys. Lett. **B206** (1988) 163, and new data presented at this conference by L. Camilleri.
9. E705 collaboration: S.W. Delchamps et al., in proceedings of the DPF Meeting, Storrs (1988) 681, K. Haller et al., eds.
10. E706 collaboration: G. Alverson et al., XXIVth Conference on High Energy Physics, R. Kotthaus and J. H. Kuhn, eds. (1988) 719.
11. L. Cormell and J. F. Owens, Phys. Rev. **D22**, 1609 (1980); E. Berger, E. Braaten and R. Field, Nucl. Phys. **B239**, 52 (1984).
12. CDF collaboration: R. Blair et al., ANL-HEP-CP-89-07 (1989); UA1 collaboration, C. Albajar et al., Phys. Lett. **B209**, (1988) 385; UA2 collaboration: R. Ansari et al., Zeit. Phys. **C41**, (1988) 395.
13. D. W. Duke and J. F. Owens, Phys. Rev. **D30** (1984) 49.
14. P. M. Stevenson and H. D. Politzer, Nucl. Phys. **B277** (1986) 758.
15. BCDMS collaboration: A.C. Benvenuti et al., Phys. Lett. **223B** (1989) 485.
16. EMC collaboration: K. J. Aubert et al., Nucl. Phys. **B259** (1985) 189.
17. E. Bonvin et al., Phys. Lett. **236B** (1990) 523.

Recent Results on Direct Photons from CDF

The CDF Collaboration*
Presented by Robert M. Harris
Fermi National Accelerator Laboratory
Batavia, IL   60510, USA

June 1, 1990

Abstract

We report on preliminary measurements of direct photons in $\bar{p}p$ collisions at $\sqrt{s} = 1.8$ TeV from the 1988-89 run of the Collider Detector at Fermilab (CDF). The inclusive direct photon cross section, measured for photon transverse momentum in the range $13 < P_t < 68$ GeV, has an excess at low P_t compared to recent Quantum Chromodynamic (QCD) calculations. The pseudorapidity distribution of the away-side jet, for events with $27 < P_t < 33$ GeV, agrees with QCD predictions. Measurements of the K_t kick in photon-jet events are also presented.

1 Physics Motivation

Measurements of photons, coming directly from the hard collision of partons, provide a test of QCD which is free from the energy measurement uncertainties associated with jets. The compton diagrams, shown in Fig. 1a, dominate the production of direct photons at the Born level, so measurements of direct photons are sensitive to the gluon distribution of the proton. The high center of mass energy of the Tevatron allows the CDF detector[1] to test QCD and probe the gluon distribution in a previously unexplored range of fractional momentum ($.015 < x < .075$).

*The collaborating institutions are listed in Appendix A

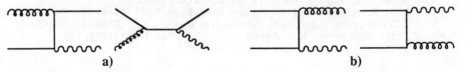

Figure 1: Lowest order diagrams for direct photons. **a)** Compton and **b)** annihilation.

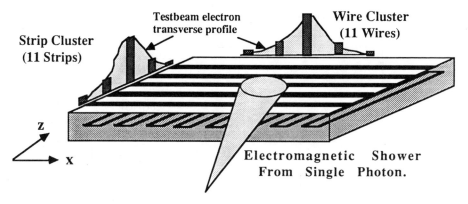

Figure 2: The transverse profile of a photon in a CES chamber.

2 Direct Photon-Background Separation

The signal for a direct photon is the deposition of isolated energy in the electromagnetic calorimeter with no associated charged track in the tracking chamber. The dominant background comes from the decays of the neutral mesons π^0 and η into multiple photons. We employ two methods for separating direct photons from neutral mesons: method I uses the transverse profile of the electromagnetic shower and method II counts the number of conversion pairs produced just outside the tracking chamber.

2.1 Method I: Transverse Shower Profiles

The central electromagnetic strip (CES) chambers[2] are multiwire proportional chambers embedded in the central electromagnetic calorimeter at shower maximum (6 radiation lengths). A calorimeter cluster consists of three projective towers spanning $\Delta\eta \times \Delta\phi \approx .3 \times .26$. The CES anode wires are perpendicular to cathode strips, as shown in Fig. 2. Within the boundaries of the calorimeter cluster are CES strip and wire clusters, consisting of eleven strips and eleven wires respectively. The highest energy strip cluster and the highest energy wire cluster are chosen to measure the transverse profile and position of photons. The electromagnetic shower from a single photon, shown schematically in Fig. 2, has a narrow transverse profile, and produced a small χ^2 when fit to an electron transverse profile. Conversely, a neutral meson decaying into multiple photons has a wider profile, and produced a larger χ^2 when fit to an electron transverse profile. Fits were done for both the wire and strip clusters and each raw χ^2 was scaled by a function of the calorimeter cluster energy to produce a corrected χ^2 which was independent of energy for electrons and (presumably) photons. Then the *average* χ^2 ((strip χ^2 + wire χ^2)/2) was the variable used to separate

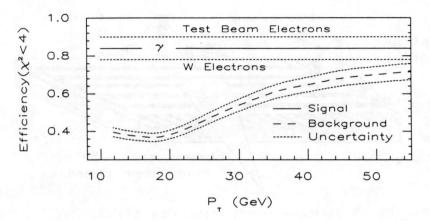

Figure 3: The efficiency for photons and background to have $\chi^2 < 4$ vs. P_t

single photons from neutral mesons.

Let ϵ_γ be the efficiency for a single photon to have $\chi^2 < 4$, and let ϵ_{π^0} be the efficiency for multiple photons from neutral meson decays to have $\chi^2 < 4$. Then the number of clusters with $\chi^2 < 4$ and $\chi^2 > 4$ is related to the number of photons and neutral mesons by:

$$\begin{bmatrix} N_{\chi^2<4} \\ N_{\chi^2>4} \end{bmatrix} = \begin{bmatrix} \epsilon_\gamma & \epsilon_{\pi^0} \\ 1-\epsilon_\gamma & 1-\epsilon_{\pi^0} \end{bmatrix} \begin{bmatrix} N_\gamma \\ N_{\pi^0} \end{bmatrix} \quad (1)$$

Inverting Eq. 1 gives the number of photons and neutral mesons:

$$\begin{bmatrix} N_\gamma \\ N_{\pi^0} \end{bmatrix} = \frac{1}{\epsilon_\gamma - \epsilon_{\pi^0}} \begin{bmatrix} 1-\epsilon_{\pi^0} & -\epsilon_{\pi^0} \\ \epsilon_\gamma - 1 & \epsilon_\gamma \end{bmatrix} \begin{bmatrix} N_{\chi^2<4} \\ N_{\chi^2>4} \end{bmatrix} \quad (2)$$

We estimated ϵ_γ using a measured testbeam electron shower for each photon in a full detector simulation which includes all analysis cuts (see section 2.1.1). Figure 3 shows the estimated photon χ^2 efficiency midway between the upper systematic bound from raw testbeam electrons and the lower systematic bound from W electrons. Similarly, we estimated ϵ_{π^0} from the simulated decays of the neutral mesons π^0, η, and K_s^0. Figure 3 shows the P_t dependence of the χ^2 efficiency of this background. Here we assumed the mesons π^0, η, and K_s^0 are produced with relative rates 1 : 0.6 : 0.25. Varying the relative meson production rates within the limits 1 : 0.75 : 0.5 and 1 : 0.45 : 0.0 produced only a small variation in efficiency. The largest systematic uncertainty on the background χ^2 efficiency, coming from the propagation of the photon simulation uncertainty, is displayed in Fig. 3. This is the dominant systematic uncertainty in the measurement of the direct photon cross section using method I. At high P_t the multiple photons from a neutral meson are so close together that the

Trigger Cuts	γ Acceptance
$P_t > 10$ GeV (77.5 nb^{-1})	$\sim$1 at 14 GeV
$P_t > 23$ GeV (2.55 pb^{-1})	$\sim$1 at 27 GeV
$E_{T,HAD}/E_{T,EM} < .125$	$\sim$1
Isolation: $E_{CONE}/E_\gamma < .15$	$.82 \to .96$
Analysis Cuts	
No Track Pointing at Photon	.98
2nd CES Strip and Wire Clusters < 1 GeV	.86
CES Strip and Average $\chi^2 < 20$	$\sim$1
14 cm $<$ |CES Z| < 217 cm	.92
|CES X| < 17.5 cm	.77
$|\eta| < 0.9$	$\sim$1
|Z Vertex| < 50 cm ($\sigma_Z = 32$ cm)	.88
Missing E_t Significance < 3	$\sim$1
	$.43 \to .50$

Table 1: The data sample, trigger, and analysis cuts for method I

χ^2 efficiency for the background is almost the same as for a single photon. Thus, for high values of P_t, Eq. 2 becomes singular and so does the systematic uncertainty in the number of direct photons. To avoid large systematic uncertainties, method I was only used up to transverse momenta of 33 GeV.

2.1.1 Data Sample and Event Selection for Method I

The data sample, trigger, and cuts for the method I analysis are summarized in table 1. The low P_t trigger was prescaled to reduce the rate throughout the run, while the high P_t trigger was not prescaled. The triggers required that 89% of the transverse energy of the photon be in the electromagnetic compartment of the calorimeter and also required the photon to be *isolated*: the extra energy inside a cone of radius $\sqrt{(\Delta\eta)^2 + (\Delta\phi)^2} = 0.7$ centered on the photon is required to be less than 15% of the photon energy. Charged background is eliminated by requiring there be no track pointing at the calorimeter cluster associated with the photon. Neutral hadron background is reduced by requiring that additional strip chamber clusters within the boundaries of the calorimeter cluster be less than 1 GeV each. Fiducial cuts are imposed to avoid uninstrumented regions at the strip chamber edges, and a cut on the z coordinate of the event vertex was imposed to maintain the calorimeter towers projective geometry. Finally, a cut on the event missing transverse energy significance ($\sqrt{(\Sigma\vec{E_x})^2 + (\Sigma\vec{E_y})^2}/\sqrt{\Sigma E_t}$) removes residual cosmic ray *bremsstrahlung*. The cuts have an acceptance ranging from 43% (at 14 GeV) to 50% (at 33 GeV). The P_t depen-

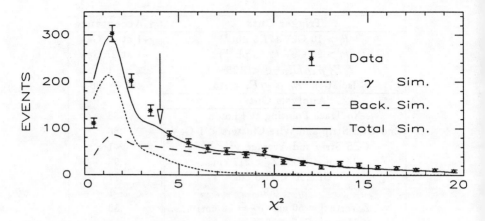

Figure 4: The χ^2 distribution for the data compared to simulation.

dence of the acceptance comes from the fractional isolation cut which eliminates more photon events at low P_t because of fluctuations in the underlying event.

The χ^2 distribution of the data after all cuts is shown in fig. 4 for the interval $14 < P_t < 20$ GeV. The peak at low χ^2 is produced predominantly by direct photons. The background and photon simulations have been normalized to the number of events using Eq. 2. For $\chi^2 > 4$ the simulation models the data quite well. For $\chi^2 < 4$ the simulated distribution has lower χ^2 than the data. Similarly, testbeam electrons have a lower χ^2 than electrons from the decay $W \to e\nu$, which gave the systematic bounds in χ^2 efficiency shown previously in Fig. 3. Only the efficiency for $\chi^2 < 4$ is needed in the method I technique of Eq. 2; knowledge of the detailed shape of the χ^2 distribution inside of $\chi^2 < 4$ is not necessary.

2.1.2 Results from Method I

Using the number of direct photons (N_γ) in a bin of transverse momentum (ΔP_t) and a bin of pseudorapidity ($\Delta \eta$), the acceptance (α) and the integrated luminosity ($\int L$), we obtain the inclusive direct photon cross section:

$$\frac{d^2\sigma}{dP_t d\eta} = \frac{N_\gamma}{\Delta P_t \cdot \Delta \eta \cdot \alpha \cdot \int L} \qquad (3)$$

which is shown twice in Fig. 5. The inner error bars are the statistical uncertainty and the outer error bars are the P_t dependent part of the systematic uncertainty combined in quadrature with the statistical uncertainty. The P_t independent component of the systematic uncertainty is shown as the normalization uncertainty. In Fig. 5 our measurement is compared to QCD calculations[3] using three different sets of parton distribution functions[4,5] for a single choice of the renormalization scale $Q^2 = P_t^2$.

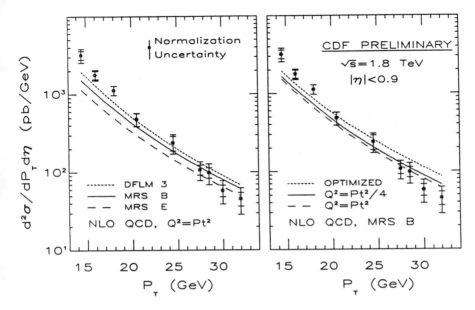

Figure 5: The inclusive direct photon cross section from method I and QCD predictions.

Also in Fig. 5, the same measurement is compared to QCD calculations[3] using three different choices of the renormalization scale for a single set of parton distribution functions[5]. The calculations are discussed in greater detail in section 3. The cross section measured with method I has a slightly steeper dependence on P_t than the QCD predictions.

2.2 Method II: Photon Conversions

The central drift tubes[6] (CDT) are three layers of gas counters just outside the central tracking chamber (CTC), as shown in Fig. 6a. The outer wall of the CTC and the inner two CDT layers provide 18% of a radiation length for the conversion of photons into electron-positron pairs. Neutral mesons decaying into two photons are twice as likely to produce a conversion pair, which is then detected as a CDT *hit* cluster. Requiring $\chi^2 < 4$ reduced the neutral meson background and preferentially selected asymmetric decays. Conversions of direct photons, or conversions of the higher energy photon from an asymmetric decay, produced no azimuthal separation of the CDT hits and CES clusters, and made the spike at zero in Fig. 6b. However, conversions of the lower energy photon of an asymmetric decay produced a measurable azimuthal separation of CDT hits and CES clusters, and made the bump shown in Fig. 6b. From this plot the production ratio γ/π^0 was estimated.

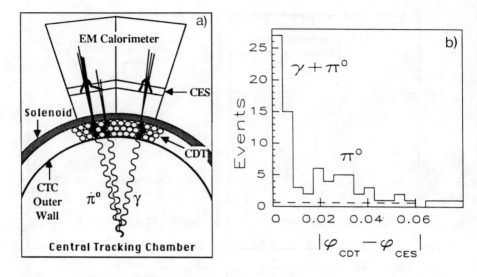

Figure 6: **a)** Conversions detected in the CDT and associated CES cluster. **b)** Azimuthal separation of CDT and CES clusters for $9 < P_t < 11$ GeV.

From the γ/π^0 ratio, and the number of neutral clusters in the calorimeter (N_C), and the total number of CDT hits ($N_{H,total}$) associated with neutral clusters, we estimated the probability of observing a conversion:

$$P_\gamma = \frac{N_{H,total}}{N_C} \left[\frac{1+\gamma/\pi^0}{2+\gamma/\pi^0}\right] \approx 0.10 \pm 0.02 \qquad (4)$$

The conversion probability expected from the amount of material and the efficiency of hit selection is about 11%, consistent with the value given in Eq. 4 within the systematic uncertainty. This systematic uncertainty, which is approximately P_t independent, dominated the uncertainty in the normalization of the direct photon cross section from method II. Finally, the number of clusters with one or more hit (N_H) is related to the number of photons and neutral mesons by:

$$\begin{bmatrix} N_H \\ N_C \end{bmatrix} = \begin{bmatrix} P_\gamma & 2P_\gamma - P_\gamma^2 \\ 1 & 1 \end{bmatrix} \begin{bmatrix} N_\gamma \\ N_{\pi^0} \end{bmatrix} \qquad (5)$$

Inverting Eq. 5 gives the number of photons and background:

$$\begin{bmatrix} N_\gamma \\ N_{\pi^0} \end{bmatrix} = \frac{1}{P_\gamma^2 - P_\gamma} \begin{bmatrix} 1 & P_\gamma^2 - 2P_\gamma \\ -1 & P_\gamma \end{bmatrix} \begin{bmatrix} N_H \\ N_C \end{bmatrix} \qquad (6)$$

2.2.1 Data Sample and Event Selection for Method II

The data sample for the method II analysis came from the trigger discussed in section 2.1.1, except the integrated luminosity was 45 nb^{-1} and 1.4 pb^{-1} for the low and high P_t thresholds respectively. The analysis cuts in section 2.1.1 were used **except**: the CES χ^2 cuts were not present, the CES 2nd cluster cut was at 2 GeV with a corresponding γ acceptance of 90%, and there was a CES strip and wire energy matching cut with a γ acceptance of 91%. In **addition**, the following cuts were introduced for the method II analysis:

- CES average $\chi^2 < 4$ to reduce the neutral meson background.
- CDT-CES $\Delta\phi < 0.1(0.07)$ for the low (high) P_t trigger.
- CDT-CES $\Delta z < 10$ cm to reduce stray track backgrounds.

The $\Delta\phi$ and Δz cuts, used in the definition of a CDT hit, had an efficiency of 100% and 75% respectively. These efficiencies are part of the observed photon conversion probability defined in Eq. 4. Background CDT hits caused by stray tracks were estimated from CDT-CES $\Delta\phi$ distributions, and were subtracted from N_H before Eq. 6 was used to find N_γ. A typical amount of random background is illustrated by the dashed line in Fig. 6b. The total direct photon acceptance was between 35% and 42% depending on P_t. Finally, the number of photons, acceptance, and integrated luminosity were used in Eq. 3 to determine the inclusive direct photon cross section from method II.

3 Inclusive Direct Photon Cross Section

The inclusive direct photon cross sections, measured separately with method I and method II, are each shown twice in Fig. 7 along with the same QCD calculations[3] that were presented in Fig. 5. The error bars on the method II points are statistical uncertainties. The outer error bars on the method I points are the statistical uncertainties and the P_t dependent component of the systematic uncertainties added in quadrature. The P_t independent component of the systematic uncertainty for each method is shown separately as the normalization uncertainty. The cross sections measured with the two methods agree within errors. The method II analysis extends to higher P_t than the more precise method I analysis. We conclude that the measured cross section has an excess at low P_t compared to the QCD predictions shown; at $P_t > 20$ GeV the measured cross section agrees with the QCD predictions.

Before drawing conclusions about the behavior of the gluon distribution at low x, the reader should consider the uncertainties in the QCD calculation. The QCD calculations approximate the experimental *isolation* cut, requiring that partons inside a cone of radius 0.7 around the photon have less than 15% of the photon energy. The

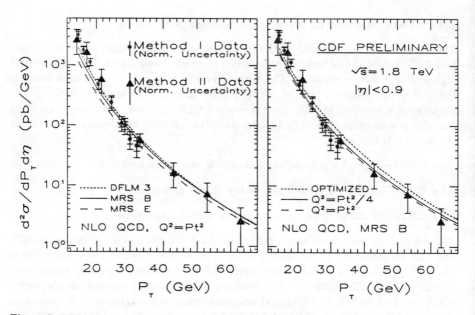

Figure 7: The inclusive direct photon cross section from method I and method II compared to QCD predictions.

calculated cross section decreases by less than 5% when the isolation condition is tightened to allow no partons in the isolation cone, and it increases by less than 7% when the isolation condition is loosened by requiring a smaller cone of radius 0.4 for the same 15% energy cut. The uncertainties in the QCD calculation associated with truncating the perturbation expansion can be roughly estimated by varying the choice of renormalization scale. The calculations presented in Fig. 5 and Fig. 7 have equal renormalization scales and factorization scales, labeled by Q^2, except for the calculation labeled *optimized* which has a renormalization scale $\mu_R^2 \approx P_t^2/50$ and a factorization scale $\mu_F^2 \approx 50 P_t^2$. The calculations are at next to leading order, that is they include all diagrams of order $\alpha\alpha_s^2$, except for final state photon *bremsstrahlung* (see Fig. 8) which is included only at an effective order of $\alpha\alpha_s$. Photon *bremsstrahlung* contributes significantly to the cross section at low P_t, and is calculated by convolving leading order diagrams with the *effective fragmentation function* for obtaining photons from partons. The effective fragmentation function of a parton into a photon has never been measured. It is modeled by a QED splitting function evolved to account for emission of soft and collinear gluons.

Considering the current uncertainties on the theoretical calculation and the preliminary nature of our measurement, it may be premature to draw conclusions about

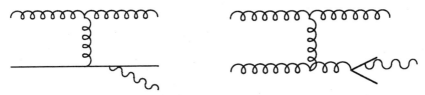

Figure 8: Examples of *bremsstrahlung* diagrams for direct photon production.

the detailed shape of the gluon distribution from Fig. 5 or Fig. 7.

4 Leading Jet in Photon Events

Direct photons are produced along with a quark or gluon which appears in the CDF detector as a jet of hadrons. From the distribution of the jet, in both polar and azimuthal angles, we can obtain additional information about the hard scattering of partons. We find jets using a fixed cone clustering algorithm[7] with radius $\sqrt{(\Delta \eta)^2 + (\Delta \phi)^2} = 0.7$. The leading jet in a direct photon event is the jet with the highest transverse momentum[7]. We measure photons and neutral mesons for $|\eta| < 0.9$ and jets for $|\eta_j| < 3.2$. For the following measurements the data sample came from the trigger discussed in section 2.1.1, except the integrated luminosity was 67 nb^{-1} and 1.4 pb^{-1} for the low and high P_t thresholds respectively. The method I analysis cuts in section 2.1.1 were used, with a few exceptions noted in section 2.2.1, and the CES average χ^2 was required to be less than 25.

4.1 Leading Jet Pseudorapidity Distribution

Pseudorapidity is related to polar angle by $\eta = \ln(\cot(\theta/2))$. The pseudorapidity of the leading jet has a distribution which depends on the convolution of parton momentum distributions, hard scattering angular distributions, and the $1/\hat{s}^2$ dependence of the parton subprocess cross section. Thus, measurements of the leading jet pseudorapidity distribution can be used to study parton momentum distributions and angular distributions for two different final states: photon-parton final states for the photon signal and parton-parton final states for the photon background.

In Fig. 9 the pseudorapidity distribution of the leading jet is shown separately for photons and neutral mesons with $27 < P_t < 33$ GeV. The error bars on the points are statistical errors only; the systematic uncertainty for photons is roughly ±40%. The leading jet pseudorapidity distribution for photons (labeled γ-jet) is compared to a full next to leading order calculation[8] and a leading order calculation. The former calculation's normalization is absolute, but the latter calculation was multiplied by 1.37 to fit the data at low $|\eta_j|$. Within statistics the data is compatible with either calculation. The leading jet pseudorapidity distribution for the background (labeled π^0-jet) is compared to a leading order QCD calculation of parton-parton scattering,

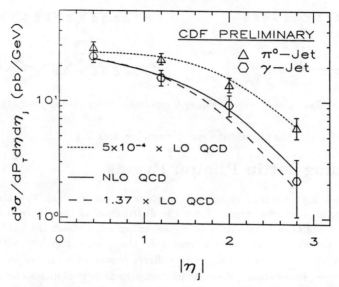

Figure 9: The pseudorapidity distribution of the leading jet, for photon and background events, is compared to QCD predictions (see text).

without fragmentation of the final state partons. This calculation was multiplied by 5×10^{-4} to fit the measured data (this number can be interpreted as an order of magnitude estimate of the probability for a gluon to fragment into a single isolated neutral meson). Measurement of the leading jet pseudorapidity in events with lower photon P_t are more difficult because of fluctuations in the underlying event, and these distributions are still being studied.

4.2 K_t kick in photon events

In lowest order QCD the jet and photon have equal and opposite transverse momentum. We have used $\Delta\phi$, the azimuthal angle between the photon and the leading jet in the transverse plane, to measure how much real events deviate from the lowest order picture. As shown in Fig. 10a, we only use the leading jet to measure $\Delta\phi$, and redefine the magnitude of the transverse momentum of the leading jet to be equal to the transverse momentum of the photon. Then the vector sum of the transverse momentum of the photon and the leading jet must lie along the perpendicular bisector of $\Delta\phi$ in the transverse plane. The magnitude of this vector

$$K_{t\perp} = 2P_t \cos\frac{\Delta\phi}{2} \qquad (7)$$

is always zero for lowest order QCD calculations in the naive parton model. Since higher order effects will cause $\vec{K}_{t\perp}$ to deviate from zero, an interesting variable is the

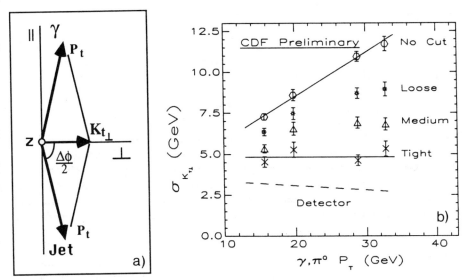

Figure 10: **a)** Definition of measured K_t kick. **b)** The RMS deviation of the K_t kick from zero vs. the photon (with background) P_t.

RMS deviation of $\vec{K_{t\perp}}$ about zero:

$$\sigma_{K_{t\perp}} = \sqrt{\langle K_{t\perp}^2 \rangle} \qquad (8)$$

In Fig. 10b we show some measurements of the RMS deviation of $\vec{K_{t\perp}}$ about zero, for direct photons with background (we have not done a background subtraction). The events in this plot were required to have a leading jet with greater than 10 GeV transverse momentum and $\Delta\phi > 90°$. With no additional cuts, the hexagons show that the RMS deviation of the K_t kick increases with increasing photon (and background) P_t, and we fit it with a solid line to guide the eye. The magnitude of this K_t kick is a significant deviation from the lowest order QCD expectations, and we hope a next to leading order calculation will be available soon. The RMS deviation of the K_t kick caused by jet angular resolution in the CDF detector was simulated and is shown as the dashed line labelled 'Detector'.

To estimate the influence of QCD radiation on the measured K_t kick we devised a cut on additional jet activity which maintains an equal fraction of events in each bin of photon (and background) P_t. Requiring that all additional jets in the event have transverse momentum $P_{t,jet} < 0.28 P_t + 6.6$ GeV is a 75% efficient cut, and the RMS deviation of the K_t kick for surviving events is shown with the asterisks in Fig 10b. Similarly, requiring that all additional jets satisfy $P_{t,jet} < 0.16 P_t + 6.2$ GeV is 50% efficient and is shown by the triangles; requiring $P_{t,jet} < 0.09 P_t + 5.0$ GeV is only 25%

efficient and is shown by the crosses. Note how the RMS deviation of the K_t kick decreases and becomes a flatter function of P_t as we tighten the cut on additional jet activity; this suggests we are removing events with QCD radiation, and that the P_t of the radiation may be correlated with the subprocess P_t. In any case, events with additional jet activity contribute significantly to the measured K_t kick.

5 Summary and Future Prospects

Preliminary results indicate that the inclusive direct photon cross section has an excess at low P_t compared to recent next to leading order QCD predictions. At $P_t > 20$ GeV the measured cross section agrees with QCD. The pseudorapidity of the leading jet, for both photons and background in the range $27 < P_t < 33$ GeV, is in good agreement with next to leading order QCD predictions. Measurements of the Kt kick in the combined photon-jet and pizero-jet final state have been made and hopefully a next to leading order calculation will be available soon.

Studies of the following topics related to photons have already begun or will soon begin: the inclusive cross section of events with two photons, the fragmentation of the leading jet in photon events, the search for $b' \to b\gamma$, the search for $W \to \pi\gamma$, and the cross section for $W\gamma$ production. All photon analysis topics will benefit from the planned installation of conversion detection chambers outside the CDF solenoid. The materials in the solenoid (1 radiation length) produce photon conversions, which will allow the new chambers to efficiently separate direct photons from background during the 1991 run.

Appendix A: CDF Collaborating Institutions

ANL - Brandeis - University of Chicago - Fermilab - INFN, Frascati - Harvard - University of Illinois - KEK - LBL - University of Pennsylvania - INFN, University of Scuola Normale Superiore of Pisa - Purdue - Rockefeller - Rutgers - Texas A&M - Tskuba - Tufts - University of Wisconsin

References

1. F. Abe et al., *Nucl. Inst. and Meth.* **A271**(1988)387.
2. L. Balka et al., *Nucl. Inst. and Meth.* **A267**(1988)272.
3. P. Aurenche, R. Baier and M. Fontannaz, FERMILAB-PUB-89/226-T(1989).
4. M. Diemoz, F. Ferroni, E. Longo and G. Martinelli, *Z. Phys.* **C39**(1988)21.
5. A. D. Martin, R. G. Roberts and W. J. Stirling, *Phys. Lett.* **B206**(1988)327.
6. S. Bhadra et al., *Nucl. Inst. and Meth.* **A268**(1988)92.
7. D. N. Brown, Ph.D. Thesis, Harvard University, 1989 (unpublished).
8. H. Baer, J. Ohnemus and J. F. Owens, FSU-HEP-900214(1990).

Comparison of direct γ production in $\bar{p}p$ and pp reactions at $\sqrt{s} = 24.3$ GeV at $4 < p_T < 6$ GeV/c $(.3 < x_T < .5)$

presented by
Leslie Camilleri
CERN
EP Division
CH-1211 Geneva 23
Switzerland

for the UA6 Collaboration

G. Ballocchi, L. Camilleri, G. von Dardel, L. Dick, F. Gaille, C. Grosso-Pilcher, J.-B. Jeanneret, W. Kubischta, EP Division, CERN, 1211 Geneva, Switzerland.
A. Bernasconi, A. Ebongué, B. Gabioud, C. Joseph, J.-F. Loude, E. Malamud, C. Morel, P. Oberson, J.L Pagès, J.-P. Perroud, D. Ruegger, G. Sozzi, L. Studer, M.-T. Tran, M. Werlen, Inst. de Physique Nucléaire, Université de Lausanne, Dorigny, 1015 Lausanne, Switzerland.
E. C. Dukes, D. Hubbard, O. Overseth, G. R. Snow, G. Valenti, Physics Dept. University of Michigan, Ann Arbor, Michigan USA.
R. Breedon, R. L. Cool, P. T. Cox, P. Cushman, P. Giacomelli, P. Mélèse, R. Rusack, V. Singh, A. Vacchi, Dept. of Physics, The Rockefeller University, 1230 York Ave., 10021 New York USA.

Abstract

Direct photons have been studied in both $\bar{p}p$ and pp interactions at $\sqrt{s} = 24.3$ GeV, in the transverse momentum range 4 to 6 GeV/c (0.3 < x_T < 0.5) for -0.4 < y < 1.2. The experiment was performed using an internal hydrogen jet target in the CERN Sp$\bar{p}$S Collider. Preliminary cross-sections in $\bar{p}p$ and in pp are presented. The $\bar{p}p$-pp difference has been measured and is compared to theoretical predictions. A preliminary determination of the gluon structure function is obtained. Prospects for improvements in the UA6 data are also discussed.

1. Introduction

The study of direct photons in hadronic collisions has many advantages over the study of jets and single particles deriving from jets: the em vertex is well understood, the observed photon arises from the elementary interaction itself without fragmentation, only a few diagrams are important, and the next to leading logarithm calculations[1] in perturbative QCD have been performed. This latter point allows incisive comparisons between theory and experiment.

Our ability to study both $\bar{p}p$ and pp interactions in the same experimental apparatus[2] is unique. At leading order two diagrams are important in the production of direct photons: gluon Compton scattering and quark-antiquark annihilation. In pp collisions the first one dominates, in $\bar{p}p$ collisions both diagrams are important, the quark-antiquark process being dominant at large p_T because quarks have a harder structure function then gluons.

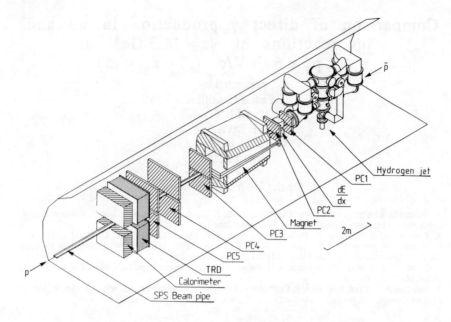

Fig. 1. The UA6 setup.

The difference of cross sections ($\bar{p}p$ - pp) selects the annihilation process which is insensitive to the gluon structure functions. It depends primarily on the quark distributions and on Λ QCD. Since the quark distributions are well measured in DIS the $\bar{p}p$ - pp difference yields a measurement of Λ QCD. Having obtained Λ QCD the $\bar{p}p$ and pp data can then yield the gluon s.f. This method avoids the correlation between the gluon s.f. and Λ present when only one cross section is available.

Direct photon x_T values up to 0.5 can be reached. Such large values are observable only when the initial state partons themselves carry a large fraction of the parent hadron momentum. That is, the direct photons measured in UA6 are sensitive to the shape of the quark and gluon distributions at high values of Bjorken x ($0.25 < x < 0.5$), not reachable by the collider experiments. In constrast the collider direct γ experiments are limited to $x < 0.2$ for UA1 and UA2 and $x < 0.1$ for CDF. Moreover, at high Q^2, structure functions evolve to a similar shape, at low x. Hence, the direct γ of UA6, at low Q^2, constrains the gluon structure function.

2. Apparatus

Figure 1 shows the UA6 setup in the CERN SPS in its configuration to observe $\bar{p}p$ collisions. It consists of a H_2 cluster jet used as an internal target, followed by a double arm spectrometer. An arm consists of five multiwire proportional chambers, two in front and three behind a 2.3 T·m dipole magnet and an em calorimeter. To observe pp collisions, the spectrometer has to be reassembled on the other side of the jet target. The jet is 0.8 cm long along the beam and has a density of 4×10^{14} nucleons/cm^3.

High instantaneous luminosities of 5.5×10^{29} cm^{-2} s^{-1} in $\bar{p}p$ can be achieved because the particle bunches traverse the jet at the SPS revolution frequency.

The two em calorimeters[3] are of the lead/proportional tube type. The lead plates are interleaved with alternating layers of horizontal and vertical tubes of 10 mm transverse dimension and 5 mm depth, giving a fine transverse segmentation. Each calorimeter is divided longitudinally into three identical modules of $8X_0$ each.

Not shown in Fig. 1 is a set of solid-state counters[4], placed near 90° in the lab, within the jet target vacuum chamber, to monitor the number of recoil protons from elastic scattering. Using the optical theorem, this determines the luminosity to $\pm 4\%$.

3. Analysis

This section describes the analysis leading to the direct γ cross-section. The $\bar{p}p$ cross-section from an integrated luminosity of 0.458 pb^{-1} has already been published[2]. The pp cross-section is from an integrated luminosity of 1.175 pb^{-1}. The data were collected in 1985 and 1986.

A simulation is necessary to compute the acceptance for γ, π^0's and η's and to compute the background from π^0's and η's in the γ sample. The γ, π^0's and η's are simulated with a WA70 distribution[5]. For each γ a shower is taken from a bank of real showers obtained by placing our calorimeter in an electron test beam. The generated events are then passed through the same analysis chain as is used for the real data. Figure 2 shows how well the Monte Carlo reproduces the π^0 mass and asymmetry distributions obtained from data. Note that the Monte Carlo is only used to give the fractions of the π^0's and η's that result in background to the direct photon sample. The absolute rates for the production of these particles are measured in the experiment at the same time as the γ's.

An em shower was taken to be a direct photon candidate if it did not reconstruct as a π^0 or η with any other shower in the same calorimeter, no charged track pointed to it within a radius of 1 cm, and its r.m.s width in the first module was less that 1.35 cm. It is important to note the absence of isolation cuts for our direct photon candidates, making our photon sample totally inclusive and easy to compare with full next-to-leading-order (NLO) predictions.

The largest contribution to the background is from π^0 decays in which one of the photons is not reconstructed because its energy is too small. The other sources of background - π^0's in which the two photons coalesce and those in which one photon is outside the detector - contribute at a lower level. The contribution from η's (in which one γ is outside the detector) is only 10% of the total background.

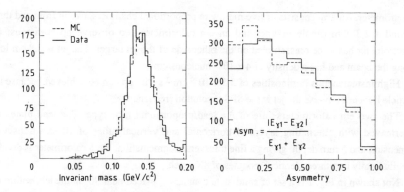

Fig. 2. Comparison between distributions of data and Monte Carlo: γγ invariant mass and energy asymmetry.

The estimation of systematic errors in the photon cross sections is still in progress but the most significant ones are (i) ± 7% coming from the number of reconstructed π^0's (ii) ± 6% due to a possible ± 0.6% non-linearity in the calorimeter response, (iii) ± 10% due to a ±.4% uncertainty in the p_T scale, (iv) ± 4% from the luminosity. When added in quadrature this adds to a total systematic uncertainty of ± 15%..

4. Results and discussions

The invariant cross-section for direct photons is shown in Fig. 3.. Also shown in Fig. 3 are predictions by Aurenche et al.[1] with Q^2 scales obtained via a principle of minimal sensitivity[6]. The solid lines are the predictions obtained by the Dukes - Owens structure functions[7] which are computed using expressions calculated at leading order whereas the dashed lines are the predictions obtained using the ABFOW structure functions[8] calculated beyond leading order.

Figure 4 shows the first preliminary measurement of the difference in cross-sections between the two reactions $\sigma(\bar{p}p \to \gamma X)$, $\sigma(pp \to \gamma X)$.

In principle, if the jet recoiling against the γ were also measured, the x's of the two partons involved in the interaction could be calculated for each event. This would allow a gluon x-distribution to be plotted and fitted. However, the acceptance of this experiment is too small to measure jets efficiently and also jets of 4-5 GeV/c are difficult to define and to measure. Without knowledge of the recoil jet, each γ p_T-bin is fed from a range of x-gluon of ~ 0.2.

Instead, a global approach, the ABFOW method, has been used. The method is: (i) use the muon Deep Inelastic Scattering data of BCDMS[9] on hydrogen and deuterium; (ii) assume the form for the gluon to be $xG(x, Q_0^2 = 2\,\text{GeV}^2.) = A_g(1-x)^\eta$; (iii) from BCDMS, for a range of η values extract, at each η, a set of valence and sea structure functions and the corresponding

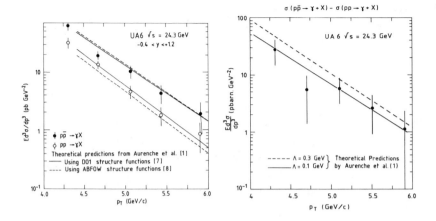

Fig. 3. Invariant cross-sections for $\bar{p}p \to \gamma X$ and for $pp \to \gamma X$

Fig. 4. First measurement of the cross sections $(\bar{p}p \to \gamma X - pp \to \gamma X)$.

value of Λ using NLO expressions; (iv) for the UA6 conditions compute the predictions of the Aurenche et al. NLO calculations with optimized scales for each of the sets of {η, valence, sea, Λ} obtained above, (v) find the χ^2 value between each set and the data; (vi) find which set gives the minimum χ^2.

Results are shown in Fig. 5. UA6 gives $\eta = 3.5 \pm 0.3 \pm 0.5$ (statistics and an evaluation of the systematics), giving a harder gluon than BCDMS ($\eta = 6.5 \pm 1.7$) but compatible with a similar direct γ experiment, WA70.

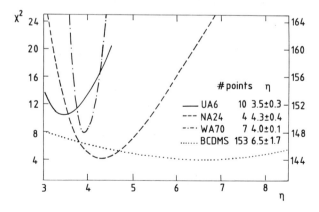

Fig. 5. The χ^2 of the fit to DIS data [9] and direct γ cross-sections.

5. Prospects

With the data collected since 1987 the total luminosity on tape amounts to $L(\bar{p}p) = 3.7$ pb^{-1}, $L(pp) = 4.0$ pb^{-1}. Furthermore, during 1990 we expect to collect $L(\bar{p}p) = 5.0$ pb^{-1}. The overall uncertainties in our direct photon cross-sections at high x_T are dominated by statistical uncertainties, as can be seen from the error bars in Fig. 4. For this reason, analysis of the data taken after 1986 will reduce the errors significantly. We hope to determine Λ accurately from the difference in cross-sections.

The systematics are also expected to decrease since (i) the addition of a new calorimeter module with 5 mm segmentation instead of 10 mm allows the meassurement of showers along two extra projections inclined at $\pm$ 60° to the original ones thus removing ambiguities and enhancing our ability to distinguish two close showers (ii) a new analysis has been developed which allows looser cuts and will decrease the background in the γ signal.

With the expected improvements we will be able to determine more precisely the value of Λ from the cross-section difference and then provide a good estimate of the gluon structure function in a range of x complementary to that of the colliders.

6. References

[1]   P. Aurenche, R. Baier, A. Douiri, M. Fontannaz and D. Schiff, Phys. Lett. **B140** (1984) 87; Nucl. Phys. **B297** (1988) 661; Nucl. Phys. **B286** (1987) 509.

[2]   A. Bernasconi et al., Phys. Lett. **B206** (1988) 163.

[3]   L. Camilleri et al., Nucl. Instr. and Meth. **A286** (1990) 49.

[4]   R. E. Breedon et al., Phys. Lett. **B216** (1989) 459.

[5]   WA70 collab., M. Bonesini et al., Z. Phys. **C37** (1988) 535; **C38** (1988) 371.

[6]   P. M. Stevenson, Phys. Rev. **D23** (1981) 2916; P. M. Stevenson and H. D. Politzer, Nucl. Phys. **B277** (1986) 758.

[7]   D. W. Duke and J. F. Owens, Phys. Rev. **D30** (1984) 49.

[8]   P. Aurenche, R. Baier, M. Fontannaz, J. F. Owens and M. Werlen, Phys. Rev. **D39** (1989) 3275

[9]   BCDMS Collab., A. C. Benvenuti et al. Phys. Lett. **B223** (1989) 485; Phys. Lett. **B237** (1990) 592.

DIRECT PHOTON CROSS SECTION MEASURED BY UA2

The UA2 Collaboration
Bern - Cambridge - CERN - Heidelberg - Milano - Orsay (LAL) -
Pavia - Perugia - Pisa - Saclay (CEN)

presented by
Gary F. Egan
University of Melbourne, Australia

ABSTRACT

Preliminay results of the direct photon cross-section measured with the UA2 detector are presented, corresponding to an integrated luminosity of 7.4 pb^{-1} at $\sqrt{s}$ = 630 GeV. Comparison to next to leading order QCD calculations show generally good agreement.

1. INTRODUCTION

Measurement of the direct production of high p_T photons in hadron-hadron collisions is a convenient way to study the hadronic constituents and their interactions. Direct photons are produced in $\bar{p}p$ collisions from leading order processes with the gluon compton effect being the dominant process at $\sqrt{s}$ = 630 GeV. One of the advantages of this study is that the measurement of the photon energy is unaffected by fragmentation effects which considerably reduces the systematic uncertainty on the cross section.

A measurement of the single photon inclusive cross section was already published by UA2[1] using a data sample corresponding to an integrated luminosity of 786 nb^{-1}. The present analysis refers to data collected during the 1988 and 1989 runs, corresponding to an integrated luminosity of 7.4 pb^{-1}.

A description of the UA2 detector is given in section 2. The data sample selection is then described followed by a determination of the background present in the sample, due principally to multi-photon events from isolated $\pi°$ and $\eta°$ decays. The preliminary direct photon cross-section is presented and compared to recent next to leading order (N.L.O.) calculations.

2. THE UA2 APPARATUS

The UA2 detector was substantially upgraded between 1985 and 1987. A quadrant of the detector is shown in Figure 1. The pseudorapidity coverage of the central calorimeter[2] ($|\eta| < 1$) has been extended to $|\eta| < 3$ with new end cap calorimeters[3]. The same construction technique of lead absorber plates with scintillator and wavelength shifter readout is used in the electromagnetic compartment of both the central and endcap calorimeters, with the lead absorber plate thickness varying from 17.0 to 24.4 radiation lengths depending on polar angle. This compartment is followed by the hadronic compartment constructed from iron absorber plates, scintillator and wavelength shifter readout.

Energy deposition in the calorimeter is defined using clusters constructed by joining all cells with an energy greater than 400 MeV which share a common edge. Those clusters with a small lateral size and a small energy leakage into the hadronic compartments are marked as electromagnetic clusters and are subsequently examined as potential electron candidates.

The layout of the central detector is shown in Figure 1. Around the beam pipe, at radii of 3.5 cm (inner) and 14.5 cm (outer), are two arrays of silicon counters used for tracking and ionization measurements[4]. Between the two is a cylindrical drift chamber with jet geometry (the Jet Vertex Detector or JVD)[5]. Outside of the inner tracking detectors is the Transition Radiation Detector (TRD)[6], consisting of two sets of radiators and proportional chambers. The outermost of the central detectors is the Scintillating Fibre Detectors (SFD)[7], which consists of approximately 60 000 fibres arranged on cylinders into 8 stereo triplets.

The last elements before the calorimeters are "preshower detectors", used to localize the early development of electromagnetic showers initiated in a lead converter. In front of the central calorimeter, this function is served by the SFD. For the end cap region, the preshower detection is accomplished by the End Cap Proportional Tubes (ECPT)[8]. For the direct photon analysis the preshower detector provides a means of statistically estimating the background in the final data sample and enabling a p_T dependent correction for the background to be made, as discussed below.

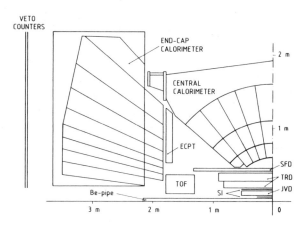

Figure 1. Schematic longitudinal view of the UA2 detector showing one quadrant.

3. DATA SAMPLE

The initial data sample is taken from the W trigger stream[9] and requires an electromagnetic calorimeter cluster in the pseudorapidity range $|\eta| < 0.76$. The cluster is required to have a longitudinal and lateral profile consistent with that expected for an electron or photon. In addition isolation cuts are imposed in order to remove background from QCD jets. The dominant background is due to hadronic jets since they often contain high p_T $\pi°$ and η mesons decaying to unresolved photon pairs. The ratio between the prompt photon and the jet cross-sections is approximately 10^{-4}, but direct photons are expected to be more isolated from other particles than $\pi°$ and η.

The isolation cuts are applied in a cone of opening angle $\sqrt{\Delta\eta^2 + \Delta\phi^2} < 0.265$, centred on the cluster centroid. Photon candidates were accepted if there were no charged tracks in the cone and at most one SFD preshower signal was observed in the cone. The final data sample was obtained after applying a transverse momentum cut on each photon candidate of $p_T > 15$ GeV/c, resulting in 25226 events and an estimated efficiency for photons of 44%. As a consequence of the isolation cuts, the contribution to photon production from quark bremsstrahlung, which becomes less important at high p_T [10], was suppressed.

4. DETERMINATION OF $\pi°$, $\eta°$ BACKGROUND

The background due to multi-photons from $\pi°$ and η was measured by considering the fractions of events in which the photon has begun showering in the converter. A photon candidate is identified as "converted" if there is one preshower signal and as "unconverted" for no signal. The measured-fraction (α) of converted photons is given by

$$\alpha = N_{\gamma 1} / (N_{\gamma 1} + N_{\gamma 0}) \tag{1}$$

where $N_{\gamma 0}$ and $N_{\gamma 1}$ are the number of unconverted and converted photon candidates respectively. The conversion fraction is shown as a function of the photon energy in Figure 2.

The conversion probability (ε_γ) for a single photon was then computed using the EGS shower simulation program [11] taking into account all the material in front of the SFD preshower detector. The simulation was tuned to reproduce UA2 results observed in testbeam studies and from $W \rightarrow e\nu$ decays. The multiphoton conversion probability (ε_π) was than determined by using ε_γ for each photon and assuming the ratio of $\eta°/\pi°$ production to be 0.6 [12]. The fraction of multi-photon contamination in the sample ($b(p_T)$) was then computed from the values of α, ε_γ and ε_π

$$b(p_T) = (\alpha - \varepsilon_\gamma) / (\varepsilon_\pi - \varepsilon_\gamma) \tag{2}$$

The average contamination for $p_T > 15$ GeV/c was found to be $<b> = 0.19 \pm 0.01$ where the error is statistical only.

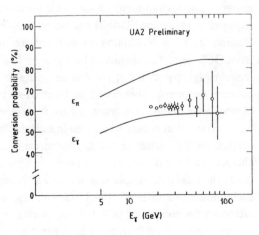

Figure 2. The photon conversion probability in the preshower. For the data (points), simulated single photons (lower curve) and simulated double photon events (upper curve).

5. PRELIMINARY DIRECT PHOTON CROSS SECTION

The inclusive cross section for direct photon production is evaluated from

$$E d\sigma / dp^3 = [N_\gamma(p_T) \cdot (1 - b(p_T))] / [2\pi \, p_T \, \Delta p_T \, \mathscr{L} \, \varepsilon_c \, A(p_T)]$$

where $N_\gamma(pt)$ is the number of photon candidates in a p_T bin of width Δp_T, $\mathscr{L}$ is the integrated luminosity corresponding to the data sample, ε_c is the efficiency of the selection criteria and $A(p_T)$ is the geometrical acceptance. Figure 3 shows the preliminary UA2 results compared to a N.L.O. calculation [13] performed using different sets of structure functions [14] and different Q^2 scales. The upper two curves use the structure functions from Ref. 14 with the optimised Q^2 scale while the lower two curves use the structure functions from Ref. 15 with the optimised Q^2 scale and $Q^2 = p_T^2$ as indicated. Only the statistical errors are shown. Experimental data agree well with the QCD calculation but do not distinguish between the different structure functions

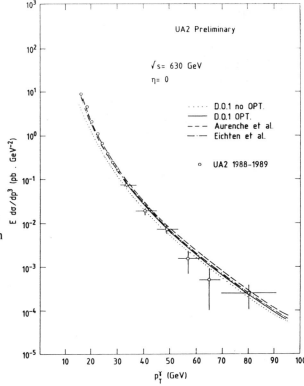

Figure 4. The preliminary direct photon cross section compared to N.L.O. calculations.

REFERENCES

1. UA2 Collaboration, J.A. Appel et al., *Phys. Lett.* **176B** (1986) 239; UA2 Collaboration, R. Ansari et al., *Z. Phys.* **C41** (1988) 395.
2. A. Beer, et al., *Nucl. Inst. and Meth.* **A224** (1984) 360.
3. F. Alberio et al., The Electron, Jet and Missing Transverse Energy Calorimetry of the Upgraded UA2 Experiment at the CERN $\bar{p}p$ Collider, in preparation for Nucl. Instr. Meth..
4. R. Ansari et al., *Nucl. Instr. Meth.* **A263** (1989) 388.
5. F. Bosi et al., CERN-EP/89-82 (1989).
6. R. Ansari et al., *Nucl. Instr. Meth.* **A265** (1988) 51.
7. R. E. Ansorge et al., *Nucl. Instr. Meth.* **A265** (1988) 33; J. Alitti et al., *Nucl. Instr. Meth.* **A279** (1989) 364.
8. K. Borer et al., *Nucl. Instr. Meth.* **A 286** (1990) 128.
9. G. Blaylock et al., The Multi-level Trigger and Data Acquisition System of the Upgraded UA2 Experiment at the CERN $\bar{p}p$ Collider, in preparation for Nucl. Instr. Meth..
10. E. L. Berger et al., *Nucl. Phys.* **B239** (1984) 52; J. F. Owens, *Rev. Mod. Phys.* **59** (1987) 485.
11. R. Ford and W. Nelson, SLAC-210 (1978).
12. UA2 Collaboration, M. Banner et al., *Z. Phys.* **C27** (1985) 329.
13. P. Aurenche et al., *Phys. Lett.* **140B** (1984) 87; P. Aurenche et al., *Nucl. Phys.* **B297** (1988) 661.
14. P. Aurenche et al., *Phys. Rev.* **D39** (1989) 3275; E. Eichten et al., *Rev. Mod. Phys.* **56** (1984) 579, **58** (1986) 1065.
15. D. W. Duke and J. F. Owens, *Phys. Rev.* **D30** (1984) 49.

GLOBAL FITS

A GLOBAL ANALYSIS OF RECENT EXPERIMENTAL RESULTS: HOW WELL DETERMINED ARE THE PARTON DISTRIBUTION FUNCTIONS?

Jorge G. Morfín

Fermi National Accelerator Laboratory, Batavia, Illinois[*]
and
Lab. de Física d'Altes Energies, Uni. Autònoma de Barcelona,
Barcelona, Spain

ABSTRACT

Following is a brief summary of the results of an analysis of experimental data performed to extract the parton distribution functions. In contrast to other global analyses, this study investigated how the fit results depend on:

1. Experimental Systematic Errors
2. Kinematic Cuts on the Analyzed Data
3. Choice of Initial Functional Forms,

with a prime goal being a close look at the range of low-x behavior allowed by data. This is crucial for predictions for the SSC/LHC, HERA, and even at Tevatron Collider energies. Since all details can be found in the just released Fermilab preprint Parton Distributions from a Global QCD Analysis of Deep Inelastic Scattering and Lepton-Pair Production by Wu-Ki Tung and J.G.M. (Fermilab-Pub-90/74 and IIT-Pub-90/11), this summary will be only a brief outline of major results.

NOTE: Due to a formatting problem in the preparation of the tables of the Fermilab/IIT Preprint, a negative sign was inadvertently dropped in some of the coefficients of the t-quark distribution in several tables. All other quark parametrizations are correct. The corrected parametrization of the t-quark distributions can be found in the tables of this summary.

---

[*] Permanent address

There are several widely used sets of parton distribution functions currently available. The following table briefly summarizes the characteristics of these other global analyses:

COMPARISON OF EXISTING PDF'S: METHODS USED

	Dk-Ow[1]	EHLQ[2]	MRS[3]	DFLM[4]
Experimental Factors				
Systematic Errors	NO	NO	NO/(YES HMRS)	YES
Correlated Errors	NO	NO	NO	YES
EMC Effect	NO	NO	YES	NO
Vary Kinematic Cuts	NO	NO	NO	NO
Theoretical Factors				
Study of small-x behavior	NO	NO	YES	NO
N-L-O QCD	NO	NO	MS-bar	DIS
Study of Functional Form	NO	NO	NO	NO
Experimental Data Used				
Latest ν Data	NO	NO	YES	YES
Latest μ Data	NO	NO	YES	NO

The objective of the present analysis has been to contain or cover ALL points listed in the above table.

I) INPUT:

The only currently available neutrino scattering results with point-to-point systematic errors is;

      CDHSW[6] F_2 and xF_3

The CCFRR collaboration will soon release their values of F_2 and xF_3 based on the largest sample of neutrino events to date. We will refit the parton distributions when this data becomes available.

From muon scattering;

      EMC[6] F_2 from H_2 and D_2
      BCDMS[7] F_2 from H_2 and D_2

Drell-Yan scattering is represented by the two FNAL experiments;
      E288[8]   Drell - Yan
      E605[9]   Drell - Yan

II) Kinematic Cuts:

  Until we have a better understanding of how non-perturbative (**higher twist**) effects contribute to measured values, we try to avoid the kinematic region which they could dominate. It is, however, still not entirely clear exactly what this region is. Our procedure to try to fix this region involves performing the analysis with Q^2 cuts of 20, 15, 10, 5, and 2 GeV^2 and noting the stability of the fit. We determined the safest cut to be;

$$Q^2 > 10 \text{ GeV}^2 \text{ and } W > 4 \text{ GeV.}$$

Within the data sets mentioned as input above, there are 776 measured points in this kinematic range.

  Results of a combined SLAC-BCDMS analysis presented[10] at this workshop indicate that as long as $x < 0.4$, one can go down to very low values of Q^2 without having to account for higher twist in the fit. We will include these new twist-4 results in our subsequent analyses.

III) DATA POINT MEASUREMENT ERRORS:

It is clearly **INCORRECT** to neglect systematic errors in a global analysis of recent high statistics experiments! As the following two figures show, the systematic errors are at least as large as, if not larger than, the statistical errors.

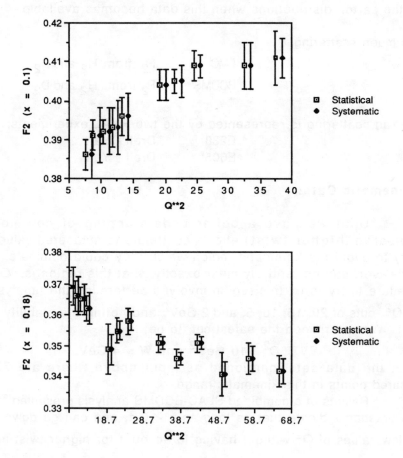

There are several ways of incorporating both systematic and statistical errors in the overall fit which can be summarized as follows;

1. <u>The statistically rigorous method</u> would be to define an overall extra "loop" in the fit and perturb the central measured value by each of the quoted systematic errors of a given measured point (eg. 4 in the case of BCDMS data) in turn. This perturbed value is then used in the statistical fit.

2. <u>Redefine the "statistical" error</u> as:

$$D^2_{p\text{-}st} = d^2_{st} + d^2_{sy_{uncor}}$$

and define

$$D^2_{p\text{-}sy} = \sum_i^{algebraic} d^2_{sy_{cor}}$$

Then, in this representation:

$$\chi^2 = c^2 + \sum_i^{npts} (T_i - M_i - c\, D_{p\text{-}sy})^2 / D^2_{p\text{-}st}$$

3. <u>Combine statistical and point-to-point systematic errors in quadrature</u> (much less CPU-thirsty than method 1 and 8 fewer parameters than method 2)

IV) FUNCTIONAL FORM:

We fit to:

$$F^a(x,Q) = e^{A_0^a} x^{A_1^a} (1-x)^{A_2^a} \log^{A_3^a}(1 + \frac{1}{x})$$

which, with the ln term, is particularly well suited to investigate small-x extrapolations.

V) PROCEDURE:

We fit only those variables which are **sensitive to the data being examined** eg. only flavor-blind valence quark distributions fit to xF_3. In addition to the shape parameters we have floating relative normalizations between data sets. With the use of sum rules, the maximum number of parameters in the fit is 11 shape parameters, 4 relative normalizations and λ.

Various combinations of data sets were used to investigate the consistency of fits and data. The various groupings are referred to as;

E-fit -- CDHSW + D-Y + EMC (472 points)
B-fit -- CDHSW + D-Y + BCDMS (647 points)
S-fit --CDHSW + D-Y + EMC + BCDMS (776 points).

VI) RESULTS:

χ^2

The B-fit is the best with χ^2/d.o.f. = 0.8 - (remember errors are the quadratic sum of systematic + statistical!), see Figs. 1-3.
The results of the E-fit are about 10% worse while the χ^2/d.o.f. for the S-fit is about 20% higher than the B-fit, see Figs. 4-7.

If, when combining BCDMS and EMC results, one takes the following points into consideration;
> 1. Include systematic errors.
> 2. An overall relative normalization brings the data sets in-line at small x where errors are small.
> 3. Reasonable Q^2-cut.

The results for the combined fit are not unacceptable, however the χ^2 for the EMC data set is still high compared to all others in the combined fit.

λ (2, 4)
>E-fit, 130-150 MeV
>B-fit, 190-225 MeV
>S-fit, 175-225 MeV

Valence Quarks
> NO surprises.

Sea Quarks
The fits are NOT sensitive to composition of the sea.

Gluon
D.I.S. data alone tends to favor a relatively hard gluon (A_2 = 3.5 - 4.5). However, D.I.S. is not particularly sensitive to the gluon.

Upon adding D-Y results, the fit favors a softer gluon distribution (A_2 = 6.5 - 7.5).

We do NOT determine the gluon distribution from direct-photon production since the current results are limited both in accuracy and x range, and there are still open questions on the correct interpretation in N-L-O QCD.

If we fix the value of A_2 for the gluon to be 4.0, this causes an increase in χ^2/d.o.f. for both the xF3 and D-Y data sets of $\approx$ 40%!

Systematic Errors

If we do NOT include systematic errors in the fit, the χ^2/d.o.f. increases to 2.5(!), λ drops by 20 %, and the valence quarks at x=0.5, 0.3 are reduced by 40%, and 20% respectively. The sea and gluon also become slightly softer.

Q^2 cut

If we vary the Q^2 cut from 20 to 5 GeV^2, we find that the value of λ (20 : 10 : 5 GeV^2) = 240 : 210 : 195 MeV. We find no other significant differences.

VII) PARAMETRIZATION OF THE PARTON DISTRIBUTION FUNCTIONS:

The A_n coefficients exhibit well-behaved Q-dependence (see Fig. 9) so that we can parametrize them as:

$$A_i = C_0^i + C_1^i\, T(Q) + C_2^i\, T(Q)^2, \quad i = 0,3$$

$$T(Q) = \ln \frac{\ln(Q^2/\Lambda^2)}{\ln(Q_0^2/\Lambda^2)}$$

This parametrization accurately describes the fit results for the kinematic range:

$$10^{-5} < x < 1.0$$
$$4.0 < Q < 10^4$$

Furthermore, the PDF's are always positive definite and are smoothly varying in both x and Q even outside of the original range. The C_i needed to determine the shape parameters (A_i) for ALL of the parton distribution functions are given in Table 1 and the Appendices. **NOTE that the t-quark parametrization in these tables has been corrected for the missing minus sign.**

VIII) IMPORTANCE OF RE-NORMALIZATION SCHEME (MS-BAR VS DIS)

There are several renormalization schemes used in next-to-leading order QCD formulations, the most common being the MS-bar and DIS schemes. Depending on which scheme has been used to calculate the hard scattering cross section, the corresponding parton distribution functions must be employed. In general, the gluon distribution will be much softer when expressed in the DIS scheme as compared to the MS-bar scheme. We provide the parton distribution functions in BOTH schemes in the tables. To help visualize the scheme dependent difference in the parton distribution functions, Fig 8. displays xg(x) and xu(x) at Q^2 = 10 GeV^2.

IX) RANGE OF ALLOWED LOW-X BEHAVIOR

Our chosen form is ideal for exploring the range of small-x behavior allowed by present data and, thus, emphasize our severe **lack** of knowledge. In Fig. 10 we plot the structure function F_2 and the gluon distribution at Q^2 = 10 GeV^2 and 10000 GeV^2 while Fig. 11 shows the ratio of predictions for three typical values of Q^2. It is obvious that the parton densities and, therefore, the physical cross sections derived from them, can differ by factors of 3 at the highest Q^2 and by an order of magnitude at more moderate Q^2.

X) WHAT DO WE NEED TO FURTHER RESTRICT THE PDF'S?

1. Statistically and systematically accurate measurements of F_2 with $x < 10^{-3}$ at reasonable Q.

2. Measurements of **Drell-Yan** cross section with small mass lepton pairs at high s (i.e. Q = 20 GeV at Tevatron) see Fig. 13.

3. Measurement of **direct photon** production over an **extended x range** and resolution of questions regarding the theoretical interpretation of the results. The direst photon results should then be part of the fit, NOT an artificially imposed constraint on the gluon distribution.

4. Measurement of **W/Z ratio at high y**, see Fig. 12.

5. Measurement of the integrated and differential production **cross sections of B-pairs** over a wide kinematic range (as proposed by, for example, BCD).

REFERENCES

[1] D. Duke and J. Owens, Phys. Rev. **D30** 49 (1984)

[2] E. Eichten et al., Rev. Mod. Phys. **56** 579 (1984) and Erratum **58** 1065 (1986).

[3] A.D. Martin, R.G. Roberts and W.J. Stirling, Phys. Rev. **D37** 1161 (1988), Mod. Phys. Lett. **A4** 1135 (1989).

[4] M. Diemoz et al., Z. Phys. **C39** 21 (1988);

[5] J.P.Berge et al., Preprint CERN-EP/89-103 (1989). Abramowicz et al, Z. Phys. **C17**, 283 (1984); Z. Phys. **C25**, 29 (1984); Z. Phys. **C35**, 443 (1984).

[6] J.J.Aubert et al., Nucl. Phys. **B293** 740 (1987).

[7] A.C. Benvenuti et al., Phys. Lett. **B223** 485 (1989) and CERN-EP/89-170,171, December, 1989.

[8] A.S.Ito et al., Phys. Rev. **D23**, 604 (1981).

[9] C.N.Brown et al., Phys. ReV. Lett. **63** 371 (1988).

[10] A. Milsztajn, *A QCD Analysis of High Statistics F_2 Data on Deuterium and Hydrogen with Determination of "Higher Twists"*. Presented at the Workshop on Hadron Structure Functions and Parton Distributions, April, 1990.

Tables

Table Ia - Fit S - DIS scheme $\Lambda(2,4) = 0.212 GeV$								$Q_0^2 = 4 GeV^2$	
	d(val)	u(val)	gluon	u(sea)	d(sea)	s	c	b	t
				A_0					
C_0	1.34	1.62	1.88	-0.99	-0.99	-0.99	-3.98	-6.28	-13.08
C_1	-0.57	-0.33	-2.78	-1.54	-1.54	-1.54	0.72	2.62	8.54
C_2	-0.08	-0.10	0.13	0.10	0.10	0.10	-0.63	-1.18	-2.70
				A_1					
C_0	0.15	0.11	-0.33	-0.33	-0.33	-0.33	-0.15	-0.18	-0.40
C_1	0.16	0.14	0.10	0.03	0.03	0.03	-0.06	0.02	0.31
C_2	-0.02	-0.01	-0.04	-0.03	-0.03	-0.03	0.00	-0.03	-0.12
				A_2					
C_0	5.30	3.68	7.52	8.53	8.53	8.53	7.46	6.56	15.35
C_1	0.43	0.53	-1.13	-1.08	-1.08	-1.08	0.96	1.40	-11.83
C_2	0.06	0.03	0.04	0.39	0.39	0.39	-0.30	-0.38	4.16
				A_3					
C_0	-1.96	-1.94	-1.34	-1.55	-1.55	-1.55	0.35	0.65	-0.43
C_1	1.08	0.87	2.92	2.02	2.02	2.02	0.89	1.13	3.18
C_2	-0.03	0.02	-0.49	-0.39	-0.39	-0.39	-0.04	-0.16	-0.82
Equivalent "Conventional Parametrization" Coefficients at $Q_c^2 = 5.0 GeV^2$ $f(x, Q_c^2) = e^{B_0} x^{B_1} (1-x)^{B_2} (1 + B_3 x)$									
B_0	-0.49	-0.31	0.48	-2.65	-2.65	-2.65	0.00	0.00	0.00
B_1	0.43	0.36	-0.15	-0.14	-0.14	-0.14	0.00	0.00	0.00
B_2	5.36	3.70	8.02	9.58	9.58	9.58	0.00	0.00	0.00
B_3	10.68	11.82	8.20	13.60	13.59	13.59	0.00	0.00	0.00

Table Ib - Fit S - MS-bar scheme									
$\Lambda(2,4) = 0.212 GeV$						$Q_0^2 = 4 GeV^2$			
	d(val)	u(val)	gluon	u(sea)	d(sea)	s	c	b	t
A_0									
C_0	1.75	2.03	1.09	-0.14	-0.14	-0.15	-2.36	-2.19	-24.77
C_1	-1.02	-0.78	-2.41	-1.98	-1.98	-1.98	-1.42	-3.86	-23.00
C_2	0.05	0.03	-0.12	0.23	0.23	0.23	0.21	1.57	34.44
A_1									
C_0	0.11	0.06	-0.24	-0.49	-0.49	-0.49	-0.49	-1.07	7.52
C_1	0.26	0.24	0.08	0.02	0.02	0.02	0.44	1.56	0.48
C_2	-0.06	-0.04	0.02	-0.02	-0.02	-0.02	-0.22	-0.73	-6.26
A_2									
C_0	6.20	4.43	5.97	10.24	10.24	10.23	9.00	11.30	-99.51
C_1	-0.41	-0.18	-0.90	-1.43	-1.44	-1.44	-0.46	-7.20	-16.45
C_2	0.29	0.22	-0.35	0.44	0.45	0.45	0.29	3.85	97.19
A_3									
C_0	-2.35	-2.35	-0.64	-2.57	-2.57	-2.57	-1.74	-4.85	36.02
C_1	1.68	1.52	2.71	2.32	2.32	2.32	3.93	10.51	16.51
C_2	-0.24	-0.19	-0.20	-0.47	-0.47	-0.47	-1.34	-4.36	-40.40
Equivalent "Conventional Parametrization" Coefficients at $Q_c^2 = 5.0 GeV^2$									
$f(x, Q_c^2) = e^{B_0} x^{B_1} (1-x)^{B_2} (1 + B_3 x)$									
B_0	0.03	0.23	0.68	-2.26	-2.26	-2.26	0.00	0.00	0.00
B_1	0.53	0.46	-0.14	-0.10	-0.10	-0.10	0.00	0.00	0.00
B_2	6.08	4.33	5.88	10.12	10.11	10.11	0.00	0.00	0.00
B_3	7.96	8.93	0.43	11.40	11.39	11.39	0.00	0.00	0.00

Figure Captions

Fig. 1 Results of B1-fit (solid) and E-fit (dashed) compared to BCDMS H2 measurement of $F_2(x,Q)$.

Fig. 2 Results of B1-fit (solid) and E-fit (dashed) compared to EMC H2 measurement of $F_2(x,Q)$.

Fig. 3 Results of B1-fit (solid) and E-fit (dashed) compared to CDHSW Iron measurement of $F_2(x,Q)$.

Fig. 4 Results of the S-fit compared to BCDMS D2 measurement of $F_2(x,Q)$.

Fig. 5 Results of the S-fit compared to EMC D2 measurement of $F_2(x,Q)$.

Fig. 6 Results of the S-fit compared to the E605 Drell-Yan cross section measurement.

Fig. 7 Comparison of the EHLQ, Duke-Owens-2, MRS-B, DFLM, and present parametrizations of the parton distribution functions to BCDMS H2 measurements of F_2 at four representative x values.

Fig. 8 Comparison of $xG(x)$ and $xu(x)$ as fit in the DIS scheme and as converted to the MSbar scheme.

Fig 9. The shape parameters A_i for the different partons as a function of $T(Q)$.

Fig 10. Predicted values of $F_2(x)$ and $xG(x)$ from fits B1 and B2 at ultra low x for $Q^2 = 10 GeV^2$ and $Q^2 = 10^4 GeV^2$.

Fig 11. The ratio of predictions (the uncertainty) for $F_2(x)$ and $xG(x)$ as a function of x for three typical values of Q^2.

Fig 12. The predictions for W-production, Z production, the W-production asymmetry, and the W/Z production ratio at the Tevatron Collider using the parton distribution function of MRS-B, DFLM, and the B1- and B2-fits from the present analysis.

Fig. 13. Prediction for low mass ($Q = 20 GeV$) Drell-Yan pair production at the Tevatron Collider for parton distribution functions as in Fig. 12.

Figure 1

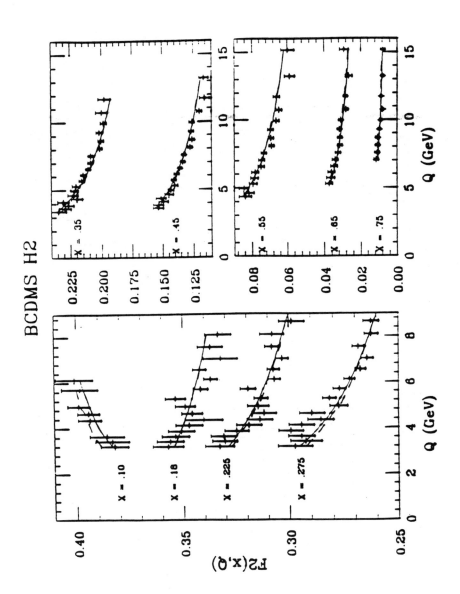

Figure 2

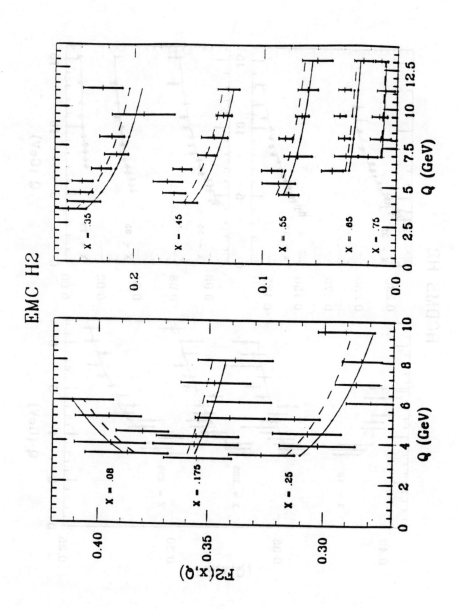

Figure 3

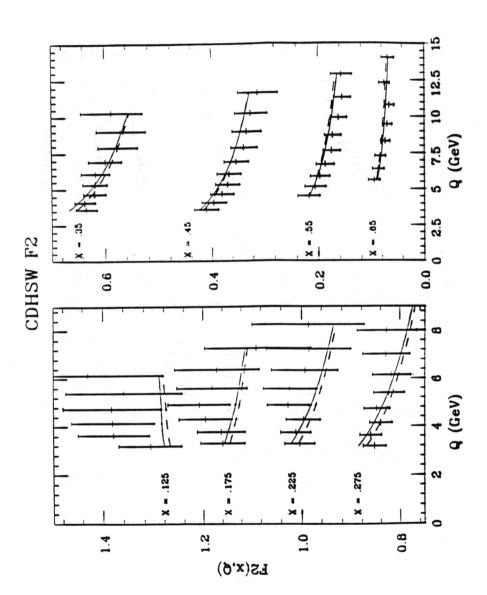

Figure 4

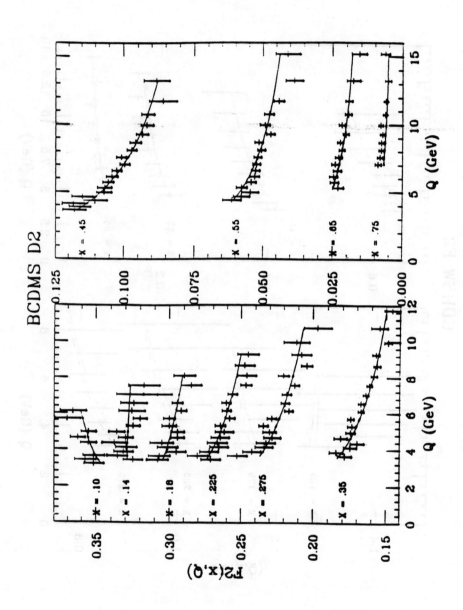

Figure 5

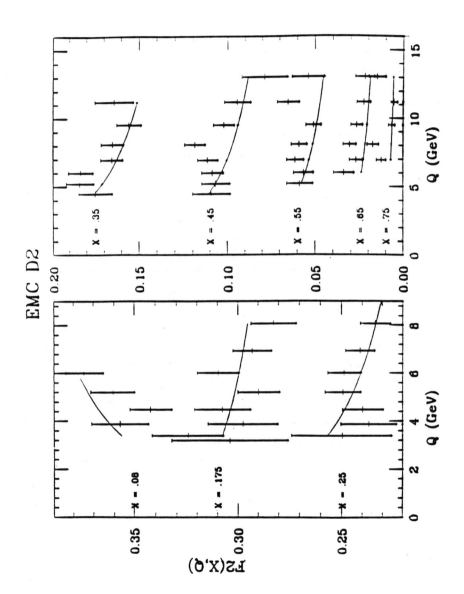

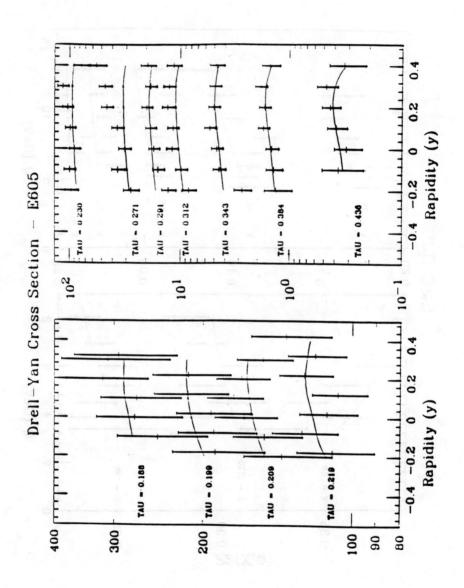

Figure 6

Figure 7

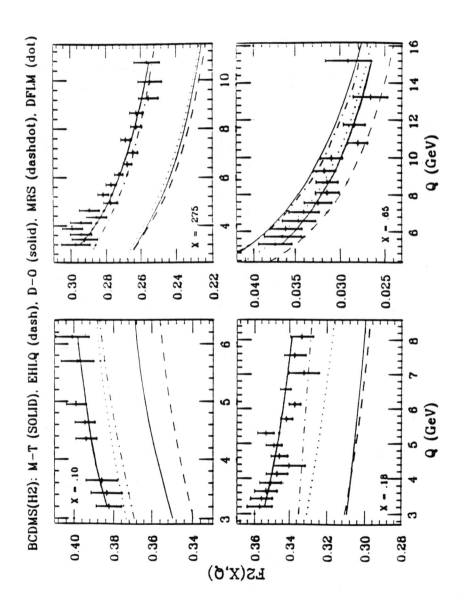

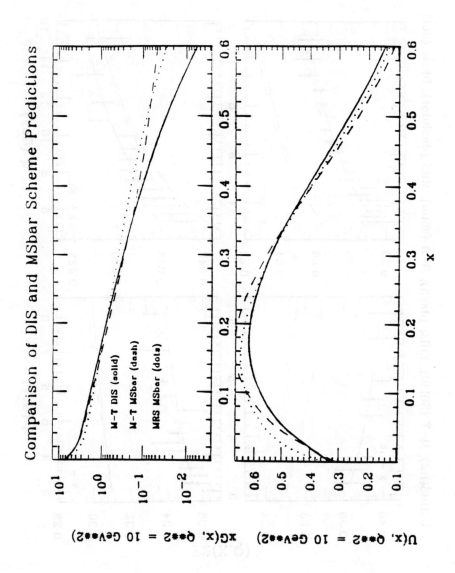

Figure 8

Figure 9

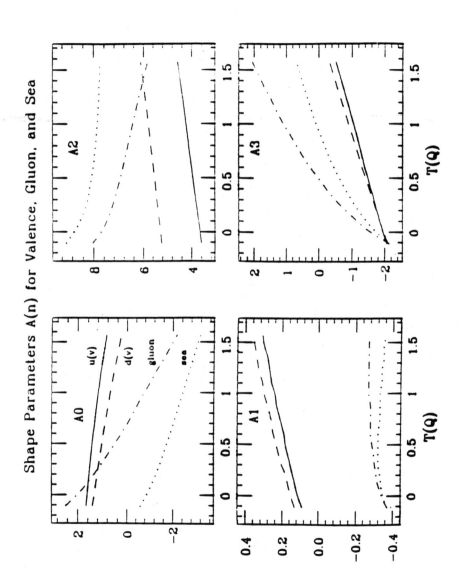

Figure 10

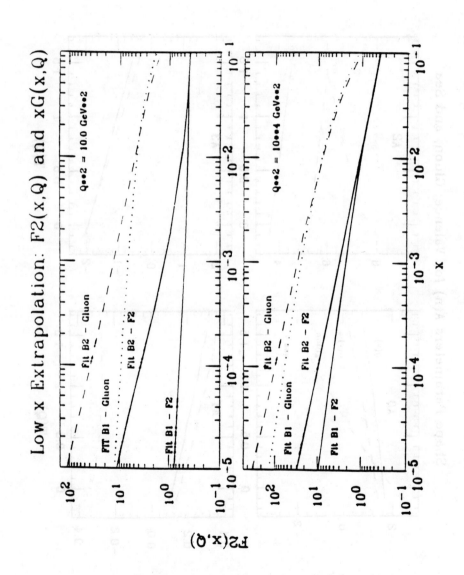

329

Figure 11

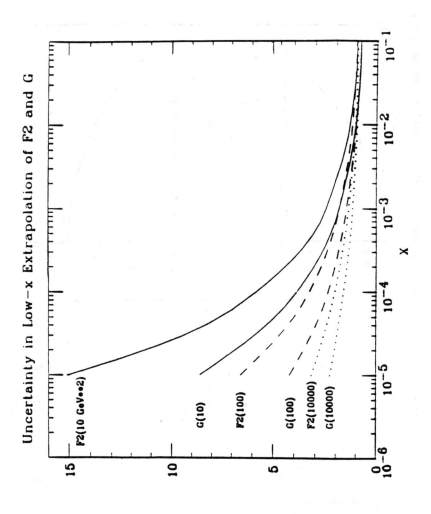

Figure 12

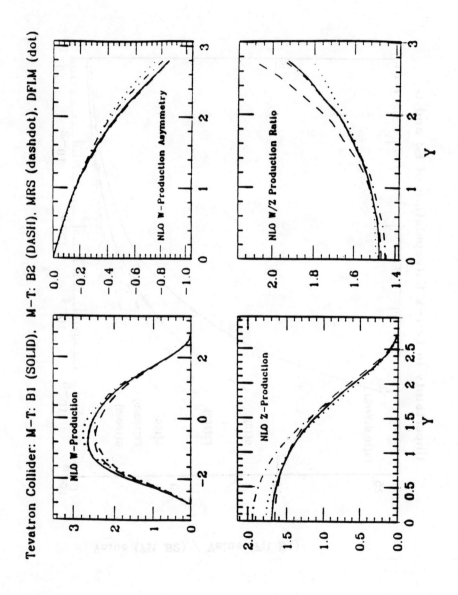

Figure 13

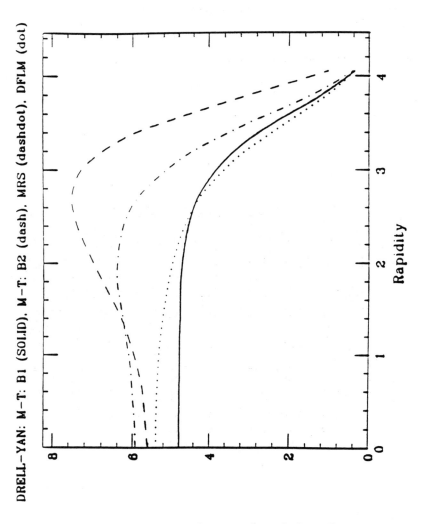

Appendix I

Following are the parton distribution function sets E, B1, and B2 discussed in the main body of this report. The parton distributions are given both in the DIS and MS-bar renormalization schemes. Note that Fit B1 and B2 are representative of the variation in low-x extrapolation allowed by the currently available data.

As a reminder, the general expression for each parton flavor is:

$$f^a(x,Q) = e^{A_0^a} x^{A_1^a} (1-x)^{A_2^a} \ln^{A_3^a}(1+\frac{1}{x}) \qquad (A1)$$

where the shape parameters are defined as:

$$A^i(Q) = C_0^i + C_1^i T(Q) + C_2^i T(Q)^2 \qquad (A2)$$

with $i = 0 - 3$, and

$$T(Q) = \ln \frac{\ln \frac{Q}{\Lambda}}{\ln \frac{Q_0}{\Lambda}} \qquad (A3)$$

Table I1 - Fit E - DIS Scheme								
$\Lambda(2,4) = 0.155 GeV$					$Q_0^2 = 4 GeV^2$			
	d(val)	u(val)	gluon	u(sea)	d(sea)	s	c	b	t
				A_0					
C_0	1.43	1.69	2.11	-0.84	-0.84	-0.84	-3.87	-6.09	-12.56
C_1	-0.65	-0.33	-3.01	-1.65	-1.65	-1.65	0.85	2.81	8.69
C_2	-0.08	-0.11	0.18	0.12	0.12	0.12	-0.73	-1.34	-2.93
				A_1					
C_0	0.16	0.11	-0.33	-0.32	-0.32	-0.32	-0.15	-0.17	-0.38
C_1	0.16	0.14	0.10	0.02	0.02	0.02	-0.07	0.01	0.30
C_2	-0.02	-0.01	-0.04	-0.03	-0.03	-0.03	0.00	-0.03	-0.12
				A_2					
C_0	6.17	3.69	7.93	8.96	8.96	8.96	7.83	6.75	14.62
C_1	0.43	0.54	-1.40	-1.24	-1.24	-1.24	1.00	1.74	-11.27
C_2	0.06	0.03	0.09	0.45	0.45	0.45	-0.36	-0.56	4.29
				A_3					
C_0	-1.94	-1.99	-1.51	-1.70	-1.70	-1.70	0.21	0.54	-0.41
C_1	1.12	0.90	3.14	2.15	2.15	2.15	0.93	1.15	3.19
C_2	-0.02	0.02	-0.55	-0.43	-0.43	-0.43	-0.03	-0.16	-0.87

Table I2 - Fit E - MS-bar Scheme
$\Lambda(2,4) = 0.155 GeV$ $Q_0^2 = 4 GeV^2$

	d(val)	u(val)	gluon	u(sea)	d(sea)	s	c	b	t
A_0									
C_0	1.79	2.12	1.58	-0.10	-0.10	-0.11	-2.53	-3.91	-6.57
C_1	-1.05	-0.85	-2.68	-2.29	-2.29	-2.29	-1.16	-0.19	1.15
C_2	0.03	0.07	0.01	0.35	0.35	0.35	0.12	-0.24	-0.48
A_1									
C_0	0.12	0.02	-0.28	-0.43	-0.43	-0.43	-0.35	-0.44	-0.90
C_1	0.24	0.32	0.05	0.09	0.09	0.09	0.26	0.38	0.95
C_2	-0.04	-0.08	0.00	-0.06	-0.06	-0.06	-0.15	-0.17	-0.33
A_2									
C_0	7.03	4.46	6.84	10.43	10.43	10.43	8.67	6.85	7.27
C_1	-0.38	-0.28	-0.93	-2.14	-2.14	-2.14	-0.10	2.15	-0.28
C_2	0.27	0.29	-0.26	0.73	0.73	0.73	0.27	-0.74	0.28
A_3									
C_0	-2.29	-2.57	-1.08	-2.49	-2.49	-2.48	-1.24	-1.56	-5.07
C_1	1.63	1.82	2.76	2.80	2.80	2.80	3.26	4.07	9.02
C_2	-0.18	-0.33	-0.32	-0.67	-0.67	-0.67	-1.06	-1.24	-2.75

Table I3 - Fit B1 - DIS Scheme
$\Lambda(2,4) = 0.194 GeV$ $Q_0^2 = 4 GeV^2$

	d(val)	u(val)	gluon	u(sea)	d(sea)	s	c	b	t
A_0									
C_0	1.30	1.59	1.48	-1.08	-1.08	-1.08	-4.22	-6.42	-12.92
C_1	-0.57	-0.34	-2.49	-1.33	-1.33	-1.33	0.88	2.67	8.33
C_2	-0.09	-0.10	0.04	-0.03	-0.03	-0.03	-0.69	-1.21	-2.68
A_1									
C_0	0.19	0.14	-0.14	-0.13	-0.13	-0.13	-0.02	-0.09	-0.36
C_1	0.15	0.13	-0.11	-0.21	-0.21	-0.21	-0.17	-0.03	0.32
C_2	-0.02	-0.01	0.03	0.06	0.06	0.06	0.03	-0.02	-0.13
A_2									
C_0	5.24	3.65	6.75	8.40	8.39	8.39	7.29	6.47	15.74
C_1	0.44	0.53	-0.54	-0.51	-0.50	-0.50	1.08	1.39	-12.73
C_2	0.05	0.03	-0.15	0.07	0.07	0.07	-0.39	-0.42	4.51
A_3									
C_0	-1.81	-1.81	-0.50	-0.88	-0.88	-0.88	0.90	1.03	-0.30
C_1	1.06	0.86	2.13	1.18	1.18	1.18	0.50	1.00	3.35
C_2	-0.02	0.02	-0.24	-0.05	-0.05	-0.05	0.08	-0.14	-0.91

Table I4 - Fit B1 - MS-bar Scheme
$\Lambda(2,4) = 0.194 GeV$ $Q_0^2 = 4 GeV^2$

	d(val)	u(val)	gluon	u(sea)	d(sea)	s	c	b	t
A_0									
C_0	1.66	2.00	0.92	-0.60	-0.60	-0.60	-2.94	-2.95	-3.88
C_1	-0.94	-0.81	-2.28	-1.76	-1.76	-1.76	-1.12	-3.21	-1.59
C_2	0.03	0.05	-0.07	0.13	0.13	0.14	0.15	1.38	-0.05
A_1									
C_0	0.18	0.09	-0.07	-0.13	-0.13	-0.13	-0.19	-0.62	-0.78
C_1	0.18	0.24	-0.16	-0.27	-0.27	-0.27	0.16	0.99	-0.07
C_2	-0.03	-0.05	0.06	0.09	0.09	0.09	-0.13	-0.51	0.40
A_2									
C_0	6.04	4.40	5.79	9.31	9.31	9.31	7.94	9.97	3.80
C_1	-0.25	-0.20	-0.68	-0.94	-0.94	-0.94	-0.05	-6.33	2.13
C_2	0.23	0.25	-0.23	0.21	0.21	0.21	0.27	3.71	0.96
A_3									
C_0	-2.09	-2.24	-0.01	-1.18	-1.18	-1.18	-0.46	-3.00	-2.37
C_1	1.42	1.53	1.93	1.31	1.31	1.31	2.93	8.42	0.48
C_2	-0.14	-0.23	-0.11	-0.10	-0.10	-0.10	-1.05	-3.61	2.30

Table I5 - Fit B2 - DIS Scheme
$\Lambda(2,4) = 0.191 GeV$ $Q_0^2 = 4 GeV^2$

	d(val)	u(val)	gluon	u(sea)	d(sea)	s	c	b	t
A_0									
C_0	1.38	1.64	1.52	-0.85	-0.85	-0.85	-3.74	-6.07	-12.08
C_1	-0.59	-0.33	-2.71	-1.43	-1.43	-1.43	0.21	2.33	7.31
C_2	-0.08	-0.10	0.15	-0.03	-0.03	-0.03	-0.50	-1.15	-2.35
A_1									
C_0	0.18	0.09	-0.72	-0.82	-0.82	-0.82	-0.58	-0.52	-0.73
C_1	0.16	0.14	0.45	0.35	0.35	0.35	0.24	0.22	0.54
C_2	-0.02	-0.01	-0.15	-0.09	-0.10	-0.10	-0.07	-0.07	-0.18
A_2									
C_0	5.40	3.74	7.75	9.19	9.19	9.19	9.63	8.33	21.14
C_1	0.42	0.54	-1.56	-0.92	-0.92	-0.92	-1.13	0.28	-19.17
C_2	0.06	0.03	0.16	0.12	0.12	0.12	0.25	-0.28	6.64
A_3									
C_0	-1.91	-2.02	-2.18	-2.76	-2.76	-2.76	-1.09	-0.52	-1.92
C_1	1.11	0.88	3.75	2.56	2.56	2.56	2.10	1.91	4.59
C_2	-0.03	0.02	-0.76	-0.40	-0.40	-0.40	-0.33	-0.31	-1.25

	Table I6 - Fit B2 - MS-bar Scheme								
	$\Lambda(2,4) = 0.191 GeV$					$Q_0^2 = 4 GeV^2$			
	d(val)	u(val)	gluon	u(sea)	d(sea)	s	c	b	t
A_0									
C_0	1.77	2.04	0.74	-0.43	-0.43	-0.43	-3.07	-4.44	-7.03
C_1	-0.98	-0.75	-2.44	-1.96	-1.96	-1.96	-1.03	-0.13	1.10
C_2	0.03	0.02	0.07	0.20	0.20	0.20	0.04	-0.23	-0.41
A_1									
C_0	0.13	0.03	-0.59	-0.86	-0.86	-0.86	-0.66	-0.68	-1.13
C_1	0.23	0.26	0.42	0.43	0.43	0.43	0.45	0.50	1.07
C_2	-0.04	-0.05	-0.15	-0.14	-0.14	-0.14	-0.17	-0.18	-0.35
A_2									
C_0	6.28	4.48	6.31	10.16	10.16	10.16	8.57	6.90	8.56
C_1	-0.34	-0.15	-1.62	-1.91	-1.91	-1.91	-0.32	1.46	-2.33
C_2	0.26	0.21	0.18	0.53	0.53	0.53	0.17	-0.53	0.87
A_3									
C_0	-2.30	-2.47	-1.37	-3.14	-3.14	-3.14	-1.68	-1.82	-5.47
C_1	1.60	1.52	3.56	3.14	3.14	3.14	3.48	4.11	9.08
C_2	-0.18	-0.19	-0.77	-0.68	-0.68	-0.68	-0.98	-1.16	-2.66

Appendix II

Table II1 (DIS scheme) and II2 (MS-bar scheme) represent a next-to-leading order fit (SN-fit) of the combined data which assumes a *non-SU(3)-symmetric sea* as suggested by some neutrino di-muon studies. The ratio of $2s/(u+d)$ is set at 0.50 for the input distribuitions. Table II3 represents a *leading order* fit of the combined data (SL-fit) which should be used in applications where leading order hard scattering matrix elements are employed.

Table II1 - Fit SN - Non-Symmetric Sea - DIS								
$\Lambda(2,4) = 0.237 GeV$						$Q_0^2 = 4 GeV^2$		
	d(val)	u(val)	gluon	u(sea)	d(sea)	s	c	b	t
				A_0					
C_0	1.42	1.68	0.90	-1.48	-1.48	-2.26	-4.68	-6.83	-14.41
C_1	-0.59	-0.33	-1.86	-0.89	-0.89	-0.90	0.92	2.68	9.65
C_2	-0.08	-0.10	-0.09	-0.12	-0.13	-0.06	-0.62	-1.13	-2.98
				A_1					
C_0	0.16	0.08	-0.17	-0.13	-0.13	-0.15	-0.06	-0.12	-0.28
C_1	0.17	0.15	-0.10	-0.19	-0.19	-0.10	-0.12	-0.01	0.15
C_2	-0.02	-0.01	0.02	0.04	0.04	0.01	0.01	-0.03	-0.06
				A_2					
C_0	5.40	3.75	5.27	7.83	7.83	7.47	5.55	5.24	11.48
C_1	0.41	0.53	0.43	-0.06	-0.05	-0.61	1.16	1.14	-7.50
C_2	0.06	0.03	-0.26	0.01	0.00	0.28	-0.26	-0.24	2.54
				A_3					
C_0	-1.99	-2.09	-0.20	-0.38	-0.38	-0.23	1.13	1.19	0.65
C_1	1.12	0.89	1.67	0.68	0.68	1.22	0.50	0.93	1.99
C_2	-0.03	0.02	-0.14	0.05	0.05	-0.16	0.03	-0.13	-0.43

Table II2 - Fit SN - Non-Symmetric Sea - MS-bar								
$\Lambda(2,4) = 0.237 GeV$						$Q_0^2 = 4 GeV^2$		
	d(val)	u(val)	gluon	u(sea)	d(sea)	s	c	b	t
				A_0					
C_0	1.84	2.08	0.31	-1.13	-1.13	-1.82	-3.69	-5.06	-9.92
C_1	-0.97	-0.66	-1.84	-1.26	-1.26	-1.40	-0.47	0.39	4.60
C_2	0.03	-0.02	-0.06	-0.01	-0.01	0.09	-0.10	-0.35	-1.53
				A_1					
C_0	0.12	0.02	-0.10	-0.15	-0.15	-0.18	-0.15	-0.25	-0.38
C_1	0.22	0.19	-0.10	-0.16	-0.16	-0.06	0.04	0.16	0.24
C_2	-0.04	-0.01	0.01	0.03	0.03	-0.01	-0.05	-0.08	-0.08
				A_2					
C_0	6.34	4.53	4.18	8.43	8.43	7.94	5.72	4.42	-1.27
C_1	-0.34	-0.04	0.05	-0.39	-0.39	-0.82	0.93	2.38	9.17
C_2	0.25	0.15	-0.12	0.05	0.05	0.30	-0.11	-0.63	-2.88
				A_3					
C_0	-2.40	-2.51	0.34	-0.64	-0.64	-0.56	0.26	-0.14	-1.60
C_1	1.53	1.24	1.64	1.01	1.01	1.65	1.85	2.72	4.40
C_2	-0.16	-0.05	-0.16	-0.06	-0.06	-0.31	-0.50	-0.75	-1.08

	Table II3 - Fit SL - Leading Order								
	$\Lambda(1,4) = 0.144 GeV$					$Q_0^2 = 4 GeV^2$			
	d(val)	u(val)	gluon	u(sea)	d(sea)	s	c	b	t
	A_0								
C_0	1.38	1.67	1.52	-0.81	-0.81	-0.81	-3.62	-6.16	-12.68
C_1	-0.62	-0.33	-3.17	-1.13	-1.13	-1.13	0.03	2.37	8.36
C_2	-0.10	-0.13	0.25	-0.26	-0.26	-0.26	-0.48	-1.24	-2.89
	A_1								
C_0	0.16	0.08	-0.25	-0.07	-0.07	-0.07	-0.06	-0.11	-0.35
C_1	0.19	0.17	-0.01	-0.46	-0.46	-0.46	-0.21	-0.05	0.28
C_2	-0.02	-0.01	0.00	0.16	0.16	0.16	0.05	-0.02	-0.12
	A_2								
C_0	5.40	3.75	7.01	9.19	9.19	9.19	8.30	6.49	14.87
C_1	0.59	0.70	-0.90	0.35	0.35	0.35	-0.60	1.28	-12.56
C_2	0.03	0.00	-0.08	-0.49	-0.49	-0.49	0.25	-0.41	4.75
	A_3								
C_0	-1.97	-2.09	-0.79	-0.89	-0.89	-0.89	0.16	0.71	-0.17
C_1	1.24	0.98	2.90	0.33	0.33	0.33	1.26	1.37	3.39
C_2	-0.05	0.02	-0.54	0.40	0.40	0.40	-0.15	-0.26	-0.96

PARTON DISTRIBUTIONS FROM A GLOBAL ANALYSIS OF DATA

P N Harriman, A D Martin and W J Stirling
Department of Physics, University of Durham,
Durham DH1 3LE, England
and
R G Roberts[1]
Rutherford Appleton Laboratory,
Chilton, Didcot, Oxon, OX11 OQX, England.

Abstract

We present a next-to-leading order QCD structure function analysis of deep-inelastic muon and neutrino scattering data. In particular, we incorporate new $F_2^{\mu n}/F_2^{\mu p}$ data and take account of a recent re-analysis of SLAC data. The fit is performed simultaneously with next-to-leading-order fits to recent prompt photon and Drell-Yan data. As a result we are able to place tighter constraints on the quark and gluon distributions. Two definitive sets of parton distributions are presented according to whether the EMC or BCDMS muon data are included in the global fit. Comparisons with distributions obtained in earlier analyses are made and the consistency of data sets is investigated.

---

[1] Talk presented by R.G. Roberts

The aim of the exercise is to produce sets of Q^2-dependent quark and gluon distributions which are constrained by existing experimental data. In this talk I concentrate on the analysis itself and leave it to Alan Martin to present you with some of the implications of these distributions; especially at high energy and low x.

The framework for the analysis is the perturbative QCD description with next-to- leading corrections. As explained by Wu-Ki Tung[1] we must specify the scheme for the definitions of the parton distributions and in our case this is the $\overline{MS}$ scheme. When we quote a value of Λ, this corresponds to $\Lambda_{\overline{MS}}$ with 4 flavours. To take account of thresholds we match α_s as N_f changes from 4 to 5.

Here I shall suppress details of the analysis, the full description can be found in ref (2). We begin by specifying the parametrisation of the initial distributions at $Q_0^2 = 4 GeV^2$. We have

$$x[u_v + d_v] = N_{ud} x^{\eta_1} (1-x)^{\eta_2} (1 + \gamma_{ud} x) \qquad (1)$$
$$x d_v = N_d x^{\eta_3} (1-x)^{\eta_4} (1 + \gamma_d x) \qquad (2)$$
$$2x[\bar{u} + \bar{d} + \bar{s}] = A_s x^{\delta_s} (1-x)^{\eta_s} \qquad (3)$$
$$xg = A_g (1-x)^{\eta_g} \qquad (4)$$

where, in the valence quark distributions, q_V, the coefficients N_q are fixed in terms of the appropriate η_i's and γ_q's so as to reproduce the flavour content of the proton. At $Q^2 = Q_0^2$ we assume

$$\bar{u} = \bar{d} = 2\bar{s}$$

where the fraction of the strange sea is chosen so as to reproduce the observed ratio of neutrino-induced dimuon ($\mu^+ \mu^-$) to single muon events. The charm distribution is generated through the evolution equations assuming that the charm quark is massless and that $c(x, Q_0^2) = 0$. This procedure has been shown to give a good description of the measured charm structure function $F_2^{c\bar{c}}$. The coefficient A_g of the gluon distribution is determined by the momentum sum rule and so, including $\Lambda_{\overline{MS}}$, there is a total of eleven free parameters to be determined by the data.

The gluon distribution may also be allowed a factor x^{δ_g} to probe the small x behaviour - this is discussed in Alan Martin's talk.

This analysis is the latest in a series by the authors and I should pause for a moment to explain the ancestry of the present version:

1987:

MRS1, 2, 3[3] were fits to deep inelastic data; μN EMC, νN CDHSW, CCFRR. The 1,2,3 referred to the shape of the gluon at Q_0^2 and represented 'soft', 'hard' and 'singular' ($1/\sqrt{x}$ behaviour) gluons. The soft version MRS1 seemed to be the preferred set and this gives rise to:

1988:

MRSE, MRSB[4] which corresponded to soft gluon type fits where the μN data was in turn EMC or BCDMS.

1989:

The previous analysis was extended to include Drell-Yan data and prompt photon production data, as well as F_2^n/F_2^p measurements from NMC. The results were labelled MRSE', MRSB'.[5]

1990:

Present analysis.
Input data: Renormalised EMC, $F_2^{\mu p}$ data,[6] renormalised BCDMS $F_2^{\mu p}$ data[7], WA70 prompt photon data[8], E605 Drell-Yan data[9], as well as the neutrino deep inelastic data of CDHSW[10] and CCFR[11]. Finally the n/p data of BCDMS, EMC and NMC[12].

The renormalisation of the μp data is important. The famous discrepancy between the EMC and BCDMS data is at least as 10% at low x. The detailed re-analysis[13] of the SLAC data indicates that consistency between the SLAC and EMC data requires a shift upwards of 8% of the latter and a similar exercise with SLAC and BCDMS suggests a shift downwards of 2% of the BCDMS data. However because of the difference in *shape* of EMC and BCDMS the discrepancy is then primarily at large x. In fact there are other reasons for moving the EMC data by 8-10%, see ref (2).

First we perform a combined fit to the DIS data and to the WA70 pp prompt photon data. The latter are important in helping to pin down the shape of the gluon distribution. We use the NLO analysis of prompt γ production by Aurenche et al[14] and find that the parameter η_g is then constrained to be $4 \leq \eta_g \lesssim 5$ but

$\Lambda_{\overline{MS}}$ is not particularly constrained by the prompt γ analysis. Combining with the DIS data we find

$$\eta_g = 4.4, \quad \Lambda_{\overline{MS}} = 100 \pm 20 MeV \quad \text{if } F_2^{\mu p} \equiv EMC$$
$$\eta_g = 5.1, \quad \Lambda_{\overline{MS}} = 190 \pm 20 MeV \quad \text{if } F_2^{\mu p} \equiv BCDMS$$

In both fits the neutrino data and n/p data are used. Fig. 1 shows the resulting fit to the WA70 data, and the comparison with the n/p data is shown in fig. 2. Next we include the Drell-Yan data of E605 at Fermilab with a view to constraining the shape of the sea. The cross-section is

$$\frac{d^2\sigma}{dydM^2} = \frac{8\pi\alpha^2}{9M_s} K(y, M^2) \sum_q e_q^2 q(x_1, M^2) \bar{q}(x_2, M^2) + (q \leftrightarrow \bar{q})$$

and writing

$$K(y, M^2) = K_0(y, M^2) K'$$

allows us to separate the known $0(\alpha_s)$ correction, K_0, to the cross-section from the not yet fully determined $0(\alpha_s^2)$ correction, K'. It turns out that the Drell-Yan data does not change the nature of the fits got from the previous step of the analysis, the resulting values of η_s (which governs the shape of the sea) and K' are

$$\eta_s = 10.5, \quad K' = 1.15 \quad EMC - \text{type fit}$$
$$\eta_s = 9.75, \quad K' = 0.94 \quad BCDMS - \text{type fit}$$

Fig. 3 shows the fit to the E605 data.

The output is two sets of Q^2-dependent parton distributions, HMRS(B) and HMRS(E). Fig. 4 shows how these compare with other sets and fig. 5 shows our two sets at $Q^2 = M_W^2$. If you would like a subroutine which computes these distributions down to $x = 10^{-5}$ and up to $Q^2 = 10^6 \ GeV^2$ contact:

BITNET: WJS @ HEP.DUR.AC.UK
DECNET: 19681.

Finally, we can inspect how the measurement of F_2 at HERA may help to discriminate between the two sets. Fig. 6 shows the extrapolation of F_2 to the HERA region.

References

[1] W.K. Tung, talk at this workshop.

[2] P.N. Harriman, A.D. Martin, W.J. Stirling and R.G. Roberts, RAL-90-007, 1990.

[3] A.D. Martin, R.G. Roberts and W.J. Stirling, Phys. Rev. D37 (1988) 1161.

[4] A.D. Martin, R.G. Roberts and W.J. Stirling, Phys. Lett. D206 (1988) 327.

[5] A.D. Martin, R.G. Roberts and W.J. Stirling, Mod. Phys. Lett. A4 (1989) 1135.

[6] EMC: J.J. Aubert et al., Nucl. Phys. B259 (1985) 189.

[7] BCDMS: A.C. Benvenuti et al., Phys. Lett. 223B (1989) 485.

[8] WA70: M. Bonesini et al., Zeit Phys. C38 (1988) 371.

[9] E605: C.N. Brown et al., Phys. Rev. Lett. 63 (1989) 2637.

[10] CDHSW: J.P. Berge et al., CERN-EP/89-103.

[11] CCFR: F. Sciulli, report at 19th Symposium of Multiparticle Dynamics, Arles 1988.

[12] BCDMS: A.C. Benvenuti et al., CERN-EP/89-171 (1989).
EMC: J.J. Aubert et al., Nucl. Phys. B293 (1987) 740.
NMC: J. Nassalski, Madrid Conference 1989.

[13] SLAC: L.W. Whitlow et al., SLAC-PUB-5100 (1989).

[14] P. Aurenche et al., Phys. Rev. D39, (1989) 3275.

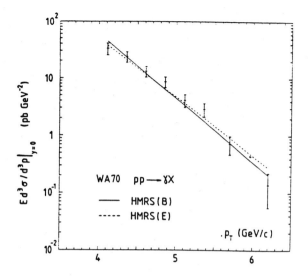

Figure 1: Data on the prompt photon transverse momentum distribution in pp collisions at $\sqrt{s} = 23$ GeV from the WA70 collaboration [8] (corrected to $y = 0$), together with the predictions using the HMRS(B) (continuous line) and HMRS(E) (dashed line) parton distributions.

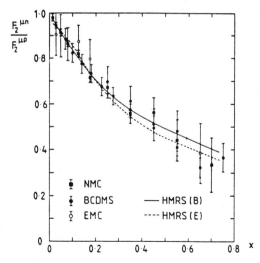

Figure 2: Data on the structure function ratio $F_2^{\mu n}/F_2^{\mu p}$ from the EM collaboration [12] (open circles), the BCDMS collaboration [12] (solid circles), and the NM collaboration [12] (solid squares), together with the HMRS(B) (continuous line) and HMRS(E) (dashed line) fits.

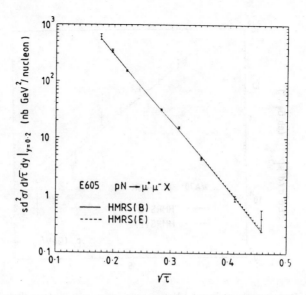

Figure 3: Drell-Yan data from the E605 collaboration [9] in pN collisions at $\sqrt{s}$ = 38.8 GeV, together with the predictions using the HMRS(B) (continuous line) and HMRS(E) (dashed line) parton distributions.

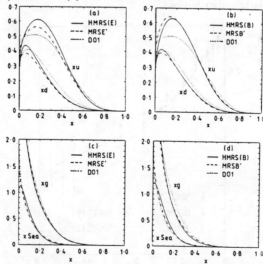

Figure 4: The continuous and dashed curves are the HMRS and MRS' parton distributions $xf_i(x, Q^2 = 20 GeV^2)$ respectively. The left-(right-)hand plots are the parton distributions obtained using data sets which include the EMC(BCDMS) $F_2^{\mu p}$ measurements. In each case, we show the distributions of Duke and Owens Set 1 for comparison.

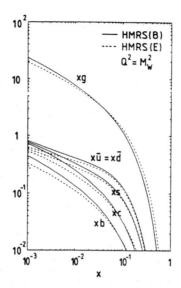

Figure 5: The HMRS gluon and sea quark distributions at $Q^2 = M_W^2$.

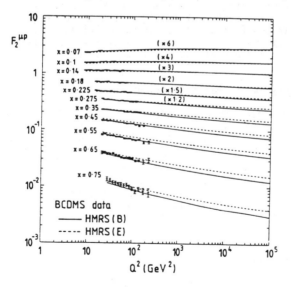

Figure 6: Predictions for $F_2^{\mu p}(x, Q^2)$ (electromagnetic part only) extrapolated to the HERA kinematic region. The data are from the BCDMS collaboration [7] and the dashed and continuous curves correspond to the HMRS(E) and HMRS(B) parton distributions respectively.

Phenomenological Consequences of HMRS Partons

P.N. HARRIMAN,[1] A.D. MARTIN,[1] R.G. ROBERTS[2] and W.J. STIRLING[1]

(presented by A.D. Martin[†])

[1]Department of Physics, University of Durham, DH1 3LE, England
[2]Rutherford Appleton Laboratory, Chilton, OX11 0QX, England

Abstract

We extend the HMRS global parton analysis to obtain a band of allowed gluon distributions with qualitatively different small x behaviour and study the implications for various processes at high energy colliders. In particular we study W and Z production as a function of collider energy. We compare predictions for the $W^{\pm}$ charge asymmetry with recent CDF measurements. Proceeding up to SSC energies we find a sizeable component of W and Z production at large rapidity which arises from a valence quark colliding with a parton with extremely small x ($x < 10^{-4}$). Finally we discuss how HERA may probe the gluon distribution at $x \sim 10^{-3}$.

1. Uncertainty in the gluon distribution

We have heard[1] that the HMRS parton distributions are obtained from a next-to-leading order analysis of data from deep-inelastic lepton-nucleon scattering, the Drell-Yan and prompt photon processes. These data cover the x region $0.03 \lesssim x \lesssim 0.7$ and tightly constrain the quark distributions for values of x in this interval. On the other hand the gluon has much more freedom. The prompt photon data determine its value at $x \sim 0.4$ and this together with the momentum sum rule are sufficient to determine the two-parameter "starting" form

$$xg(x) = A(1-x)^{\eta} \qquad (1)$$

at $Q^2 = 4$ GeV^2, which is then evolved to higher Q^2.

Clearly Eq. 1 is too restrictive if we wish to study very small x. Other possible forms of small x behaviour range from $xg \sim x^{1/2}$, appropriate to a "valence-like" gluon, to the more singular $xg \sim x^{-1/2}$ which approximates[2] the leading logarithmic resummation of the $\ell n(1/x)$ higher order contributions.[3] In ref. 4 the freedom at very small x was investigated using, in addition to the HMRS(E) partons, two

---

[†]Talk presented at the Workshop on Hadron Structure Functions and Parton Distributions, 26-28th April 1990, Fermilab, Batavia, Illinois.

alternative sets of partons, $E_{\pm}$, which were obtained by repeating the global analysis of the structure function and related data using gluon distributions which behave (at $Q^2 = 4$ GeV^2) like $xg \sim x^{\delta}$ at small x where $\delta = \pm 1/2$ respectively. To maximize the possible spread of allowed gluons, the sets $E_{\pm}$ were selected so that the experimental errors on the prompt photon measurements were approximately spanned rather than to give the optimum fit to these data. Fig. 1 shows the evolution of the gluons of the three parton sets: $E \equiv \text{HMRS(E)}$ and $E_{\pm}$. The differences are most striking at low x and low Q^2. The relative behaviour of the gluons in Fig. 1 is reflected in the predictions shown in Fig. 2 for heavy quark ($Q\bar{Q}$) and Higgs production; processes

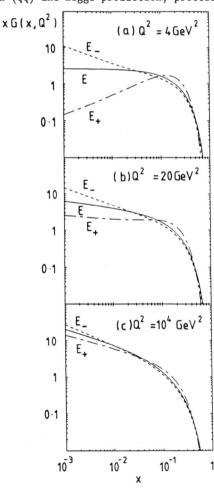

Fig. 1

The gluon distributions, $xg(x,Q^2)$ of the three sets of partons $E, E_{\pm}$ for three different Q^2 values.

which are mediated by gluon-gluon fusion. We note a large spread in the predictions for $b\bar{b}$ production at $\sqrt{s}$ = 40 TeV since $x \gtrsim 2m_b/\sqrt{s} \sim 3 \times 10^{-4}$ and $Q^2 \sim m_b^2$, as compared to the small spread for $\sigma(t\bar{t})$ which samples $x \gtrsim 2m_t/\sqrt{s} \sim 10^{-2}$. Also we see that there is a larger uncertainty in $\sigma(H)$ from the lack of knowledge of m_t (which occurs in the virtual quark loop) than of the gluon distribution.

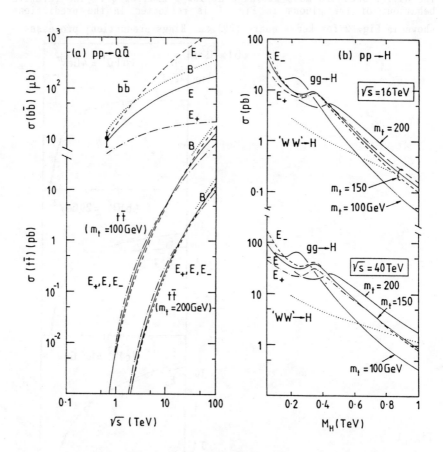

Fig. 2 Predictions for $b\bar{b}, t\bar{t}$ and Higgs production in pp collisions using $E, E_\pm$ parton distributions. Also shown are the predictions using the HMRS(B) ≡ B partons. The figure is taken from ref. 4.

2. Parton information from W and Z data at $p\bar{p}$ colliders

The $p\bar{p}$ collider measurements of W and Z production serve as a valuable independent check on the parton densities. Here the incoming partons, which are taken to carry momentum fractions x_1 and x_2 of the colliding beam particles, must satisfy a kinematic constraint which, for $p_T^2(W) \ll M_W^2$, is given by

$$x_1 x_2 \simeq M_W^2/s \qquad (2)$$

and so the parton densities are sampled for $x \simeq M_W/\sqrt{s}$ and $Q^2 \simeq M_W^2$; and similarly for Z. Thus the measurements[5] at the CERN and FNAL

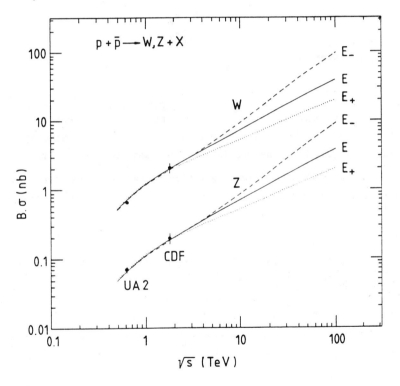

Fig. 3   The cross sections for W and Z production in $p\bar{p}$ collisions as a function of collider energy as predicted by the three parton sets $E, E_\pm$. The values obtained using the HMRS(B) set are slightly larger than those using the HMRS(E) $\equiv$ E set. The UA2 and CDF measurements[5] are shown. At LHC and SSC energies $\sigma_{W,Z}(pp) = \sigma_{W,Z}(p\bar{p})$ to within less than 1%.

colliders, with $\sqrt{s}$ = 0.63 and 1.8 TeV, probe the sea and valence quarks in the regions $x \simeq 0.13$ and $x \simeq 0.05$ respectively. The calculated and measured values of the W and Z cross sections (times the appropriate leptonic branching ratios) are compared in Fig. 3. Within the errors the agreement is excellent. The calculated cross sections include the $O(\alpha_s)$ QCD contributions but there is $O(10\%)$ uncertainty associated with the neglect of the $O(\alpha_s^2)$ and higher order corrections. A major experimental uncertainty concerns the precise value of the integrated luminosity. These uncertainties are considerably reduced if we consider ratios of observables such as the (acceptance corrected) observed number of W's to Z's

$$R \equiv \frac{B_{e\nu}\sigma_W}{B_{ee}\sigma_Z} , \qquad (3)$$

or the $W^{\pm}$ asymmetry

$$A(y) \equiv \frac{d\sigma(W^+)/dy - d\sigma(W^-)/dy}{d\sigma(W^+)/dy + d\sigma(W^-)/dy} . \qquad (4)$$

The ratio R has had a long and fruitful history in predicting the number N_ν of light neutrinos. The recent CDF measurement[6] R = 10.2 ± 0.9 gives N_ν = 2.9 ± 1.5. However this way of determining N_ν is now clearly overtaken by the precision of the measurements at LEP.

The asymmetry $A(y)$ is a sensitive probe of the valence (and sea) quark distributions. We may represent $W^{\pm}$ at $p\bar{p}$ colliders naively as follows

$$p \Rightarrow \begin{array}{c} u \rightarrow W^+ \leftarrow \bar{d} \\ d \rightarrow W^- \leftarrow \bar{u} \end{array} \Leftarrow \bar{p} .$$

Since relatively speaking the u distribution is weighted more to larger x than the d distribution there is a preference for the W^+ to follow the incoming proton direction and the W^- the incoming antiproton. As may be expected there is an intimate connection between the deep inelastic structure function ratio $F_2^{\mu n}/F_2^{\mu p}$ and the $W^{\pm}$ asymmetry[7]. A very approximate rule of thumb is that the (negative) slope of $F_2^{\mu n}/F_2^{\mu p}$ between x_1 and x_2 is proportional to $A(y)$ where $x_{2,1} = M_W \exp(\pm y)/\sqrt{s}$.

The recent NMC small x measurements[8] for $F_2^{\mu n}/F_2^{\mu p}$ are shown in Fig. 4 together with the values obtained using various sets of partons. The strong preference for the HMRS partons is not surprising since only this structure function analysis was able to include these recent data in the fit. The CDF measurements[12] of the $W^{\pm}$ asymmetry,

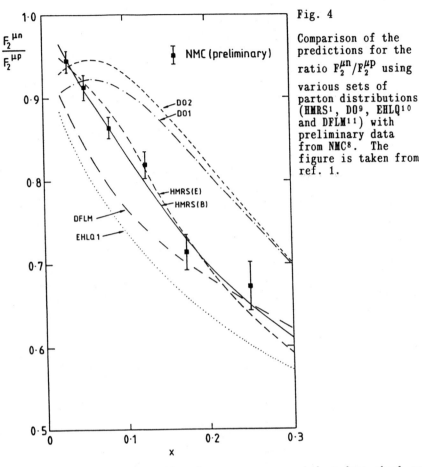

Fig. 4

Comparison of the predictions for the ratio $F_2^{\mu n}/F_2^{\mu p}$ using various sets of parton distributions (HMRS[1], DO[9], EHLQ[10] and DFLM[11]) with preliminary data from NMC[8]. The figure is taken from ref. 1.

however, are even more recent and so serve as an independent check on the HMRS partons. Experimentally it is the asymmetry A_e of the decay $e^{\pm}$ that is measured by CDF as a function of the $e^{\pm}$ rapidity y_e, rather than the parent $W^{\pm}$ asymmetry. The V-A coupling means that the e^+ from $W^+ \to e^+\nu_e$ decay is emitted preferentially opposite the direction of the W^+, whereas the e^- tends to favour the W^- direction. The $e^{\pm}$ asymmetry $A_e(y_e)$ is therefore expected to be smaller than the parent $W^{\pm}$ asymmetry $A(y)$. The preliminary CDF measurements of the asymmetry are compared with the predictions obtained using the two sets of HMRS partons of ref. 1 in Fig. 5. Both sets are in agreement with the measured asymmetry as well as with the $F_2^{\mu n}/F_2^{\mu p}$ data. Clearly an increase in the experimental precision of these two observables will provide a non-trivial constraint on the valence and sea quark distributions.

3. W and Z production at the SSC as a probe of very small x

At first sight it might appear that W and Z production at the SSC will probe the sea quark (q_s) and gluon (g) densities in the x range

$$x \sim \frac{M_{W,Z}}{\sqrt{s}} \sim 2 \times 10^{-3} \qquad (5)$$

via the subprocess $q_s\bar{q}_s \to W, Wg$ and $q_s g \to W q_s$. However the production mechanisms are more subtle. A hint of this follows from the observation that 30-40% of $\sigma_{W,Z}$ at $\sqrt{s}$ = 40 TeV arises from (i) one of the incoming partons having x < 10^{-4} and (ii) one of the partons being a valence quark. The explanation follows readily from Fig. 6. There is a rapid rise of the gluon at very small x, which through evolution immediately feeds a similar behaviour into the sea quark densities. A major contribution to $\sigma_{W,Z}$ therefore arises from the parton

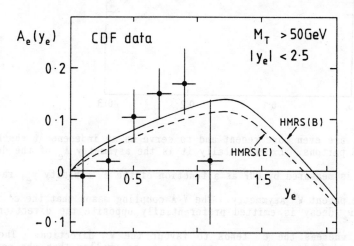

Fig. 5 Comparison of the predictions obtained using HMRS partons[1] with the preliminary CDF data[12] on the charge asymmetry A_e of $e^\pm$ originating from $p\bar{p} \to W^\pm \to e^\pm \nu$ as a function of the $e^\pm$ rapidity y_e.

convolution
$$q_s(x_1)q_v(x_2)$$
with $x_2 \sim 0.1$ in the peak of the valence q_v distribution while, due to the kinematic constraint (2), the sea quarks are sampled in the region

$$x_1 \simeq 10 \, M_{W,Z}^2/s \simeq 5 \times 10^{-5}, \qquad (6)$$

that is for x values much smaller than the expectations of Eq. 5. As x_1 and x_2 are so different the above sizeable contribution to W and Z production at SSC energies will occur at large rapidity. The predictions for $d\sigma_Z/dy$ at the SSC obtained using the three sets of partons, $E, E_\pm$ are shown in Fig. 7(a). The dramatic effects seen at large rapidities are in striking contrast to the more usual rapidity distribution of Fig. 7(b), which corresponds to Higgs production via

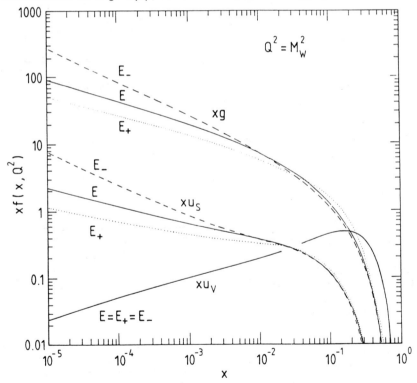

Fig. 6   The gluon, up-sea and up-valence distributions of HMRS sets $E, E_\pm$ as a function of x at $Q^2 = M_W^2$.

gluon-gluon fusion, with $M_H = M_Z$ and $m_t = 200$ GeV. The Higgs is produced much more centrally, sampling the gluon at x_1, x_2 values given by Eq. 5. The dramatic effects in $d\sigma_Z/dy$ even survive a severe cut of $|y_\mu| < 3.5$ on the rapidities of both the $\mu^\pm$ arising from Z decay, see

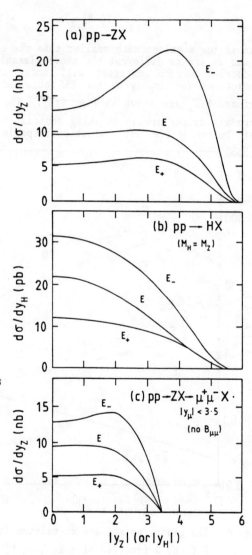

Fig. 7

The rapidity distributions of (a) the Z and (b) the Higgs (with $M_H = M_Z$) produced in pp collisions at $\sqrt{s} = 40$ TeV, as calculated using the three sets of partons $E, E_\pm$. Fig. (c) shows the effect of a rapidity cut $|y_\mu| < 3.5$ on the muons from Z decay. The figure is taken from ref. 13.

Fig. 7(c). The observation of $Z \to \mu^+\mu^-$ at large rapidity at SSC energies will therefore provide a sensitive probe of the sea quark and gluon densities at very small x values ($x < 10^{-4}$). Moreover an appreciable proportion ($\sim 30\%$) of Z production samples this interesting region.

A similar effect at large rapidity has been predicted in prompt photon[14] and in Drell-Yan[15] production and at first sight it appears that these processes could be used to probe such small x even at Tevatron energies. However to reach comparable x values would require the observation of prompt photons with $2p_T \sim 5$ GeV or Drell-Yan pairs with $M_{\mu\mu} \sim 5$ GeV at large rapidity. These events are essentially impossible to distinguish respectively from decay and bremsstrahlung photons and from muon pairs of $b\bar{b}$ and $c\bar{c}$ origin,[13] particularly in the forward (large rapidity) region.

4. Probing the gluon

We have seen that prompt photon data determine the gluon for $0.3 \lesssim x \lesssim 0.5$ and that deep inelastic scattering provides a momentum sum rule constraint, but that widely differing gluon forms are permitted at small x. Of course the trial gluons of parton sets $E, E_\pm$ are only illustrative. There are strong theoretical reasons to expect a small x behaviour similar to that contained in the E_- set ($xg \sim x^{-1/2}$) rather than the conventional behaviour (of E) shown in Eq. 1. From Fig. 7(a) we see that E_- produces the most striking effect at large rapidity and that the observation of $pp \to Z \to \mu^+\mu^-$ at SSC energies can probe the region $x \lesssim 10^{-4}$. Besides exploring the possibility of a singular behaviour of the gluon these data may also probe the effects of gluon saturation which are expected[2] to occur in the $x \lesssim 10^{-4}$ region.

In the near future the most promising methods of probing the gluon at small x appear to be the measurements of the longitudinal structure function F_L and of the inelastic cross section for J/ψ photoproduction at the HERA ep collider. The experimental simulations are presented in refs. 16,17. In summary, the cross section $d\sigma/dx$ for the inelastic photoproduction of J/ψ (via the $\gamma g \to J/\psi\, g$ subprocess) is sharply peaked at $x \sim M_\psi^2/s_{\gamma p}$. When integrated over the inelastic domain, the cross section is found[18] to be well approximated by

$$\sigma(\gamma p \to J/\psi\, X) \simeq 1.5\, \bar{x}\, g(\bar{x}, M_\psi^2)\ \text{nb}$$

with $\bar{x} \simeq 3.4\, M_\psi^2/s_{\gamma p}$. Fig. 8(a) shows the differences in the predicted values of the cross section as a function of $\sqrt{s_{\gamma p}}$ for the E,E$_\pm$ gluons. An integrated luminosity of 100 pb^{-1} should be able to distinguish the three predictions and determine the gluon in the x ~ 10^{-3} region. The second method is based on the observation that, within the framework of QCD, the small x behaviour of the longitudinal structure function F_L is dominated by the gluon contribution. In fact it has been shown that [16]

$$xg(x,Q^2) \simeq 1.77 \left[\frac{3\pi}{2\alpha_s} F_L(0.4x,Q^2) - F_2(0.8x,Q^2) \right].$$

Predictions for F_L at Q^2 = 40 GeV^2 are shown in Fig. 8(b) using the three different gluons. Again we see that measurements of the gluon in the kinematic region down to x ~ 10^{-3} should be possible at HERA.

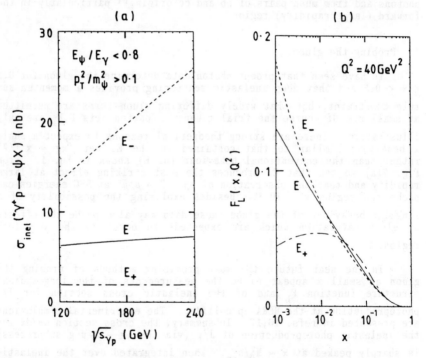

Fig. 8   Predictions for (a) the inelastic J/ψ production cross section as a function of $\sqrt{s_{\gamma p}}$ relevant to HERA, (b) the longitudinal structure function $F_L(x, Q^2 = 40$ GeV$^2)$ using the three gluons of sets E,E$_\pm$. The figure is taken from ref. 4.

References

1. R.G. Roberts, these proceedings; P.N. Harriman, A.D. Martin, R.G. Roberts and W.J. Stirling, RAL preprint 90-007 (1990), to appear in Phys. Rev.
2. L.N. Lipatov, Sov. Phys. JETP 63 (1986) 904
   J.C. Collins and J. Kwiecinski, Nucl. Phys. B316 (1989) 307
3. L.V. Gribov, E.M. Levin and M.G. Ryskin, Phys. Rep. 100 (1983) 1
4. P.N. Harriman, A.D. Martin, R.G. Roberts and W.J. Stirling, RAL preprint 90-018 (1990), to appear in Phys. Lett.
5. UA2 collaboration: J. Alitti et al., CERN preprint EP/90-20 (1990)
   CDF collaboration: presented by P. Derwent at the XXV Rencontre de Moriond, 1990
6. CDF collaboration: F. Abe et al., Phys. Rev. Lett 64 (1990) 152
7. A.D. Martin, R.G. Roberts and W.J. Stirling, Mod. Phys. Lett A4 (1989) 1135
   E.L. Berger, F. Halzen, C.S. Kim and S. Willenbrock, Phys. Rev. D40 (1989) 83
8. NMC: presented by J. Nassalski at the Europhysics Conference on HE Physics, Madrid 1989; updated data presented at this meeting by C. Peroni
9. D.W. Duke and J.F. Owens, Phys. Rev. D30 (1984) 49
10. E. Eichten, I. Hinchliffe, K. Lane and C. Quigg, Rev. Mod. Phys. 56 (1984) 579
11. M. Diemoz, F. Ferroni, E. Longo and G. Martinelli, Z. Phys. C39 (1988) 21
12. CDF collaboration, J. Hauser, these proceedings
13. A.D. Martin and W.J. Stirling, Durham preprint DTP/90/34 (1990)
14. P. Aurenche, R. Baier and M. Fontannaz, FNAL preprint - 89/226-T (1989)
15. F. Olness and W.-K. Tung, Journal Mod. Phys. A2 (1987) 1413
16. S.M. Tkaczyk, W.J. Stirling and D.H. Saxon, Proc. HERA Workshop, ed. R.D. Peccei, DESY 1988, Vol. 1, page 265
17. A.M. Cooper-Sarkar et al., Z. Phys. C39 (1988) 281
18. A.D. Martin, C.-K. Ng and W.J. Stirling, Phys. Lett. 191B (1987) 200

SPIN STRUCTURE OF THE NUCLEON

MEASUREMENTS OF NUCLEON SPIN STRUCTURE FUNCTIONS WITH POLARISED LEPTONS

KLAUS RITH

Max-Planck-Institut für Kernphysik
D-6900 Heidelberg 1, F.R.G

ABSTRACT

The EMC/SLAC results for the proton spin structure function $g_1^p(x)$ and its possible interpretations are discussed. Details of the four experimental proposals to measure the spindependent structure functions $g_1(x)$ and $g_2(x)$ for both proton and neutron in deep inelastic polarised lepton scattering are presented.

1. Introduction

The information about the internal spin structure of the nucleons is contained in the structure functions $g_1^{p,n}(x)$ and $g_2^{p,n}(x)$ which can be measured in deep inelastic scattering of longitudinally polarised charged leptons off a polarised proton or neutron target. If one defines α as the angle between the beam momentum vector $\vec{k}$ and the target polarization vector $\vec{P}$ and ϕ as the angle between the polarisation plane formed by $\vec{k}$ and $\vec{P}$ and the lepton scattering plane formed by $\vec{k}$ and by the scattered lepton momentum vector $\vec{k}'$, then the spin-dependent part of the deep inelastic cross section is given by the difference of cross sections for two opposite target polarizations[1]:

$$\frac{d^3(\sigma(\alpha) - \sigma(\alpha + \pi))}{dx\, dy\, d\phi} = \frac{e^4}{4\pi^2 Q^2}\left[\cos\alpha\left(\left[1 - \frac{y}{2} - \frac{y^2}{4}\gamma\right]g_1(x,Q^2) - \frac{y}{2}\gamma g_2(x,Q^2)\right)\right.$$
$$\left. - \sin\alpha\cos\phi\sqrt{\gamma(1 - y - \frac{y^2}{4}\gamma)}\left(\frac{y}{2}g_1(x,Q^2) + g_2(x,Q^2)\right)\right] \quad (1)$$

The kinematics is described by $\nu = E - E'$, the energy transfer by the virtual photon to the nucleon in the deep inelastic scattering process, and by the negative square of the invariant mass Q^2 of the virtual photon. The Bjorken scaling variables are defined as $x = Q^2/2M\nu$ and $y = \nu/E$. Here E and E' are the energies of the incoming and the scattered lepton and M is the nucleon mass. In the infinitive momentum frame, x can be interpreted as the fraction of the nucleon momentum carried by the struck quark. The quantity γ is defined as $\gamma = \sqrt{Q^2}/\nu$.

Experimentally, the two structure functions $g_1^{p,n}(x,Q^2)$ and $g_2^{p,n}(x,Q^2)$ can be separated by performing two measurements with different orientations α of the target polarisation vector $\vec{P}$, for instance by polarising the target longitudinally ($\alpha = 0$), and measuring the longitudinal asymmetry $A_{\parallel}$ and by polarising the target perpendicular to the beam direction ($\alpha = 90°$), and measuring the transverse asymmetry $A_{\perp}$.

Theoretically, $g_1(x,Q^2)$ has a transparent interpretation in the framework of the parton model, where it can be written in terms of polarized quark distributions $q_f^+(x,Q^2)$ and $q_f^-(x,Q^2)$ for each flavor f of quark and antiquark. $q_f^{\pm}(x,Q^2)$ is defined to be the probability to find a quark in an infinite momentum frame with momentum fraction x and a helicity which is the same as $(+)$ or opposite to $(-)$ that of a parent nucleon with positive helicity.

$$g_1(x,Q^2) = \frac{1}{2}\sum_f e_f^2 \left[q_f^+(x,Q^2) - q_f^-(x,Q^2)\right] \qquad (2)$$

where e_f is the quark charge in units of $|e|$.

The second spin dependent structure function $g_2(x,Q^2)$ does not have an equally transparent interpretation. Its knowledge is required for an unambiguous determination of $g_1(x,Q^2)$ from the cross section (1). In addition, it contains further important information. In the operator product expansion it is given by

$$g_2(x,Q^2) = -g_1(x,Q^2) + \int_x^1 \frac{dz}{z} g_1(z,Q^2) + \tilde{g}_2(x,Q^2), \qquad (3)$$

where the term $\tilde{g}_2(x,Q^2)$ arises only for massive quarks and is sensitive to quark-gluon correlations, which appear as twist-3 operators[1,2].

Experimentally the situation is very unsatisfactory up to now. Only two measurements of $g_1^p(x,Q^2)$ have been performed (with limited statistical and systematic accuracy) for the proton[3,4]; no data at all exist for $g_1^n(x,Q^2)$ of the neutron and the structure function $g_2(x,Q^2)$ is completely unexplored.

In the case of the deuteron, which is a spin-1 target, there are in principle additional leading twist spin-dependent structure functions beyond spin-$\frac{1}{2}$, namely $b_1(x)$[5] and $\Delta(x)$[6]. $b_1(x)$ can be measured by scattering an unpolarised electron beam from a polarised deuterium target with the target spin directed parallel to the direction of the incident electron beam and arranged in each of its $m_I = +1, 0, -1$ substates.

$$b_1(x, Q^2) = \frac{1}{2} \sum_f e_f^2 \left[q_f^+(x, Q^2) + q_f^-(x, Q^2) - 2q_f^0(x, Q^2) \right] \quad (4)$$

where $q_f^0(x)$ is the probability to find a quark (with positive or negative helicity) in a target with helicity 0.

$\Delta(x)$ is measured by scattering an unpolarised electron beam from a polarised deuterium target in the $m_I = 0$ substate, with the target polarisation directed perpendicular to the direction of the incident electron beam and studying the ϕ distribution of scattered electrons. It is sensitive to gluon components in the deuteron. Both $b_1(x)$ and $\Delta(x)$ are expected to be non-zero, but small, in the case of the deuteron.

2. The EMC/SLAC result for $g_1^p(x)$ and its consequences

Fig. 1 shows the presently available experimental information for $g_1^p(x)$ obtained by the EMC[3] and SLAC[4] experiments in deep inelastic scattering od muons (electrons) from frozen ammonia (butanol) targets. The kinematic range extends from about $x = 0.7$ down to $x = 0.015$, but below $x \approx 0.1$ the x dependence is only poorly determined with error bars of 100% to 200%.

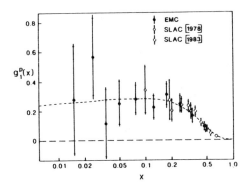

Fig. 1 The world's data on $g_1^p(x)$

It is worthwhile to point out that 110 days of data taking were required by EMC to achieve even this accuracy. The reason for the large error bars are the dilution factors entering the statistical error δA_1 of the longitudinal asymmetry A_1 from which $g_1^p(x)$ has been determined:

$$\delta A_1 \simeq (\sqrt{N^{\uparrow\downarrow} - N^{\uparrow\uparrow}} \cdot f \cdot D \cdot P^T \cdot P^B)^{-1}, \tag{5}$$

where D can be regarded as a depolarisation factor of the virtual photon (which varies between 0.85 at low x and 0.25 at large x), P^T is the target polarisation, P^B is the beam polarisation and f is the fraction of events originating from polarised free nucleons in the target ($f \simeq 3/17$ for NH_3, $f \simeq 10/74$ for butanol but $f = 1.0$ for polarized hydrogen or deuterium gas). For typical values of these quantities ($f = 0.15$, $P^B = P^T = 0.8$, $D = 0.5$) δA_1 is more than 20 times bigger than the expectation from the counting rates alone. From this discussion it is evident that the best way to decrease the statistical errors of g_1^p substantially is to use targets with pure atomic species ($f = 1$ instead of $\simeq 0.15$), while it is much more difficult to increase the luminosity by big factors compared to the previous experiments.

From these data an the integral of $g_1^p(x)$ has been evaluated:

$$I_1^p = \int_0^1 g_1^p(x)\,dx = 0.126 \pm 0.010 \pm 0.015. \tag{6}$$

The result deviates by more than two standard deviations from the theoretical prediction of the Ellis Jaffe Sum Rule[7] which including QCD corrections gives a value of $I_1^p = 0.189 \pm 0.005$.

Assuming the validity of the Bjorken Sum Rule[8] which, taking into account QCD corrections[9], reads for $Q^2 = 10.7$ GeV^2:

$$\int_0^1 (g_1^p(x) - g_1^n(x))\,dx = \frac{1}{6}\frac{g_A}{g_V}(1 - \frac{\alpha_s}{\pi}) = 0.191 \pm 0.003, \tag{7}$$

one concludes that

$$I_1^n = -0.065 \pm 0.010 \pm 0.015 \tag{8}$$

and hence that the spin dependent structure function $g_1^n(x)$ is much larger than hitherto assumed[10] and is negative over a wide range of x.

Using arguments based on SU(3) flavor symmetry[11-13] one can calculate the mean z component of the spin carried by each quark flavor in a proton with spin $s_z = 1/2$. The surprising result of this analysis is that little of the proton spin originates from the spin of the quarks

$$<s_z>_{quarks} = 0.060 \pm 0.047 \pm 0.069, \tag{9}$$

and that the strange sea has a substantial polarisation opposite to the spin of the parent proton

$$< s_z >_{s-quarks} = -0.095 \pm 0.016 \pm 0.023. \qquad (10)$$

These conclusions are supported by the results of a low energy elastic neutrino nucleon scattering experiment[14].

The above results have led to an extensive discussion about the internal spin structure of the nucleon and how quark and gluon spins and orbital angular momenta contribute to the nucleon spin. It is beyond the scope of this contribution to review the large body of literature generated by this unexpected experimental result. Rather, I shall only summarize some of the key ideas.

Examples for possible explanations are:
- the EMC result for I_1^p and/or the extrapolation for $x \to 0$ is wrong[15];
- perturbative QCD breaks down and the Bjorken Sum Rule is violated[16,17];
- isospin effect due to $m_u \neq m_d$ [18];
- SU(3) arguments are not applicable[19];
- the deviations of I_1^p from the predictions of the Ellis Jaffe Sum Rule are caused by a strong Q^2 dependence of the integral (which is, however, not visible in the experimental data available). This could, for example, be due either to a significant higher twist correction associated with the Drell-Hearn-Gerasimov Sum Rule[20], which gives a negativ value of the integral for $Q^2 = 0$, or to nonconservation of the U(1) axial current in QCD[21] as a consequence of the Adler-Bell-Jackiw anomaly[22];
- the spin of the nucleon is predominantly due to orbital angular momentum contributions of quarks[23]. Such a situation arises naturally in the Skyrme model of the nucleon in the chiral limit of massless quarks and in leading order of the $1/N_c$ expansion (where N_c is the number of colours)[11,24];
- gluons contribute substantially to the spin of the nucleon. It has been argued[25] that due to the axial anomaly the singlet axial current does not measure the quark spins alone but rathe a combination of quark spins and gluon spins. To explain the EMC/SLAC result a very large gluon polarisation Δg of about 5 units of angular momentum is required which has to be compensated by orbital angular momentum contributions of similar size but opposite sign;
- a small η'-nucleon coupling which enters the axial singlet vector current via a generalized Goldberger Treiman relation[26].

3. Proposed new experiments

In my opinion the present status of the theoretical discussion demonstrates that our understanding of the internal spin structure of the nucleon is very incomplete. This results in large parts because the focus has been on a single number, namely the value of I_1^p. For a better understanding of the underlying dynamics we not only require measurements of the x-dependence of $g_1^p(x)$ with much better statistical and systematic accuracy than that of the previous experiments but also precise measurements as well of $g_1^n(x)$ as of $g_2^{p,n}(x)$.

At present four polarised lepton scattering experiments have been proposed which want to use very different and partly innovative technologies.

3.1 *The SMC experiment at CERN*[27]

The SMC collaboration will use the external muon beam at CERN and the Forward Muon Spectrometer previously used by EMC (1978-85) and NMC (1986-89).

Compared to the setup used for the EMC polarisation experiment[3] the detector system has already been upgraded by NMC in many aspects. To further increase the reliability and minimize sources of systematic errors SMC will add more driftchamber and proportional chamber planes and the driftchambers W67, which suffer from aging effects, will be replaced by drifttubes. In the area behind the spectrometer two polarimeters will be added to measure the polarisation P^μ of the incident muon as well from the spectrum of the decay electrons as from elastic muon-electron scattering from a magnetized thin iron target. The goal in precision for measuring P^μ is 5%.

Since the muon beam has only very low intensity (about $3.1 \cdot 10^6$ muons/s) large solid polarised targets have to be used to get sufficiently high luminosity. The target material will be chemically doped butanol and deuterated butanol (possibly propanediole). It is foreseen to use in the first year of data taking (1991) the old EMC target (with butanol instead of NH_3). For later use a new target is being built with a length of 2·60 cm (about $4.5 \cdot 10^{25}$ cm^{-2}) and a 2.5 T solenoid to produce a high field homogenity of $\pm 2 \cdot 10^{-5}$ (EMC: 2·40 cm, $\pm 10^{-4}$). A dipole field of 0.5 T will produce transverse target polarisation for an exploratory measurement of the transverse asymmetry. The target polarisation is expected to be about 80 % for H and 40 % for D. It will be measured with an expected accuracy of 3 %. It is planned to reverse the spin direction once every few hours. The experiment will cover a kinematic range $0.001 < x < 0.7$ and $Q^2 > 2$ GeV^2.

The advantages of this project are that there are only few unknowns. The muon beam, generated from π, K -decay in flight, is automatically polarised to about 80 %, the spectrometer and the software chain exist since many years, the target technology is far advanced and well understood and radiative corrections are generally smaller than for electron scattering experiments.

The main disadvantage of this experiment is the small fraction f of polarisable nucleons in the target (f <10/74 for butanol), the substantial background from the vessels, the ^{3}He/^{4}He-bath, the rf-components etc. and the rather low deuteron polarisation. A huge amount of data has to be collected where only a small fraction arises from the polarised nucleons and from the experience of EMC and NMC several years will be required to process the data. To obtain the neutron asymmetry the deuteron and proton asymmetries have to be subtracted which are small numbers with large corrections.

3.2 *E-142 at SLAC.*[28] This group has produced a preliminary version of a proposal to measure the spindependent structure function $g_1^n(x)$ of the neutron with an external polarised electron beam at SLAC. The beam will be produced by a polarised GaAs source and will have the following parameters: polarisation P^e = 0.4, Intensity $5 \cdot 10^{11}$ e/pulse, pulse rate 120 Hz, energy 22.66 GeV. The beam polarisation will be reversed on a pulse to pulse basis, the beam polarisation will be determined by Moller-scattering to 5 %.

The target will be a polarised ^{3}He. As the spins of the protons in a ^{3}He nucleus are in opposite directions, polarized ^{3}He can be regarded to a good approximation as an effective polarised neutron target (the ratio of the magnetic moments is $\mu(^3He)/\mu(n) = 1.112$) with a dilution factor of f = $1/(2\sigma^p/\sigma^n + 1) \approx 1/3$, where σ^p and σ^n are the deep inelastic cross sections for proton and neutron targets. The advantages of such a target is that $A_1^n(x)$ and $g_1^n(x)$ can be obtained directly without the subtraction needed in the case of deuterium and hydrogen targets. There is however some concern about nuclear effects, which might even be different for e.g. quasielastic electron or proton scattering from such a target and for deep inelastic scattering, where the quark-gluon structure is probed.

The SLAC target will be based on the technique of ^{3}He polarisation by spin exchange with a high density laser optically pumped Rb vapor[29]. The target will be a 30 cm long cylindrical glass cell with windows of 0.1 mm thickness containing a ^{3}He density of $3 \cdot 10^{20}$ cm^{-3} ($\approx$10 atm). The expected polarisation of the target is 50 % and will be measured to an accuracy of 5 % by NMR. In addition to the ^{3}He there are $6 \cdot 10^{14}$ Rb atoms/cm^3 in the cell and about $8 \cdot 10^{18}$ N$_2$ molecules/cm^3, which are necessary to non-radiatively quench the Rb excited states populated by the absorption of laser light. The total target thickness including this extra material and the walls is $\approx 6 \cdot 10^{22} cm^{-2}$. The biggest uncertainty in the determination of the target thickness will come from the windows. This is the reason for working at such a high pressure and large target length.

Due to the extra target material the dilution factor is no longer $f \approx 1/3$ but it becomes $f \approx 1/(3.86\sigma^p/\sigma^n + 2.86)$ which is $\approx 1/0.67$ for x $\to$ 0 and $\approx 1/12.8$ for x=0.6 and therefore is even worse than in the case of solid polarized proton targets. Another drawback of the cell walls is the substantial electromagnetic background generated by the beam passing through.

The scattered electrons will be detected simultaneously by two spectrometers at 4.5° (solid angle 0.15 msr) and 7.5° (solid angle 0.45 msr). With two momentum settings the 4.5° spectrometer will cover an x range from 0.04 to 0.20 for $Q^2 > 1\ GeV^2$, the other will provide a range in x of 0.1 to 0.6 for $Q^2 > 5\ GeV^2$.

Do to the very high luminosity the experiment can achieve in its very restricted kinematic range a very good statistical accuracy for $g_1^n(x)$ within a rather short running time of 90 hours (100 % efficient). There will be however substantial systematic errors.

A 3 atm target has been tested successfully with a 10 μA electron beam at BATES this spring and the collaboration has been asked by SLAC to submit a fully worked out proposal.

3.3 *The HERMES experiment at HERA* [30]. The idea of this proposal is to use a polarized internal gas target of hydrogen and deuterium or of ^{3}He together with the high current (60 mA) longitudinally polarised beam of the HERA electron storage ring.

The target design, indicated schematically in fig. 2, is a thin-walled windowless 40 cm long storage cell which is fed by a high intensity source of polarised atoms. The cell walls are coated by special materials to inhibit depolarisation and recombination of the target atoms. A magnetic guiding field of about 0.33 prevents depolarisation by the magnetic field of the electron bunches and a system of collimators protect the target cell from synchrotron radiation.

By this technique, which to my knowledge has been proposed for the first time 10 years ago by P. Schüler[31], target densities of 10^{14} to 10^{15} nuclei/cm^2 can be achieved, which is two orders of magnitude higher than in a situation where the stored electron beam intersects directly with the particle flux from the source. The target is very thin ($\approx 10^{-10}$ radiation length) with no windows, so scattering occurs only from the polarised atoms.

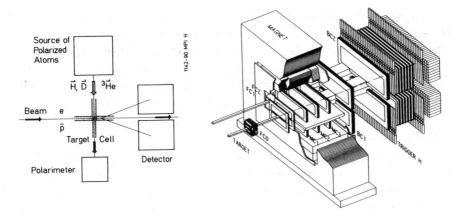

Fig. 2 Principle of the storage cell target

Fig. 3 The HERMES spectrometer

In the case of the H and D targets the polarised atoms will be deliverd by an atomic beam source based on Stern-Gerlach separation of a thermal atomic beam. The source is designed to deliver an intensity of 10^{17} atoms/s in a single substate to the cell[32]. The polarisation will be measured by rf spectroscopy to an accuracy of 2-3 %. The ^{3}He source will polarise the atoms by metastability exchange optical pumping[33]. In this case no additional material like Rb or N_2 is required. For a feed rate of 10^{17} atoms/sec the target thickness will be $9 \cdot 10^{14}$ nucleons/cm^2. This density could be increased by another factor of $\approx$3-9 by cooling the storage cell or/and increasing the feed rate by a factor of 3.

Transverse electron beam polarisation in storage rings can be produced by means of the Sokolov-Ternov effect[34] and has been observed at SPEAR, PEP, DORIS and PETRA, but presently it is not yet possible to firmly predict the degreeof polarisation which can be achieved at bigger machines like HERA and LEP. Longitudinal polarisation has to be produced by special spin rotators, a technique which will be tested at HERA for the first time at a storage ring. At present it is assumed that a polarisation of 50 % can be achieved. The degree of beam polarisation will be measured by Compton backscattering of polarised laser light. With standard techniques it should be possible to measure it with an accuracy of 2-3 %.

The experiment will be installed in the EAST HALL of HERA where the first set of spin rotators will be mounted for test purposes. The spectrometer (angular acceptance 40 - 250 mrad) is shown in fig. 3. For a beam energy of 35 GeV the accessible x range extends from 0.02 to 0.8, the Q^2 range from 1 to 20 GeV^2. The magnet is divided into two symmetrical parts by a horizontal iron plate which shields the electron and proton beams from the magnetic field. Due to this arrangement the spectrometer is divided into two identical halves. Tracking will be done by silicon strip detectors, proportional wire chambers and drift chambers, electron identification and pion rejection by a high resolution calorimeter wall and a transition radiation detector.

The main advantage of this proposal compared to other approaches is the fact that the target atoms are pure atomic species and hence the dilution factors are much larger than for other techniques ($f = 1$ for hydrogen and deuterium, $f \approx 1/3$ for ^{3}He) and high statistical accuracy can be achieved in relatively short running time. Measurements with longitudinal and with transverse holding field are planned for all the target materials which will provide very detailed information about the structure functions $g_1(x)$ and $g_2(x)$ and their integrals for both proton and neutron and also first information about the additional structure functions $b_1(x)$ and $\Delta(x)$ for the deuteron.

3.4 *The HELP proposal for LEP* [35]. This proposal is very similar to the HERMES proposal since it wants to use the longitudinally polarised circulating 50 GeV electron beam in LEP, an internal polarised jet gas target and a large solid angle Forward Spectrometer . The luminosity will be however by orders of magnitude smaller, since the beam current in LEP is only 3 mA compared to 60 mA at HERA and the (very optimistic) projection for the thickness of the polarized jet target is $9 \cdot 10^{12}$ atoms/cm^2 compared to 10^{14} atoms/cm^2. Therefore very long running times would be required to achieve reasonable accuracy even if the beam intensity could be increased by a factor of 10 by filling more bunches into the ring. The degree of longitudinal polarisation which could be achieved at LEP is similar uncertain as in the HERA case.

Much higher target densities could be achieved with unpolarised cluster targets and in my opinion the second part of the proposal which suggests to study hadronisation in electroproduction with the same detector should be considered seriously.

3.5 *Comparison of the experiments* The main parameters of the four experiments are compared in table 1. The 'Figure of Merit' is defined as

$$FM = I^B \cdot T \cdot (P^B \cdot P^T \cdot f)^2$$

It should be noted that a direct comparison of the Figure of Merit for the deuterium and ^{3}He targets is misleading. While A_1^n can be determined directly from the ^{3}He data, a subtraction of the deuterium and proton data is required in the case of the other targets. In this case the error for the neutron asymmetry is given by $\delta A_1^2 = (1 + \sigma^p/\sigma^n)^2 \delta A_1(D)^2 + (\sigma^p/\sigma^n)^2 \delta A_1(H)^2$ and the effective figure of merit for the neutron measurement is more than a factor of 5 worse than for the deuteron alone.

Table 1. The main parameters of the four proposed experiments

Proposal	Beam	P^B	E^B (GeV)	I^B (part./s)	Target	P^T	f	T (cm^{-2})	FM
SMC (CERN)	$\vec{\mu}$	0.8	120	3.1·10^6	C$_4$H$_9$OH C$_4$D$_9$OD	0.8 0.4	<0.13 <0.24	4.2·10^{25} 4.8·10^{25}	1.0 1.0
HERMES (DESY)	$\vec{e}$ (HERA)	>0.5?	35	3.6·10^{17}	H D ^{3}He open cell	0.8 0.8 0.5	1 1 <0.33	1.0·10^{14} 2.0·10^{14} 9.0·10^{14}	6.4 13.0 2.5*
E142 (SLAC)	$\vec{e}$ (GaAs)	0.4	22.7	6.0·10^{13}	^{3}He closed cell	0.5	<0.14	3.6·10^{22}	19.2* (0.14)†
HELP (CERN)	$\vec{e}$ (LEP)	>0.5?	50	1.9·10^{16}	H D jet	0.95 0.95	1 1	0.9·10^{13} 1.8·10^{13}	0.043 0.087

* should be increased by a factor of ≈ 5 in comparison to D for the measurement of neutron asymmetries.
† including solid angles of spectrometers

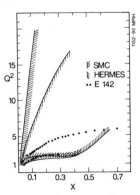

Fig 4. The x,Q^2 range covered by SMC, HERMES and E-142.

The kinematic range covered by SMC, HERMES and E-142 is compared in fig. 4. Both SMC and HERMES cover a large kinematic range and overlap in most of it. Due to the higher beam energy the mean Q^2 of SMC is about a factor of 2.5 higher than for HERMES, the x range extends down to 0.01 for SMC, to 0.02 for HERMES. The E-142 spectrometers have only a small solid angle and thus the accessible kinematic range is very limited.

4. Anticipated accuracies

4.1 Polarised structure functions. SMC has requested 220 days of beam time with a muon energy of 120 GeV with a division of the beam flux in ratio 2/1 between deuterium and hydrogen targets. With the expected statistics collected during this time the following statistical accuracies could be achieved:

$$\delta A_1^p(SMC) \simeq 0.5 \cdot \delta A_1^p(EMC), \qquad \delta A_1^n(SMC) \geq \delta A_1^p(EMC).$$

HERMES has requested 400 hrs (100 % efficient) beamtime for each of the six planned measurements (longitudinal and vertical polarisation; H,D and ^{3}He target). Realistically this corresponds to about 50 days of data taking each.

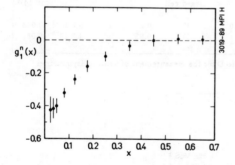

Fig. 5 Projected accuracy for a measurement of $g_1^n(x)$ by HERMES

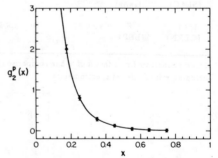

Fig. 6 Projected accuracy for a measurement of $g_2^p(x)$ by HERMES

The statistical errors for the $g_1^{p,n}$ the will be about a factor of 3 smaller than those anticipated for SMC. Fig. 5 shows the expected statistical accuracy for $g_1^n(x)$, assuming a longitudinal beam polarisation of 50 %. The shape of g_1^n has been chosen under the assumption that the EMC/SLAC result for the proton is correct and the

Bjorken Sum Rule is fulfilled. Of course many different other shapes leading to the same I_1^n are possible. These error bars are so small that the x dependence of $g_1^n(x)$, which is essential for constraining theoretical models better, will be well defined even if I_1^n would be substantially smaller than presently estimated and it will also allow to determine the integral I_1^n and the Bjorken Sum rule rather precisely.

E-142 will be able to map out the x-dependence of g_1^n between x = 0.04 and 0.6 with about the same precision as HERMES but 2-3 times more data points.

In fig. 6 the statistical uncertainties are shown for the extracted $g_2^p(x)$ assuming a 400 hour measurement of HERMES for each of the longitudinal and transverse asymmetries for a hydrogen target. The line indicates the positivity limit for $A_2 = \sqrt{R}$. It is obvious that such a measurement will largely reduce the uncertainty in the determination of $g_1^p(x)$ which otherwise would be proportional to the difference between $g_2(x) = 0$ and the positivity limit. The data will constrain the x-dependence of $g_2(x)$ rather well and therefore allow a determination of a quark-gluon correlation function in the nucleon[1].

4.2 Integrals. Table 2 shows the statistical and systematic precision of the Sum Rules which can be extracted from the errors of the structure functions $g_1^p(x)$ and $g_1^n(x)$. The absolute values of the integrals are derived from the EMC/SLAC proton result and from the Bjorken Sum Rule, the percentual errors will of course change according to the values measured in future experiments. It should be noted that the values given in this table differ partly from those quoted in the different proposals which contain several numerical errors.

Table 2. Anticipated accuracies of the Sum Rules

Integral	Value	Target	HERMES		SMC		E-142	
			stat.	sys.	stat.	sys.	stat.	sys.
I_1^p	0.126	H	2.3	5.2	5.3	8.2 (6.8)		
I_1^d	0.061	D	6.6	5.2	17.4	12.8 (8.5)		
I_1^n	-0.065	D - H ^{3}He	7.6 7.3	11.2 7.8	19.2	20.0 (15.5)	6.5	33
$I_1^p - I_1^n$	0.191	2H - D ^{3}He	3.7 4.1	7.1 4.3	9.0	11.6 (9.6)		

For the calculation of the systematic errors HERMES has assumed that $\delta P^B/P^B = \pm 2.5\%$, $\delta P^T/P^T = \delta F_2/F_2 = \pm 3\%$, while SMC has assumed $\delta P^B/P^B = \delta F_2/F_2 = \pm 5 \delta P^T/P^T = \pm 3\%$ and that the error due to the neglect of A_2 is given by $A_2 \leq sqrtR$. For better comparison to HERMES the values in brackets give the total systematic error under the assumption $\delta F_2/F_2 = \pm 3\%$ and that an exploratory measurement with transverse polarisation reduces the error due to A_2 to half its value.

The errors for HERMES are substantially smaller than for SMC, the statistical errors will be small enough for detailed studies of the systematic errors.

The systematic error for E-142, which for me looks very conservative, is dominated by the uncertainty in the beam polarisation, the neglect of A_2 and nuclear effects in ^{3}He.

5. Additional experimental information

5.1 Δg from Ψ Production? Several people have argued that the gluon polarisation Δg can be determined from the cross section asymmetry in polarized leptoproduction of Ψ's. This is in principle correct but in practice it turns out that the required accuracy can not be acchieved with the luminosities of the experiments proposed until now.

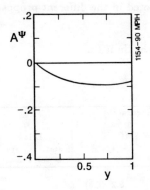
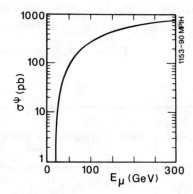

Fig. 7 Prediction [36] for A^Ψ for 200 GeV muons

Fig. 8 Cross section for Ψ leptoproduction

Fig. 7 shows the asymmetry A^Ψ calculated by Altarelli and Stirling [36] for an incident muon energy of 200 GeV under the assumption $\Delta g = 5$. Even for this large value of the gluon polarisation the absolute value of the asymmetry is smaller than 0.1 for all y.

The cross section σ^Ψ for Ψ charged leptoproduction is shown in fig. 8 as a function of incident lepton energy. It varies rather rapidly for energies below 100 GeV and increases more slowly for higher energies. Typical approximate values are 40 pb at 35 GeV, 320 pb at 120 GeV and 600 pb at 200 GeV.

Ψ events can be detected via their two muon decay (branching ratio 7%). For a muon energy of 200 GeV the SMC forward spectrometer has an acceptance of about 10% for these events. If the SMC experiment would use all the requested beam time of 220 days to run at 200 GeV then with a total luminosity of $L = 10^{39} cm^{-2}$ about 4200 Ψ's could be collected. In the framework of the photon gluon fusion model [37] only the inelastic events can be used to extract the gluon distribution. From the experience of the recent analysis of Ψ production by the NMC [38] these are about 38% of the total sample. Taking into account the dilution factors one then could achieve a statistical accuracy for the asymmetry

$$\delta A_\Psi = (\sqrt{N^\Psi_{inel.}} \cdot f \cdot D \cdot p^T \cdot p^B)^{-1} \cong 0.47$$

which is about five times bigger than the maximal expected asymmetry.

In reality the situation will be even worse since SMC is planning to run most of the time at 120 GeV where the cross section is further down by a factor of two and the spectrometer acceptance for three muon events is essentially zero.

Also the HERMES experiment has no chance to determine Δg because of the small cross section at 35 GeV.

Therefore polarised lepton scattering experiments can probably little contribute to solve the question of the magnitude of gluon polarisation compared to other processes like polarised Drell-Yan and direct photons in polarised proton-nucleon scattering which have been discussed at this conference.

5.2 Orbital angular momenta? The question whether at least part of the nucleon spin can be attributed to orbital motion of its constituents has been discussed in detail by Meng et al.[39]. He argued that such orbital momentun contributions should manifest themselves in the Φ distributions of hadrons produced in deep inelastic scattering of unpolarised electrons or muons on a transversaly polarised target. The hadrons should be produced preferentially in the plane perpendicular to the polarisation plane if the constituents of the nucleon are performing an ordered - in addition to or instead of the random - motion.

5.3 Individual polarised quark distributions. The individual polarised valence and seaquark distributions could be deduced from asymmetries in the multiplicities of the difference and the sum of π^+ and π^- produced from proton and deuterium targets.[40] At the moment particle identification is not foreseen for SMC and HERMES and hadron distributions would have to be corrected for proton and kaon contributions. Such measurements would, however, be very important to clarify the question whether the valence quark spins add up to the value expected in the nonrelativistic quark model and are just compensated by a seaquark polarisation of equal magnitude but opposite sign.

6. Conclusions

The proposed experiments use very different and partly innovative technology and are in many aspects complementary. In a few years from now they hopefully will provide us with detailed informations about the different spin structure functions and will help to substantially improve our understanding of the internal structure of the nucleon.

7. Acknowledgement

I have benefited much from discussion with numerous experimentalists and theorists. I want to thank R. Arnold, G. Baum, N. de Botton, V. Hughes, R.L. Jaffe, U. Landgraf, S. Rock, A. Schäfer, P. Schüler and all my collegues from the HERMES project, especially M. Düren and E. Steffens.

8. References

1. R.L. Jaffe, *preprint CTP 1798* (1989).
2. W. Wandzura and F. Wilczek, *Phys. Lett.* **B172** (1977) 195.
   E.V. Shuriak and A.F. Vainshtein, *Nucl. Phys.* **B201** (1982) 142.
3. J. Ashman et al., *Phys. Lett.* **B206** (1988) 364,
   Nucl. Phys. **B328** (1989) 1.
4. M.J. Alguard et al., *Phys. Rev. Lett.* **37** (1978) 1261, **41** (1978) 70.
   G. Baum et al., *Phys. Rev. Lett.* **51** (1983) 1135.
5. P. Hoodbhoy et al., *Nucl. Phys.* **B312** (1989) 571.
6. R.L. Jaffe and A. Manohar, *Phys. Lett.* **B223** ((1989) 218.
7. J. Ellis and R.L. Jaffe, *Phys. Rev.* **D9** (1974) 1444,
   erratum **D10** (1974) 1669.
8. J.D. Bjorken, *Phys. Rev.* **148** (1966) 1467, **D1** (1970) 1376.
9. J. Kodaira, *Nucl. Phys.* **B165** (1980) 129.
10. R.D. Carlitz and J. Kaur, *Phys. Rev. Lett.* **38** (1977) 38.

11. S. Brodsky et al., *Phys. Lett.* **B206** (1988) 309.
12. J. Ellis and M. Karliner, *Phys. Lett.* **B213** (1988) 73.
13. M. Glück and E. Reya, *Z. Phys.* **C39** (1988) 569.
14. L.A. Ahrens et al., *Phys. Rev.* **D35** (1987) 785.
15. F. Close and R.G. Roberts, *Phys. Rev. Lett.* **60** (1988) 1471.
16. G. Preparata and J. Soffer, *Phys. Rev. Lett.***61** (1988) 1167.
17. L.S. Celenza et al., *preprint BCCNT/89/051/193* (1989).
18. A. Schaefer, *Phys. Lett.* **B208** (1988) 175.
19. H. Lipkin, *Phys. Lett.* **B214** (1988) 429.
20. S.D. Drell and A.C. Hearn, *Phys. Rev. Lett.* **16** (1966) 908.
21. R.L. Jaffe, *Phys. Lett.* **B193** (1987) 101.
22. J.S. Bell and R. Jackiw, *Nuovo Cim.* **60A** (1969) 47.
23. L. Seghal, *Phys. Rev.* **D10** (1974) 1663.
    P.G. Ratcliffe, *Phys. Lett.* **B192** (1987) 180.
    G.P. Ramsay et al., *Phys. Rev.* **D39** (1989) 361.
24. Z. Ryzak, *Phys. Lett.* **B217** (1989) 325.
25. A.V. Efremov and O.V. Teryaev, *Dubna preprint E2-88-287* (1988).
    G. Altarelli and G,G, Ross, *Phys. Lett.* **B212** (1988) 391.
    R.D. Carlitz et al., *Phys. Lett.* **B214** (1988) 229.
26. T.P. Cheng and L.F. Li, *Phys. Rev. Lett.* **62** (1988) 1441.
    H. Fritzsch, *Phys. Lett.* **B224** (1989) 189.
27. SMC - J. Beaufays et al., *CERN/SPSC 88-47, SPSC/P242* (1988).
28. E-142 - R. Arnold et al., *draft proposal* (1989).
29. T.E. Chupp et al., *Phys. Rev.* **36C** (1987) 2244.
30. HERMES - K. Coulter et al., *DESY/PRC 90/1* (1990).
31. P. Schüler, *1980 International Symposium High-Energy Physics with Polarised Beams and Targets*, ed. C. Joseph and J. Soffer, (Birkhäuser, Basel, 1981), p. 460.
32. E. Steffens, *Nucl. Phys.* **A497** (1989) 519c.
33. F.D. Colegrove et al., *Phys. Rev.* **132** (1963) 2561.
    R.G. Milner et al., *Nucl. Instr. and Meth.* **A274** (1989) 56.
34. A.A. Sokolov and I.M. Ternov, *Sov. Phys. Dokl.* **8** (1964) 1203.
35. HELP - G. Ballochi et al., *CERN/LEPC 89-10, LEPC/M 88* (1989).
36. G. Altarelli and W.J. Stirling, *Preprint CERN- TH 5249/88* (1988).
37. J.P. Leveille and T. Weiler, *Nucl. Phys.* B147 (1979) 147.
    V. Barger et al., *Phys. Rev.* **D20** (1979) 630.
    E.L. Berger and D. Jones, *Phys. Rev.* **D23** (1981) 1521.
38. C. Peroni, *these proceedings*,
    M. de Jong, *private communication*.
39. Ta-chung Meng et al., *Phys. Rev.* **D40** (1989) 769.
40. L.L. Frankfurt et al., *Phys. Lett.* **B230** (1989) 141.

POLARIZED DRELL-YAN EXPERIMENTS

J. C. Collins

Department of Physics, Illinois Institute of Technology
Chicago, IL 60616, U.S.A.

and

High Energy Physics Division, Argonne National Laboratory
Argonne, IL 60439, U.S.A.

ABSTRACT

A theorist's view of the possibilities for polarized Drell-Yan experiments is described.

1 Introduction

I will give a brief summary of the possibilities for Drell-Yan experiments with polarized beam and target.

For the purposes of this talk I will concentrate on the conventional Drell-Yan process of dimuon (or dielectron) production in hadron-hadron collisions via a virtual photon. (Fig. 1.) The process enables one to measure polarized parton densities in different combinations from those to which deep-inelastic scattering experiments are sensitive. If parton densities are already known, then the process provides an excellent test of QCD. The process is very clean to treat theoretically: the dimuon provides a direct probe of the hard scattering unperturbed by fragmentation effects.

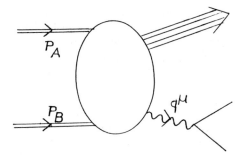

Fig. 1. Drell-Yan process; notation.

At collider energies, when one makes Ws and Zs, *single* spin asymmetries [1] are large because of the parity violation in the weak interactions. However, I will not discuss this case.

There is much structure to measure, because there are seven variables to play with: the overall center of mass energy, the mass, Q, of the pair, its rapidity y, its transverse momentum, $q_\perp$, and the polar angles θ and ϕ of one of the muons. This implies that even if one uses the process to measure parton densities, there are very nontrivial constraints to be satisfied by the data, if it is to obey QCD.

In the absence of experiments to date, there is relatively little literature on the theory of the process. When I searched on the SPIRES database for all papers with either 'spin' or 'polarized' in the title and with 'Drell', 'Yan', 'dilepton' or 'dimuon' in the title, I only found 18 papers, the relevant ones being Refs. [1-12]. Even so, the process is well understood, and phenomenological calculations can easily be made as needed.

2 Basic parton model

The lowest order QCD approximation for the cross section is

$$\frac{d\sigma}{dQ^2 dy} = \frac{\alpha_{em}}{Q^4} \times \text{constant} \times \sum_{i,s} e_i^2 f_{i,s/A}(x_A) f_{\bar{i},\bar{s}/B}(x_B). \tag{1}$$

Here, i represents the flavor of the quark coming from hadron A and s represents its helicity. The quark-antiquark annihilation only occurs for left-handed quarks on right-handed antiquarks, or vice versa, so that the process has 100% analyzing power. That is, if a single flavor of quark dominates, then the spin asymmetry is

$$A = \lambda_A \lambda_B \Delta_A \Delta_B, \qquad (2)$$

where the λs are the beam and target hadron polarizations, and the Δs are the fractional polarization of the partons, relative to the hadron polarization (i.e., the asymmetry in the parton distributions).

Clearly, if we performed the experiment in proton-antiproton collisions, then for moderately large $\tau \equiv Q^2/s$, valence quark-antiquark annihilation dominates. Since at large x we know that the valence quarks are highly polarized, representing the polarization of the constituent quarks, the asymmetry in the Drell-Yan cross section will be large (tens of per cent). Moreover, $\sigma(+-)$ will be substantially greater than $\sigma(++)$, where the $+$ and $-$ labels represent the hadron helicity.

However, if we perform a proton-proton (or proton-nucleus) experiment, then the process goes between valence and sea quarks. The asymmetry should be rather less. Moreover, the EMC results suggest a negative polarization for the sea, so that $\sigma(+-)$ will be somewhat smaller than $\sigma(++)$, if we choose a region like large x_F which forces us to have large x valence quarks annihilating small x sea quarks.

3 QCD corrections

In deeply inelastic scattering with virtual photon exchange, there are four structure functions F_1, F_2, G_1, and G_2, with the last two only existing for the polarized case. But in polarized Drell-Yan, there are 48 [5], of which only 4 are present in the unpolarized case. These are obtained by measuring the angular distribution of the leptons in the pair, and varying the beam and target polarization. Clearly to measure them all requires a Herculean effort. Most of them exist only for transversely polarized beam and target. Moreover in low order perturbation theory, most of them vanish. For example, the Born graphs just give a $1 + \cos^2 \theta$ distribution. The sole effect of spin is to give the asymmetry given in Eq. 2.

For the moment let us ignore the angular distribution, and focus on the asymmetry of the cross section $d\sigma/dydQ^2dq_\perp^2$. When $q_\perp \gg Q$, the lowest order cross section is of order α_s. The graphs have been calculated by Mani and Noman[8]. The parton level cross section for the 'annihilation' process $q + \bar{q} \to \gamma^* +$ gluon is

$$(1 - \lambda_q \lambda_{\bar{q}}) \times \text{unpolarized cross section}, \qquad (3)$$

exactly as for the Born graph that is relevant for low transverse momentum. But the result for the 'Compton' process, $q + \text{gluon} \to \gamma^* + q$ is

$$\text{constant} \left\{ \frac{\hat{s} + 3m^2}{\hat{s}} + \frac{\hat{s} - m^2}{q_\perp^2} \left[1 - \frac{2m^2(\hat{s} - m^2)}{\hat{s}^2} \right] \right.$$
$$\left. + \lambda_g \lambda_q \left[-\frac{\hat{s} + 3m^2}{\hat{s}} + \frac{\hat{s} - m^2}{q_\perp^2} \left(1 - \frac{2m^2}{\hat{s}} \right) \right] \right\}. \tag{4}$$

If $\hat{s}$, the subprocess center-of-mass energy squared, the low, then this cross section is proportional to $1 - \text{constant}\lambda_q \lambda_g$, but if $\hat{s}$ is large then the cross section is proportional to $1 + \text{constant}\lambda_q \lambda_g$. This change of sign in the Compton process can be seen in the plots given in [8].

One could further probe these effects by measuring the associated jet.

4 Intrinsic $k_\perp$

I would like to make a suggestion of my own for measurements in this process, that would help clarify, perhaps, the discussion about the EMC measurements of polarized deeply inelastic scattering.

Suppose one measures the Drell-Yan process as a function of transverse momentum, not at large $q_\perp$, but where $q_\perp \ll Q$. Then it is correct within QCD to treat the process as $q\bar{q}$ annihilation, with intrinsic transverse momentum:

$$\frac{d\sigma}{d^4q} \propto \sum_i e_i^2 \int d^2k_{A\perp} \, P_i(x_A, k_{A\perp}) P_i(x_B, q_\perp - k_{A\perp}). \tag{5}$$

Note that there is no gluon-induced process.

As is well-known, the transverse momentum distribution is substantially broadened by gluon radiation. This is absorbed by a Sudakov-like form factor into the energy-dependence of the transverse momentum distributions $P(x, k_\perp)$. The radiation is spin-independent. So the spin asymmetry of $d\sigma/d^4q$ directly probes the asymmetry of the parton densities. (There are some gluonic effects buried in non-leading parts of the Sudakov formalism, but these are being probed at a scale set by the transverse momentum, not by the dimuon mass[13].)

Thus for this case the connection to naive quark model results should be closer than for the conventional case of the parton densities that are used in deeply inelastic scattering. These are integrated over all $k_\perp$ and therefore involve an ultraviolet subtraction. These subtractions are associated with the issues discussed by Jaffe and by Mueller in their talks elsewhere in this volume.

Roughly speaking there are two extremes for the interpretation of the EMC data: First is that the quark model fails to describe the data well, and there is a substantial polarization of the sea quarks, including the s and $\bar{s}$, that is opposite to the valence polarization. The alternative is that the gluon polarization is large, and that by effects related to the anomaly this induces an extra contribution to the asymmetry.

The anomaly is a short-distance effect, so one might hope that Drell-Yan at low transverse momentum is much less sensitive to this effect. If the gluon explanation is correct, then the sea polarization should be rather small, so that there will be little spin asymmetry in the Drell-Yan cross section in proton-proton collisions. But if the sea explanation is correct, then there will be a significant asymmetry, with $\sigma(+-)$ somewhat smaller than $\sigma(++)$.

5 Conclusions

The Drell-Yan process should be a good direct probe of parton spin. However, the asymmetry in proton-proton collisions may only be a few percent. (It would be much larger in proton-antiproton collisions.)

At low transverse momentum and large x_F, the process directly probes the sea, and in particular the distribution of $\bar{u}$ quarks.

At large $q_\perp$ the process probes the gluon distribution. (Note that one need not take Q so large in this case.

Acknowledgments

This work was supported in part by the U.S. Department of Energy.

References

1. B. Pire and J.P. Ralston, Phys. Rev. D28, 260 (1983).

2. R.P. Bajpai, M. Noman and R. Ramachandran, Phys. Rev. D24, 1832 (1981).

3. C.K. Chen, Phys. Rev. D16, 1576 (1977).

4. M. Chaichian, M. Hayashi and K. Yamagishi, Z. Phys. C20, 237 (1983).

5. J.T. Donohue and S. Gottlieb, Phys. Rev. D23, 2577 (1983), and D23, 2581 (1983).

6. S. Gupta, D. Indumathi and M.V.N. Murthy, Z. Phys. C42, 493 (1989), erratum: ibid. C44, 356 (1989).

7. F.J. Gilman and T. Tsao, Phys. Rev. D21, 159 (1980).

8. H.S. Mani and M. Noman, Phys. Rev. D24, 1223 (1981).

9. H.S. Mani and S.D. Rindani, Phys. Lett .84B (1979) 104-108.

10. A.A. Pankov and I.S. Satsunkevich, Yad. Fiz. 41 395 (1985).

11. J.P. Ralston and D.E. Soper, Nucl.Phys. B152, 109 (1979).

12. R.L. Thews, Phys. Lett. 100B, 339 (1981).

13. J.C. Collins and D.E. Soper, in *Proceedings of Moriond Workshop on Lepton Pair Production, 25-31 January 1981*, ed. J. Tran Thanh Van (Editions Frontières, Dreux, 1981); J.C. Collins, D.E. Soper and G. Sterman, *Nucl. Phys.* **250** (1985) 199, and references therein.

PROMPT PHOTON EXPERIMENTS USING POLARIZED BEAMS

P. F. SLATTERY

Department of Physics and Astronomy
University of Rochester
Rochester, New York 14627, USA

ABSTRACT

Two possible experiments that could in principle be carried out at Fermilab in the next few years to measure prompt photon production by polarized protons incident on a polarized target are briefly reviewed and compared. Some questions are posed concerning the physics significance of an experimentally realizable measurement of this sort.

Introduction

It is well recognized that the study of prompt photon production provides the most direct way of investigating the gluon content of hadrons. In particular, to leading order in the strong coupling constant, the Compton process dominates the production of prompt photons in nucleon-nucleon collisions. These diagrams (Fig. 1) are directly sensitive to the gluon structure of the interacting hadrons.

Prompt photon experiments thus, at least in principle, provide an attractive means of investigating the extent to which gluons contribute to the net spin of the nucleon. Recent interest in this question has been stimulated by electroproduction measurements made by the EMC group at CERN of the spin dependent structure function of the proton $g_1^p(x, Q^2)$.[1] These results have been interpreted to indicate that, contrary to intuitive expectations, very little of the net spin of the proton appears to be carried by its constituent quarks.[2] This would imply that gluons and/or parton orbital angular momenta provide the dominant contributions to the nucleon spin.

In this paper, I will briefly review and compare two possible prompt photon experiments that could be carried out at Fermilab in the next few years using high energy polarized protons incident on a polarized target. I will then pose certain questions that I believe need to be answered before an experiment of this sort is actually carried out.

Proposal P809

At present, only one group at Fermilab has pending an explicit proposal to measure direct photon production in the polarized domain. The group involved is the E704 collaboration, and their proposal has been assigned the number P809.

This group proposes to upgrade the momentum of the present Fermilab MP beamline from its current value of about 200 GeV/c to roughly 500 GeV/c. The MP beam transports polarized protons resulting from Λ decays. The protons are initially produced with transverse polarization, which is subsequently converted to longitudinal polarization via a spin-rotator system consisting of "snake magnets". This technique has the advantage of permitting frequent alternation between positive and negative beam helicities, which has significant benefits to minimizing systematic uncertainties. The beam is equipped with a tagging system that permits the experimenters to achieve a net polarization of 45% for selected particles. The anticipated yield of these spin-tagged protons in P809 is 8×10^7/spill for 4×10^{12} protons/spill on the primary MP target.

The beamline upgrade proposed in P809 requires the addition of 14 4-inch quadrupole magnets to achieve the performance specified in the proposal. The described upgrade does not address the fact that the MP beam competes with the MC beam within the present Meson Area targeting scheme. In the past, this has significantly limited the integrated luminosity made available to experimenters in the MP line, since the MC beam serves Fermilab's high priority K decay experiments. (To take an historical perspective, the MP beam is scheduled to be off during the second half of the 1990/91 fixed target run, and ran for only a few weeks during the 1987/88 run.)

The experimenters propose to employ a polarized ^6LiD target. This material has the very distinct advantage that, to the extent that a polarized ^6Li nucleus can be treated as a polarized deuteron plus an unpolarized α-particle, then up to 50% of the total number of nucleons in the target can be polarized. This provides an enormous potential improvement relative to more conventional polarized targets, in which typically < 15% of the constituent nucleons are polarized, for experiments at high transverse momentum, where the identity of the interacting nucleus cannot be determined on an event-by-event basis. The accompanying disadvantage is that the nuclear physics of ^6Li is not well understood at present, although the P809 experimenters propose to remedy this deficiency.

The collaboration proposes to detect and measure direct photons using an augmented version of the present E704 Pb-glass calorimeter. This upgraded calorimeter is pictured in Fig. 2. The inner cells have a transverse area of $38 \times 38 \ mm^2$, while the outer cells are $114 \ mm$ on a side. The inner Pb-glass blocks are supplied by Serpukhov, and presently exist, but the ADC readout for the complete inner calorimeter depicted is not yet on hand. The outer calorimeter would be an entirely new device.

The performance anticipated for the overall calorimeter is excellent, as indicated in Fig. 3, which presents a Monte Carlo estimate of the energy asymmetry expected in the π^o and η regions. The very long plateaus reflect the capability of the P809 calorimeter to detect and measure photons down to an energy of 500 MeV. Figure 4 summarizes the views of the proposers concerning what might be expected from the experiment. The curves are from Berger and Qiu[3], and represent an attempt by these authors to place limits on the upper and lower range of reasonable values of A_{LL}, the double spin asymmetry parameter for direct photon production. The data point summarizes what the experimenters believe could be accomplished in 2500 hours of beam time.

A Possible Polarized Version of E706

The second group at Fermilab that has given serious thought to mounting an experiment of this type is the E706 collaboration. In contrast to the E704 group, for which P809 represents an extension of their ongoing investigation of polarization phenomena into the area of prompt photon experimentation, the E706 experimenters have been considering whether to extend their current investigation of prompt photon production into the polarized domain. The two groups can thus be viewed as approaching the problem from opposing perspectives.

At present, the MW beam transports high energy hadrons directly from its primary target – i.e. it is a standard secondary beam line. A study has been made of the feasibility of modifying the existing MW beam to transport longitudinally polarized protons originating from Σ^+ decays.[4] The conclusion has been reached that the existing beam line elements could be reconfigured to transport a 400 GeV/c polarized proton beam of this origin. The target region would have to be augmented by a new high field sweeping magnet, but the fact that the existing beam is aligned at a 12 mr angle relative to the direction of the primary beam is a distinct advantage for a tertiary beam originating from the decay of charged secondaries. (In contrast, the MP beam is aligned at 0 mr relative to the direction of the primary beam, since it transports polarized protons originating from neutral decays.)

The proton flux using the existing 3-inch quadrupoles in MW would be somewhat lower than that predicted for the upgraded MP beam (2×10^7/spill for 4×10^{12} primaries/spill), although a 50% increase in intensity could be obtained through the use of 4-inch quadrupoles, in analogy with those that are postulated for the high energy version of MP. Of greater significance, however, is that fact that the net polarization of the MW beam would be 63%. Since the figure of merit of a polarized beam increases as the square of its net polarization, this is equivalent to a factor of two increase in flux relative to a beam with 45% net polarization.

The primary differences between the two beams are the following. As stated earlier, protons arising from Σ^+ decay are initially longitudinally polarized. Thus, a spin-rotator system is not absolutely required, providing one is satisfied to run with a beam of fixed helicity. If, on the other hand, one elects to periodically reverse the helicity of the beam to minimize systematic uncertainties, a system of snake magnets of twice the rotating power of a set adequate for an initially transversely polarized beam would be required. The other difference concerns the primary targeting of the MW beam. The beam is at present targeted simultaneously with the MT beam, and to date has been run whenever the Tevatron has been operated in the fixed target mode. There appears to be no technical reason why the Meson Area targeting scheme could not be modified to permit the MP beam to run in parallel with MC. However, unless and until this change is made, any statistics limited experiment would gain significantly from being run in MW versus MP.

The E706 collaboration has neither an existing polarized target, nor at present any expertise in this area. In all of its thinking about possibly mounting a polarized prompt photon experiment, the group has assumed that it would expand the existing collaboration to include individuals with the necessary target and expertise. Thus, in order to evaluate a potential extension of E706 into the polarized domain, it is necessary to assume arbitrarily that the existing MW spectrometer is augmented by a suitable polarized target.

One definite advantage of the existing E706 spectrometer to any prompt photon experiment is the proven capability of its electromagnetic calorimetry. The experiment is equipped with a 3-meter diameter liquid argon calorimeter (LAC) involving r - ϕ readout. This calorimeter was first employed during the 1987-88 fixed target run, and approximately 5 M physics quality events were accumulated. Figure 5 presents representative data taken at that time: 5(a) displays the two photon mass distribution, and demonstrates the excellent signal-to-background in both the π^o and η regions; 5(b) displays the background subtracted photon energy asymmetry distribution in the π^o mass region. As can be seen, this latter distribution rolls off at lower values of asymmetry than does the corresponding distribution in Fig. 3. This is consistent with the higher photon energy threshold (of a few GeV) characterizing the LAC data collected by E706 during the 1987/88 data run, but some of the observed effect is also to be attributed to the inevitable difference between real data and Monte Carlo events, and to the fact that the distributions shown in Fig. 3 have not been corrected for the presence of background in the π^o and η mass regions. The E706 experimenters continue to make progress in understanding in detail the technical capabilities of their calorimeter, which is presently being employed in the experiment's 1990/91 data run. During this run, the LAC is being read out in its entirety, without suppression of channels containing pulse heights below a pre-established threshold. It is anticipated that this will significantly lower the effective energy threshold for the detection of low energy photons, but it nevertheless is not expected that the device will be able to reliably detect photons down to an energy as low as the 500 MeV threshold predicted for the P809 Pb-glass calorimeter.

Although the energy resolution achievable using liquid argon calorimetry is inferior to that of Pb-glass calorimetry, a very fine grained device may be constructed at acceptable cost using the former technology. Specifically, the r-defining strips in the E706 LAC are approximately 5.5 mm wide, and the inner edge of the ϕ-defining strips are 3.5 mm in width. Even at 9 meters from the experimental target, this exceptionally fine granularity provides excellent descrimination between one and two photon induced showers.

Figure 6 compares the geometric acceptances of the two calorimeters at their respective locations. This figure takes into account the fact that the P809 calorimeter is positioned 11 meters from the experimental target (a consequence of its coarser segmentation) by scaling up the dimensions of the LAC by 11/9 before projecting its outline onto the face of the P809 calorimeter. As can be seen, the LAC coverage extends significantly further backward in the interaction center-of-mass, particularly in comparison with the inner portion of the P809 calorimeter. (The increased forward coverage of the P809 calorimeter is going to be challenging to exploit fully due to the frequency of overlapping showers in this region of laboratory phase space.)

Conclusions

My primary conclusion is that, irrespective of the technical differences identified in the preceding sections, a comparable prompt photon experiment, involving polarized protons incident on a polarized target, could be mounted in either MP or MW in time to begin taking data during the 1992/93 fixed target run. The higher energy and intensity predicted for the upgraded MP beam would largely be offset by the higher average polarization and increased running time expected for a reconfigured MW beam, plus the larger geometric acceptance of the E706 LAC compared to the proposed P809 Pb-glass calorimeter. The sensitivity of the P809 calorimeter to very low energy photons is intrinsically beneficial to an experiment of this type, but this potential advantage should be balanced against the LAC's proven (and continuously improving) capability for detecting and measuring high p_T prompt photons. The incremental cost of mounting an experiment of this type in each of the two areas appears likely to be significantly different, but this issue is moot unless and until there is more than one proposal to carry out such a measurement at Fermilab.

This brings to the fore the central question: What is the physics significance of an experiment of the general precision depicted in Fig. 4? This question has several distinct aspects:

(1) What is the actual value of A_{LL}? The data point in Fig. 4 is plotted on the higher of the two theoretical estimates, but it might just as well have been plotted on the lower, in which case the ability of the measurement to distinguish reliably even the sign of A_{LL} could be questioned.

(2) How cleanly can prompt photon production be measured at p_T values near 4 GeV/c? Figure 7 displays the γ/π° ratio anticipated by the P809 experimenters. Signal and background are predicted to be approximately equal at 4 GeV/c, which makes the measured asymmetry quite sensitive to uncertainties in estimating the number of residual photons from meson decays in the final single photon sample. Moreover, the prompt photon cross section at 500 GeV is not very well understood in the 3-4 GeV/c region of p_T. Thus, the curves presented in Fig. 7 should be treated with caution.

(3) How reliably can the results of leading order perturbative QCD be applied to prompt photon production at these relatively low p_T values? The simplest interpretation of an experiment of this type is in terms of the leading order diagrams depicted in Fig. 1. However, next-to-leading order effects are also important, even at values of p_T significantly higher than 4 GeV/c. Some of these are shown in Fig. 8,[5] the purpose of which is simply to illustrate that present theoretical understanding of prompt photon production is fairly sophisticated, and subtle effects are known to be important, particularly at relatively low values of p_T. Such higher order effects have not been included in calculating the curves shown in Fig. 4.

Posing these sorts of questions is often easier than decisively answering them. At the present time, the E704 collaboration has decided that the experiment is worth carrying out at Fermilab in the next few years, while the E706 collaboration has recently reached the opposite conclusion. It is perhaps interesting to speculate (as R. Jaffe seemed to suggest at the time of the Workshop) that this difference of opinion may arise, at least in part, because the former group tends to compare the experiment to other sorts of polarized experimentation, whereas the E706 experimenters most naturally compare the experiment to other contemporary prompt photon experiments.

Acknowledgements

This paper is based in part on research supported by the U.S. Department of Energy, the National Science Foundation, and the University Grants Commission of India.

References

1. J. Ashman et al., *Phys. Lett.* **B206** (1988) 364; *Nucl. Phys.* **B328** (1989) 1.
2. S. J. Brodsky, J. Ellis and M. Karliner, *Phys. Lett.* **B206** (1988) 309.
3. E. L. Berger and J. W. Qiu, *Phys. Rev.* **D40** (1989) 778; *Phys. Rev.* **D40** (RC) (1989) 3128.
4. W. Baker and D. Carey, private communication.
5. P. Aurenche et al., *Phys. Lett.* **B140** (1984) 87.

Figure Captions

Fig. 1 The leading order QCD Compton diagrams.
Fig. 2 The proposed P809 Pb-glass calorimeter.
Fig. 3 The expected photon energy asymmetry distributions for the P809 Pb-glass calorimeter in the π^o and η mass regions.
Fig. 4 The experimental result expected from P809. The quantity A_{LL} is the double spin asymmetry parameter, and the error bar represents the statistical uncertainty in its determination. The curves represent theoretical limits on reasonable values of A_{LL}.
Fig. 5 Representative data from the 1987/88 run of E706:
(a) The two photon mass distribution (note the semilog scale);
(b) The photon energy asymmetry distribution in the π^o mass region.
Fig. 6 The geometric acceptance of the E706 LAC projected onto the location of the P809 Pb-glass calorimeter.
Fig. 7 The experimental γ/π^o ratio expected for P809.
Fig. 8 Next to leading order contributions to QCD Compton scattering:
(a) Real emission diagrams;
(b) Virtual exchange diagrams.

QCD Compton Scattering

Born Diagrams

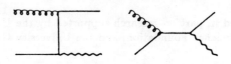

Figure 1

P809

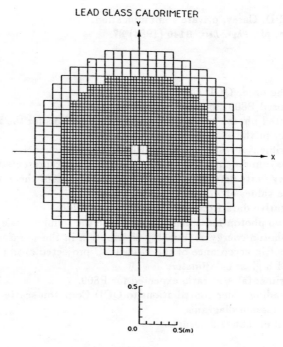

Figure 2

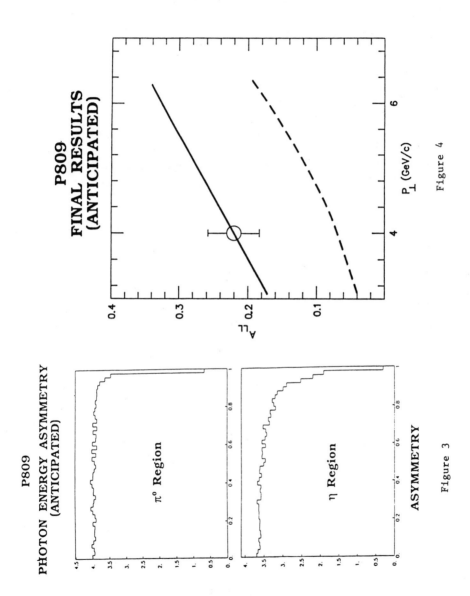

Figure 3

Figure 4

π^0 & η **DECAYS**

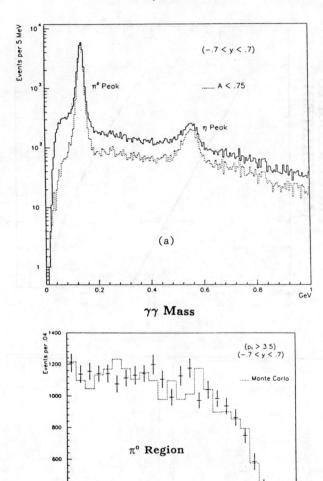

Figure 5

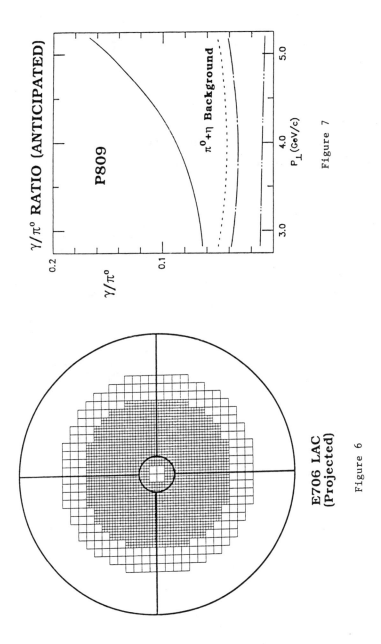

Figure 7

Figure 6

QCD Compton Scattering
Next-to-Leading Diagrams

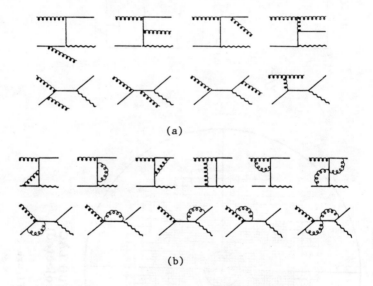

Figure 8

PHYSICS WITH TRANSVERSELY SPINNING QUARKS

X. ARTRU
*Institut de Physique Nucléaire de Lyon, IN2P3-CNRS et Université Claude Bernard
43 boulevard du 11 novembre 1918, F-69622 Villeurbanne Cedex, France*

and

M. MEKHFI
Laboratoire de Physique Théorique, Université d'Oran-Es Senia, 31, Algérie

ABSTRACT

A short introduction to transverse polarisation of quarks in the Björken limit is given. Some wrong prejudices about the existence, the observability and the evolution of transverse spin asymmetry are dissipated. A toy model where quarks remember the transverse polarisation of the baryon is presented. The low sensitivity of conventional Deep Inelastic Lepton Scattering to transverse spin is explained. Some "clean" and feasible experiments, which measure the quark transverse polarisation at leading order in α_s and $1/Q$, are proposed. The evolution of the transversely polarized quark density with Q^2 is shown to obey a simple Altarelli-Parisi equation, without coupling to the gluon density.

1. Introduction

The polarization of a quark inside a polarized baryon is specified by the three distributions[1-3] in the Björken variable x:

the unpolarized quark distribution $q(x)$,
the helicity asymmetry $\Delta q(x) \equiv q_+(x) - q_-(x)$,
the transverse spin asymmetry $\Delta_\perp q(x) \equiv q_{+\hat{n}}(x) - q_{-\hat{n}}(x)$,
where

$q_\pm(x)$ =distribution of helicity $\pm\frac{1}{2}$ quarks in a baryon of *helicity* $+\frac{1}{2}$,
$q_{\pm\hat{n}}(x)$ = distribution of quarks polarised in the *transverse* direction $\pm \hat{n}$ in a baryon polarised along $+\hat{n}$ $(q(x) = q_+(x) + q_-(x) = q_{+\hat{n}}(x) + q_{-\hat{n}}(x))$.

It is *a priori* as important to know $\Delta_1 q(x)$ as $\Delta q(x)$. If you want to describe a fish you don't know the name, you specify its length, its vertical size and its transverse size. Ignorance of the later may result in confusing a moon-fish with a globe-fish... However transverse spin in deep inelastic reactions has not been, up to now, a popular subject. This may be due to the following prejudices:

- "In the massless limit a quark cannot have but a *longitudinal* spin". Accordingly, $\Delta_1 q(x)$ would be a *higher twist* quantity.
- "Even if $\Delta_1 q(x)$ were sizeable, it would give no *observable* effect in the lowest order in α_s and $1/Q$".
- "Transverse spin is a complicated thing; its evolution with Q^2 is tricky".

We want to show here that these prejudices are wrong. The details of our calculations may be found in Ref.[4]

2. Nonvanishing of $\Delta_1 q(x)$

To the first prejudice, we reply that no rule forbids even a massless quark to be in the coherent superposition of helicity states

$$|\pm \hat{x}> = \frac{|+> \pm |->}{\sqrt{2}} ; \qquad (1)$$

which is to be interpreted as the state spinning toward $\pm\hat{x}$. For example, electrons and positrons in a storage ring can acquire a large transverse polarization despite of their small mass (in fact, the smaller the mass, the faster this process). It could be however that a quark does not remember the transverse polarisation of its parent baryon; to convince you that this is not necessarily so, let us consider the simplest toy model, where the baryon is made of a quark and an elementary *scalar* diquark; the baryon-quark-diquark vertex coupling is assumed to be

$$g\ <\hat{s}_q|V|\hat{s}_B> = g\ \bar{u}(x\vec{p}+\vec{k}_T,\hat{s}_q)\ u(\vec{p},\hat{s}_B)\ . \qquad (2)$$

The polarised quark distibutions in x are obtained by the Weisszäker-Williams method[2,5]:

$$dN_q = \frac{g^2}{16\pi^3} \times \frac{x\,dx}{1-x} d^2\vec{k}_T \ |<\hat{n}_q|V|\hat{n}_B>|^2 \left(\frac{1}{k^2-m_q^2}\right)^2, \qquad (3)$$

where the quark four-momentum k is off-mass-shell:

$$k^2 = x\,m_B^2 - (k_T^2 + x\,m_{qq}^2)/(1-x)\ .$$

We get
$$\Delta_1 q(x) = q_+(x) = (x\, m_B + m_q)^2\, f_0(x)\, ,$$
$$q_-(x) = f_2(x)\, ,$$
$$f_n(x) = \frac{1}{16\pi^2}\frac{1}{(1-x)} \int_0^\infty k_T^n\, dk_T^2 \left[\frac{g(k^2)}{k^2 - m_q^2}\right]^2 , \qquad (4)$$

where we have replaced g by a covariant vertex function $g(k^2)$ which suppresses the pole at $k^2 = m_q^2$ (for confinement) and the k_T divergence (softness assumption). Thus, in this model, $\Delta_1 q(x)$ is nonnegligible (unless $q_+(x)/q_-(x)$ is small, i.e., the quark helicity is most of the time opposite to the proton helicity; this would however imply an unlikely large $<k_T^2>$).

A similar conclusion holds if we choose a *pseudovector* diquark (corresponding to the triplet qq system). In this case, the formulas are more complicated; taking, for instance, $m_q = 0$, $m_{qq} = m_B = M$, we get

$$q_+(x) = (1-x)^{-2}[x^2 M^2 (1+x)^2 f_0(x) + 2(1+x^2)f_2(x)]\, ,$$
$$q_-(x) = f_2(x) + 2x^2 M^2 f_0(x)\, ,$$
$$\Delta_1 q(x) = -(1-x)^{-2}[x^2 M^2 (1+x)^2 f_0(x) + 4f_2(x)] \qquad (5)$$

Note that $\Delta_1 q(x)$ is positive if the diquark is scalar, negative if the diquark is pseudovector. We may take a mixture of these two models, the relative weights being chosen according to $SU(6)$.

3. Observability of quark transverse spin at leading twist

3.1. A simple example. The second prejudice can be cast in doubt once we know that, in the case of $e^+ e^- \to \mu^+ \mu^-$, transverse polarisation leads to an *azimuthal* asymmetry of the form

$$\frac{d\sigma}{d\omega} = \left(\frac{d\sigma}{d\omega}\right)_{nonpolar.} \times\, [\,1 - PP'\, A_{NN}(\theta)\, \cos 2(\phi - \phi_0)\,]\, , \qquad (6)$$

where $\vec{P}$ and $\vec{P'}$ are the transverse polarization vectors of e^+ and e^- respectively, ϕ is the azimuth of the scattering plane, ϕ_0 the azimuth of the bisector of $\vec{P}$ and $\vec{P'}$ and A_{NN} the normal-normal spin correlation parameter,

$$A_{NN}(\theta) = -\frac{\sin^2\theta}{1+\cos^2\theta}\, . \qquad (7)$$

The same type of asymmetry should exist in the Drell-Yan reaction $q\bar{q} \to \mu^+ \mu^-$, replacing $\vec{P}$ and $\vec{P'}$ for instance by $\vec{P}_q = \vec{P}_{target} \times \Delta_1 q(x)/q(x)$ and $\vec{P}_{\bar{q}} = \vec{P'}_{projectile} \times \Delta_1 \bar{q}(x')/\bar{q}(x')$.

Before looking for other possible reactions, let us have some more insight into the physics of transverse spin.

3.2. Transverse spin asymmetries and helicity conservation.

Let us first remark that, in the *helicity basis*, transverse polarization effects allway appear as *interferences* between different helicity amplitudes. Indeed, for fixed $\hat{s}_a$,

$$|<f|T|\hat{s}_a,\hat{s}_b=+\hat{x}>|^2 - |<f|T|\hat{s}_a,\hat{s}_b=-\hat{x}>|^2$$

$$=<f|T|\hat{s}_a,+><f|T|\hat{s}_a,->^* +c.c.$$

$$=<\hat{s}_a,-|T^\dagger|f><f|T|\hat{s}_a,+> +c.c. \quad . \tag{8}$$

This is an *helicity-flip* term of $ImT\#T^\dagger\ T$. It corresponds to the the exchange of a $b\bar{b}$ state of total helicity ± 1 in the *t-channel* of the unitarity diagram.

What happens then if we want to measure $\Delta_1 q(x)$ with deep inelastic lepton scattering? The corresponding unitarity ("hand bag") diagram is shown in Fig.1. The (approximate) conservation of helicity at the electromagnetic vertices suppresses helicity exchange in the t-channel, so the transverse asymmetry is only of order m_q/Q or k_T/Q. Higher twist contribute the same order. This kind of probe is therefore not very sensitive. Furthermore, transversely polarized D.I.L.S. measures the *structure function* $g_1(x) + g_2(x)$ which is related in a nonsimple way to the polarized *parton distribution* $\Delta_1 q(x)$. Thus, although $g_2(x)$ is an interesting quantity by itself, such an experiment is not a clean measure of transverse quark spin.

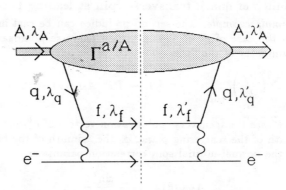

Fig.1 - Unitarity diagram for deep inelastic electron scattering. $\Gamma^{a/A}$ represents the quark density matrix.

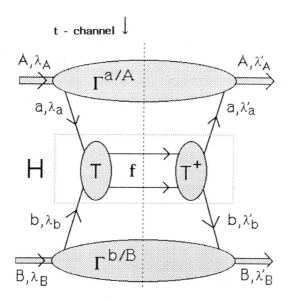

Fig.2 - Unitarity diagram for hard hadron-hadron collision; the subprocess $a+b \to f$ is framed in the rectangle H.

3.3. A first class of "clean" probes. Replace the lepton by an other quark or antiquark; the unitarity diagram is then Fig.2. The following principles will guide us:

- H conserves the t-channel helicity: a and b must be transversely polarized.
- The quark lines must connect the upper and lower part of the diagram. This can be done either in annihilation process (Fig.3a) or with the cross term in scattering of identical quarks (Fig.3b).
- If we integrate over the azimuthal angle ϕ of f, rotationnal invariance forces H to conserve J_z. On the other hand, if the t-channel helicity is ± 1, J_z changes by two units. Therefore, to get a nonzero result, we must *not* integrate over ϕ.

Under these conditions, the differential cross section writes, in a way analogous to Eq.6,

$$d\sigma = \int dx\, dy\, d\hat{\sigma}_{\text{unpol}}(\hat{s},\hat{\theta})$$

$$\times [\, a(x)\, b(y) - P_A\, P_B\, \Delta_1 a(x)\, \Delta_1 b(y)\, \hat{A}_{NN}(\hat{\theta})\, \cos 2(\phi - \phi_0)\,]\,, \qquad (9)$$

where $\hat{A}_{NN}(\hat{\theta})$ is the normal-normal spin correlation parameter of the subprocess.

The Drell-Yan mechanism already mentionned belong to the case of Fig.3a. It has $|\hat{A}_{NN}| = 1$ at $\hat{\theta} = \pi/2$, therefore $p\bar{p} \to \mu^+ \mu^- + X$ should be a good probe of the transverse quark spin. Comparable asymmetries exist for $q\bar{q} \to gluon + gluon$, $gluon + \gamma$, $\gamma + \gamma$ and $heavy\ quark\ pair$.

For identical quark scattering (case of Fig.3b)[6,7],

$$\hat{A}_{NN} = -\frac{\sin^4 \hat{\theta}}{11 + 34\cos^2 \hat{\theta} + 3\cos^4 \hat{\theta}} . \qquad (10)$$

This asymmetry should be best measured in $p + p \to two\ \pi^+\ of\ large$ and $opposite\ p_T + anything$, to enhance the valence u-quark contribution. It could also be seen in $p + p \to high\ p_T\ \pi^0$ or jet, but this would require more statistics. The smallness of $\hat{A}_{NN}$ here is partly due to a color factor which expresses the necessity of matching the colors of the two quarks.

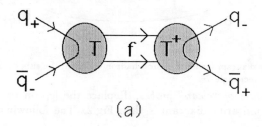

(a)

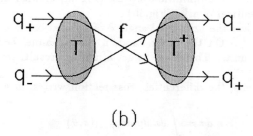

(b)

Fig.3 - Subprocesses allowing the exchange of the helicity states $|\pm 1 >$ in the t-channel; a) quark-antiquark annihilation, b) cross term in elastic scattering of two identical quarks.

3.4. A second class of "clean" probes. Let us come back to lepton scattering on transversely polarized nucleons, but *looking at a final polarized baryon.* Indeed, we can overcome the helicity argument presented in §3.2 by measuring the transverse polarisation of the ejected quark, i.e., taking $\lambda_f = -\lambda'_f$ in Fig.1. It can be done with the fragmentation into a Λ baryon; the *spin transfer parameter* B_{NN} will be proportionnal to $\Delta_1 q(x) \times \Delta_1 F_{\Lambda/q}(z)$, the second factor being the *transverse spin asymmetry of the fragmentation function*. This experiment does not require a polarized lepton beam. $\Delta_1 F_{\Lambda/q}(z) \times \Delta_1 F_{\bar{\Lambda}/\bar{q}}(z')$ could be independently measured in $e^+ e^- \to \Lambda_{\text{pol}} + \bar{\Lambda}_{\text{pol}} + X$.

3.5. Precaution. In both classes of probes, the two polarized baryons must be linked by the hard-interacting quark line and nothing else. This requires the reaction to have at least four external baryons, two of which are undetected. In particular, the mechanism in which the two baryons share the same *junction* line[8] should be avoided. This mechanism may be dominant at medium invariant mass or transfer between the two baryons, but is suppressed by an extra power of this mass or transfer.

4. Evolution of $\Delta_1 q(x, Q^2)$

Even for flavor-singlet t-channel, $\Delta_1 q(x, Q^2)$ evolves uncoupled with the gluon distributions, unlike $q(x, Q^2)$ and $\Delta q(x, Q^2)$. This is due to the fact that a two-gluon state in the t-channel would be a barrier for helicity ± 1 exchange. The Gribov-Lipatov-Altarelli-Parisi kernel and its moments are

$$\Delta_1 P(z) = \frac{4}{3}\left[\frac{2}{(1-z)_+} - 2 + \frac{3}{2}\delta(z-1)\right] \tag{11}$$

$$\Delta_1 P_n = \frac{4}{3}\left(\frac{3}{2} - 2\sum_{j=1}^{n}\right). \tag{12}$$

All the moments of $\Delta_1 q$ are decreasing; we see no obvious conserved quantity, despite the fact that sum rules have been proposed for $g_2(x)$ [9] or $\Delta_1 q(x)$ [10] (However, we do not find agreement with the $\Delta_1 P(z)$ of Ref. [10]).

5. Conclusion

The transversely polarised quark distribution, not yet explored, should bring as much information on hadronic structure as the longitudinal one, without beeing redundant. It is *a priori* sizeable, theoretically simple, and not coupled to gluon distributions. The ordinary polarized deep inelastic lepton scattering experiments are not sensitive to it at first order in α_s and $1/Q$, but we have the following clean experimental probes:
- $p_{\text{pol}} + \bar{p}_{\text{pol}} \to 2\,\pi^+$ at opposite side,
- $p_{\text{pol}} + \bar{p}_{\text{pol}} \to l^+\,l^-$, *gamma + gluon jet, 2 gluon jets, heavy quark pair*,
- $l^\pm + p_{\text{pol}} \to l^\pm + \Lambda_{\text{pol}} + anything$,
- $e^+\,e^- \to \Lambda_{\text{pol}} + \bar{\Lambda}_{\text{pol}} + anything$.

In the cases *polarized beam + polarized target*, the asymmetry will be in $\cos(\phi_A + \phi_B)$, where ϕ_A and ϕ_B are the spin orientations with respect to the scattering plane.

Let us mention that quite analogous effects exist for gluons, replacing *"transverse spin"* by *"linear polarisation"* [4,11]

6. References

1. R.P. Feynman: *Photon-hadron interactions*, New York, Benjamin 1972.
2. G. Altarelli, G. Parisi: *Nucl. Phys.* **B126** (1977) 298, and references therein.
3. Yu L. Dokshitzer, D.I. Dyakonov, S.I. Troyan: *Phys. Rep.* **58** (1980), and references therein.
4. X. Artru, M. Mekhfi: Z. Phys. C - *Particles and Fields* **45** (1990) 669.
5. P. Kessler, *Nuovo Cimento* **17** (1960) 809.
6. C.K. Chen: *Phys. Rev. Lett.* **41** (1978) 1440.
7. K. Hidaka, E. Monsay, D. Sivers: *Phys. Rev.* **D19** (1979) 1503.
8. X. Artru, M. Mekhfi: *Phys. Rev.* **D22** (1980) 751.
9. H. Burkhardt, W.W. Cottingham: *Ann. Phys.(NY)* **56** (1970) 453.
10. I. Antoniadis, C. Kounas: *Phys. Rev.* **D24** (1981) 505.
11. R. L. Jaffe, A. Manohar: *Nucl. Phys.* **B321** (1989) 343.

THEORY

Parton Distribution Functions and Higher Order Corrections

R. K. Ellis
Fermi National Accelerator Laboratory
P.O. Box 500, Batavia, Illinois 60510, U.S.A.

Abstract

The higher order contributions to various QCD processes are considered. The terms due to the exchange of an eikonal gluon between the incoming partons are studied. These terms dominate the high energy behaviour. This allows the extraction of the impact factor for each of these processes and hence the strength of their lowest order couplings to the QCD pomeron.

1. Introduction

In this paper I describe work[1] done in collaboration with D. A. Ross, the aim of which is to extract common features of next to leading order corrections in the high energy limit. I shall consider three processes of phenomenological interest which occur in lowest order without gluon exchange in the t channel. For these processes gluon exchange in the t channel occurs only as a higher order correction in the first non-leading order. For large values of s this gluon exchange is the dominant part of the higher order contribution (at least at the parton level). This term is the most important at high energies since, unlike the leading order expression, it does not fall off like $1/s$. The dimensions of the cross-section are carried by some other quantity, such as the transverse momentum or heavy quark mass, which remains fixed as s becomes large.

I first review the hard scattering formalism to define notation. In calculating the contribution from gluon exchange and integrating over the phase space of the outgoing partons, one encounters a collinear divergence when the exchanged gluon goes on its mass shell. This collinear divergence is subtracted in the normal fashion[2] and absorbed into the parton distribution functions. Thus the perturbatively calculated cross section σ for a process initiated by partons of type i and j may be written as,

$$\sigma_{ij}(s,m^2) = \sum_{i'j'} \int dx_1 dx_2 \, \hat{\sigma}_{i'j'}(x_1 x_2 s, m^2) \, \Gamma_{i'i}(x_1,\epsilon) \Gamma_{j'j}(x_2,\epsilon) \tag{1.1}$$

The short distance cross section $\hat{\sigma}$ is free from collinear singularities and is calculable as a perturbation series in the running coupling constant. The collinear singularities have been regulated by continuation of the dimension of space-time to $n = 4 - 2\epsilon$ dimensions. For example, in the $\overline{MS}$ factorisation scheme[5] the factorisation piece Γ is defined to be,

$$\Gamma_{ij}(x,\epsilon) = \left[\delta_{ij}\delta(1-x) - \frac{1}{\bar{\epsilon}}\frac{\alpha_S}{2\pi}P_{ij}(x) + O(\alpha_S^2)\right] \tag{1.2}$$

where

$$\frac{1}{\bar{\epsilon}} = \frac{1}{\epsilon} + \ln 4\pi - \gamma_E. \tag{1.3}$$

P_{ij} is the normal Altarelli-Parisi function[6] and γ_E is the Euler constant. For a process with an observed massless parton in the final state, $p_1 + p_2 \to p_3 + X$, the factorisation formula becomes,

$$\frac{E_3 d\sigma_{ij}^k}{d^{n-1}p_3} = \sum_{i'j'k'} \int dx_1 dx_2 \frac{dx_3}{x_3^2} \, \Gamma_{kk'}(x_3,\epsilon) \frac{\hat{E}_3 d\hat{\sigma}_{i'j'}^{k'}}{d^{n-1}\hat{p}_3} \Gamma_{i'i}(x_1,\epsilon) \Gamma_{j'j}(x_2,\epsilon) \tag{1.4}$$

The short distance cross section $\hat{\sigma}$ is evaluated at rescaled values of the parton momenta, $\hat{p}_1 = x_1 p_1, \hat{p}_2 = x_2 p_2, \hat{p}_3 = p_3/x_3$.

The finite parts of the factorisation piece, Γ, are subject to a prescription ambiguity. One has the freedom to subtract any finite part along with the collinear divergence. At high energy we find that most of the leading power higher order corrections to the processes considered here can be removed by a suitable choice of this finite part. What this means, in effect, is that factorisation prescription dependence allows one to reduce considerably the coupling of the perturbative pomeron[3,4] to these hard scattering processes. This leads to a gluon distribution which is different from the distribution defined in the $\overline{MS}$ factorisation scheme at small values of x.

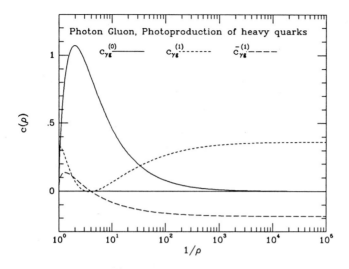

Figure 1: Photon-gluon coefficients defined by Eq. (2.1)

2. Results of explicit calculations

In this section I present the results of explicit calculations of the high energy behaviour.

2.1 High energy behaviour of heavy quark photoproduction

I shall begin by studying the case of the photoproduction of heavy quarks. In perturbation theory the short distance cross section for the photoproduction of heavy quarks of mass m and electric charge e_Q may be written as,

$$\hat{\sigma}_{\gamma j} = \frac{e_Q^2 \alpha \alpha_S(\mu^2)}{m^2} \left\{ c_{\gamma j}^{(0)}(\rho) + 4\pi \alpha_S \left[c_{\gamma j}^{(1)}(\rho) + \ln\left(\frac{\mu^2}{m^2}\right) \bar{c}_{\gamma j}^{(1)}(\rho) \right] + \ldots \right\}. \qquad (2.1)$$

where $\rho = 4m^2/s$ and s is the square of the total photon-parton centre of mass energy. The label j indicates the type of the incoming parton and μ is the factorisation and renormalisation point. Analytic expressions for the functions $c^{(0)}$ and $\bar{c}^{(1)}$ can be found in ref. [7]. An analytic expression for the function $c^{(1)}$ is not available, although in ref. [7] a fit to numerical results in the $\overline{MS}$ scheme is provided. Numerical results for $c^{(1)}$ for the photon-gluon process are shown in Fig. 1 as a function of ρ. As expected the non-leading correction has a completely different high energy behaviour than the leading term.

2.2 High energy behaviour of heavy quark production

I now consider the high energy behaviour of the hadroproduction of heavy quarks. The short distance cross section for the hadroproduction of heavy quarks may be written

$$\hat{\sigma}_{ij} = \frac{\alpha_S^2(\mu^2)}{m^2} \left\{ f_{ij}^{(0)}(\rho) + 4\pi\alpha_S \left[f_{ij}^{(1)}(\rho) + \ln\left(\frac{\mu^2}{m^2}\right) \overline{f}_{ij}^{(1)}(\rho) \right] + \ldots \right\} \quad (2.2)$$

The labels i and j indicate the type of the incoming partons and μ is again the factorisation and renormalisation scale. Analytic expressions for the functions $f^{(0)}$ and $\overline{f}^{(1)}$ can be found in ref. [8]. A plot of the numerical results for gluon-gluon scattering in the $\overline{MS}$ scheme is shown in Fig. 2 taken from ref. [8]. Note that the results of ref. [8] for the gluon-gluon initiated process shown in Fig. 2 have been confirmed by Beenakker *et al.*[9]. As can be seen the leading order contributions, ($f^{(0)}$) vanish in the high s limit, so for sufficiently large energies the parton cross sections are dominated by the higher order diagrams which involve the exchange of a spin one gluon in the t-channel.

It is clear that the high-energy behaviour of heavy quark production is dominated by these terms at the parton level. I now address the question of whether these terms are of importance in physical hadron hadron scattering, after the convolution with the rapidly falling gluon distribution functions. In order to make a crude estimate I take the gluon distribution to be given by the simple scaling form,

$$g(x) = \frac{3}{x}(1-x)^5 \quad (2.3)$$

As an example I shall consider bottom quark production at high energy. Using the gluon distribution in Eq. (2.3) and the results of 8, I obtain the total bottom production cross section shown in Fig. 3. The $O(\alpha_S^2)$ and $O(\alpha_S^3)$ estimates give vastly disparate results for the value of the cross section. I shall now show that the majority of the difference between the two orders is due to the high energy region

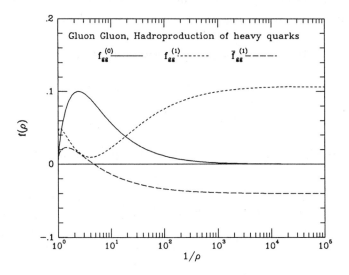

Figure 2: Gluon-gluon coefficients defined by Eq. (2.2)

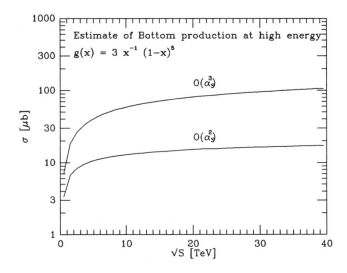

Figure 3: Estimate of the growth of the bottom cross-section with energy

and the differing high energy behaviour of the parton level cross-sections. To assess the importance of the various regions I define a parton flux function.

$$\Phi(\tau,\mu) = \tau \int_0^1 dx_1 dx_2 \, g(x_1,\mu) g(x_2,\mu) \delta(x_1 x_2 - \tau) \tag{2.4}$$

In terms of this flux the result for the hadronic cross section is written

$$\sigma(S,m^2) = \int_{\rho^H}^1 \frac{d\tau}{\tau} \Phi(\tau,\mu) f\left(\frac{\rho^H}{\tau}, \frac{\mu^2}{m^2}\right) \tag{2.5}$$

where $\rho^H = 4m^2/S$. The contributions of the leading and next-to-leading order terms are,

$$L(\tau) = \Phi(\tau,\mu) f^{(0)}\left(\frac{\rho^H}{\tau}\right) \tag{2.6}$$

$$H(\tau) = 4\pi\alpha_S \Phi(\tau,\mu) f^{(1)}\left(\frac{\rho^H}{\tau}\right) \tag{2.7}$$

This allows the total cross section to be written as

$$\sigma(S,m^2) = \frac{\alpha_S^2}{m^2} \int_{\rho^H}^1 \frac{d\tau}{\tau} \Big[L(\tau) + H(\tau) \Big] \tag{2.8}$$

where for simplicity I have taken $\mu = m$. The contribution of these two terms to the cross section can be read off from Fig. 4. The high energy region in the parton cross section is responsible for a large fraction of the hadronic cross section even at Tevatron energies. This provides sufficient motivation to study the high energy behaviour of parton cross sections.

I stress that Fig. 3 is not meant to give a good estimate of the bottom cross-section at supercollider energies, but rather to indicate that theory of heavy quark production as it is currently formulated is inadequate to describe this energy regime.

2.3 High energy behaviour of direct photon production

The total cross section for the production of direct photons contains a mass singularity, so we consider the cross section for the production of direct photons with transverse momentum $\vec{k}_T$ larger than some fixed value $\vec{p}_T$. The short distance cross

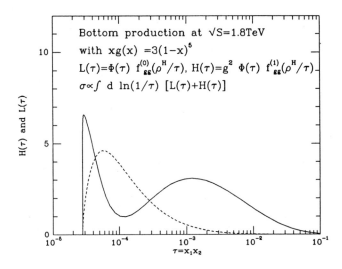

Figure 4: Leading and next-to-leading contributions to bottom production

section for this process is given in perturbation theory by,

$$\hat{\sigma}_{ij}(\vec{k}_T > \vec{p}_T) = \frac{\alpha \alpha_S(\mu^2)}{\vec{p}_T^2} \left\{ e_{ij}^{(0)}(\rho) + 4\pi\alpha_S \left[e_{ij}^{(1)}(\rho) + \ln\left(\frac{\mu^2}{\vec{p}_T^2}\right) \bar{e}_{ij}^{(1)}(\rho) \right] + \ldots \right\} \quad (2.9)$$

where

$$\rho = \frac{4\vec{p}_T^2}{s} \quad (2.10)$$

In non-leading order, $O(\alpha \alpha_S^2)$, the terms giving the leading power behaviour in s involve the exchange of a gluon in the t channel. The complete form of the functions $e_{qg}(\rho)$ for all values of ρ is shown in Fig. 5. The quark charge has been set equal to unity. This figure is obtained by numerical integration of the results of Aurenche et al.[13].

In the following section we describe the calculation of the high energy asymptote of the graphs with gluon exchange in the t-channel.

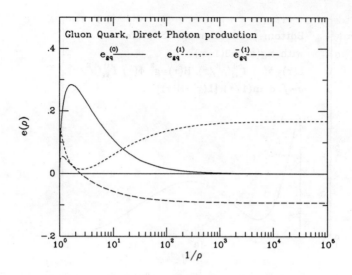

Figure 5: Plot of direct photon production cross-section

3. Impact factors

The general result for the diagrams with one gluon exchange in the t-channel can be written as,

$$\sigma = \frac{1}{2s} \int d^{(n)} \text{PS}_3 \sum \left| \frac{U_\mu g^{\mu\lambda} L_\lambda}{q^2} \right|^2 \tag{3.1}$$

where U_μ and L_λ are the upper and lower parts of the graph and the factor $g^{\mu\lambda}/q^2$ is the propagator of the exchanged eikonal gluon. In the high energy limit the dominant contribution to the polarisation sum of the exchanged gluon is given by the replacement,

$$g^{\mu\lambda} \to \frac{2p_2^\mu p_1^\lambda}{s}, \quad s = 2p_1 \cdot p_2. \tag{3.2}$$

where p_1 and p_2 are the momenta of the upper and lower lines respectively. The calculation of the asymptotic terms in diagrams with t-channel gluon exchange has already been described in a somewhat different context in refs. [10,11]. Following ref. [11] we perform a Sudakov decomposition of the momenta in the phase space integral.

$$p_3^\mu = \alpha_3 p_1^\mu + \beta_3 p_2^\mu + t_3^\mu$$

$$q^\mu = \alpha p_1^\mu + \beta p_2^\mu + q_T^\mu \qquad (3.3)$$

where $p_1 \cdot q_T = p_2 \cdot q_T = p_1 \cdot t_3 = p_2 \cdot t_3 = 0$. The dominant contribution to the cross-section at large s comes from the region $\alpha \approx m^2/s, \beta \approx m^2/s$ and $q^2 \approx -\vec{q}_T^2$.

We shall define the function I as follows,

$$I = \frac{1}{16\pi^3 s} \int d(p_1 \cdot q) \int d^4 p_3 d^4 p_4 \delta(p_3^2 - m^2) \delta(p_4^2 - m^2) \delta^4(p_1 + q - p_3 - p_4)$$
$$\times \sum \left[\frac{p_2^\mu p_2^\nu}{p_1 \cdot p_2} U_\mu U_\nu^* \right]. \qquad (3.4)$$

The momenta of the final state heavy quarks are denoted by p_3 and p_4. I describes the coupling of the eikonal gluon to the upper vertex at which the heavy quark pair is produced. It is closely related to the function known in the literature as the impact factor[12]. In this paper we will refer to the function I defined by Eq. (3.4) as the impact factor.

We now calculate the heavy quark photoproduction impact factor I_γ, as defined by Eq. (3.4). The calculation is performed using the Sudakov decomposition, Eq. (3.3), to define integration variables. The upper limit of the transverse momentum integral may be extended to infinity, if we neglect terms which are power suppressed at high energy. In this approximation we may combine denominators using a Feynman parameter x and shift the transverse momentum integral to obtain,

$$I_\gamma^{AB}(\vec{q}_T^2, m^2) = \frac{1}{2} \delta^{AB} e_Q^2 \alpha \alpha_S \int d\vec{t}^2 \int_0^1 d\alpha_3 A(\frac{\vec{q}_T^2}{m^2}, \alpha_3, \vec{t}^2). \qquad (3.5)$$

The indices A and B are the color indices of the exchanged gluon. The function A is defined as follows

$$A(\frac{\vec{q}_T^2}{m^2}, \alpha_3, \vec{t}^2) = \int_0^1 dx \left\{ \frac{\vec{q}_T^2[1 - 2\alpha_3(1-\alpha_3)(1-2x(1-x))]}{[m^2 + x(1-x)\vec{q}_T^2 + \vec{t}^2]^2} \right.$$
$$\left. -4\alpha_3(1-\alpha_3) \left[\frac{m^2 + x(1-x)\vec{q}_T^2}{[m^2 + x(1-x)\vec{q}_T^2 + \vec{t}^2]^2} - \frac{m^2}{[m^2 + \vec{t}^2]^2} \right] \right\} \qquad (3.6)$$

and $\vec{t}$ is a Euclidean vector in the transverse space, ($\vec{t}^2 > 0$). Note that the function A vanishes for small $\vec{q}_T^2$.

For compactness of notation we define the scaled impact factor j_γ as follows,

$$\overline{I}_\gamma^{AB}(\vec{q}_T^2, m^2) = \delta^{AB} \frac{e_Q^2 \alpha \alpha_S}{2} \frac{\vec{q}_T^2}{m^2} j_\gamma(\frac{\vec{q}_T^2}{m^2}) \tag{3.7}$$

The impact factor of the photon is obtained from Eq. (3.5) after performing two simple integrations.

$$j_\gamma(a) = \frac{2}{3} \int_0^1 dx \frac{1 + x(1-x)}{[1 + x(1-x)a]}. \tag{3.8}$$

Eq. (3.8) is in agreement with the results of ref. [10]. After integration over the Feynman parameter x, the impact factor of the photon in heavy quark production can be written as,

$$j_\gamma\left(\frac{\vec{q}_T^2}{m^2}\right) = \frac{2}{3} \frac{m^2}{\vec{q}_T^2} \left[\frac{\vec{q}_T^2 - m^2}{m^2} F\left(\frac{4m^2}{\vec{q}_T^2}\right) + 1\right]. \tag{3.9}$$

The function F is defined as

$$F(\rho) = \int_0^1 dx \frac{\rho}{\rho + 4x(1-x)} = \frac{\rho}{2\sqrt{1+\rho}} \ln\left(\frac{\sqrt{1+\rho}+1}{\sqrt{1+\rho}-1}\right). \tag{3.10}$$

The power series expansion of F for large and small ρ is,

$$F(\rho) \to 1 - \frac{2}{3\rho} + \frac{8}{15\rho^2} - \frac{16}{35\rho^3} + O(1/\rho^4)$$

$$F(\rho) \to -\frac{\rho}{2} \ln\left(\frac{\rho}{4}\right) + O(\rho^2) \tag{3.11}$$

It is convenient to subtract the impact factor at $\vec{q}_T = 0$ to remove the collinear singularity. After subtraction the weighted integral of j_γ is given by,

$$\int_0^\infty \frac{da}{a} \left[j_\gamma(a) - j_\gamma(0) \theta(\mu^2 - am^2)\right] = \left[\frac{41}{27} - \frac{7}{9} \ln\left(\frac{\mu^2}{m^2}\right)\right] \tag{3.12}$$

It turns out that the subtraction defined above is exactly the same as the $\overline{MS}$ bar subtraction. The final result for the short distance cross section for the photoproduction of a heavy quark is therefore given by,

$$\hat{\sigma}_{\gamma g} = \frac{e_Q^2 \alpha \alpha_s^2}{m^2} N \left[\frac{41}{27} - \ln\left(\frac{\mu^2}{m^2}\right)\frac{7}{9}\right] + O(\frac{1}{s}) \tag{3.13}$$

$N = 3$ is the number of colors and $V = 8$ is the number of gluons. This result holds in the high energy limit in the $\overline{MS}$ scheme. Eq. (3.13) is in agreement with the results for $c_{\gamma g}^{(1)}, \bar{c}_{\gamma g}^{(1)}$ as defined in Eq. (2.1) and plotted in Fig. 1 in the high energy limit ($\rho \to 0$). The corresponding result for the γq process is simply related by a factor of $V/(2N^2)$

$$\hat{\sigma}_{\gamma q} = \frac{e_Q^2 \alpha \alpha_s^2}{m^2} \frac{V}{2N} \left[\frac{41}{27} - \ln\left(\frac{\mu^2}{m^2}\right) \frac{7}{9} \right] + O(\frac{1}{s}) \qquad (3.14)$$

We now turn to the calculation of the impact factor of the gluon which is slightly more complicated because of the presence of the three gluon coupling. The calculation proceeds in a similar way to the photoproduction calculation described above. The result for the impact factor of the gluon, as defined by Eq. (3.4), can be written in terms of the Sudakov variables as follows,

$$I_g^{AB}(\vec{q}_T^2, m^2) = \frac{1}{2V} \delta^{AB} \alpha_s^2 \int d\vec{t}^2 \int_0^1 d\alpha_3 B(\frac{\vec{q}_T^2}{m^2}, \alpha_3, \vec{t}^2) \qquad (3.15)$$

where B is related to the function A defined in Eq. (3.6) above

$$B(\frac{\vec{q}_T^2}{m^2}, \alpha_3, \vec{t}^2) = \left[NA(\frac{\alpha_3^2 \vec{q}_T^2}{m^2}, \alpha_3, \vec{t}^2) - \frac{1}{2N} A(\frac{\vec{q}_T^2}{m^2}, \alpha_3, \vec{t}^2) \right]. \qquad (3.16)$$

We define the scaled impact factor by the function j_g,

$$I_g^{AB}(\vec{q}_T^2, m^2) = \frac{\delta^{AB} \alpha_s^2}{2V} \frac{\vec{q}_T^2}{m^2} j_g(\frac{\vec{q}_T^2}{m^2}) \qquad (3.17)$$

After integration j_g may also be written in terms of the function F defined in Eq. (3.10),

$$j_g(a) = \frac{2}{3a} \left\{ N \left[\frac{(a-2)(a+4)}{a} F\left(\frac{4}{a}\right) + \frac{8}{a} - \frac{10}{3} \right] - \frac{1}{2N} \left[(a-1) F\left(\frac{4}{a}\right) + 1 \right] \right\}. \qquad (3.18)$$

In the limit of small argument a we obtain for these impact factors,

$$j_\gamma(a) = \frac{7}{9} + O(a)$$

$$j_g(a) = \left[\frac{4N}{15} - \frac{1}{2N} \frac{7}{9} \right] + O(a). \qquad (3.19)$$

The weighted integral of j_g appropriate for the calculation of the cross section after subtraction of the collinear singularity is,

$$\int_0^\infty \frac{da}{a}[j_g(a) - j_g(0)\,\theta(\mu^2 - am^2)] =$$
$$N\left[\left(\frac{154}{225} - \frac{1}{2N^2}\frac{41}{27}\right) - \ln\left(\frac{\mu^2}{m^2}\right)\left(\frac{4}{15} - \frac{1}{2N^2}\frac{7}{9}\right)\right] \quad (3.20)$$

This subtraction is exactly equivalent to the $\overline{MS}$ scheme. The result for the gluon-quark short distance cross section is,

$$\hat{\sigma}_{gq} = \frac{\alpha_S^3}{m^2}\left[\left(\frac{77}{225} - \frac{1}{2N^2}\frac{41}{54}\right) - \ln\left(\frac{\mu^2}{m^2}\right)\left(\frac{2}{15} - \frac{1}{2N^2}\frac{7}{18}\right)\right] \quad (3.21)$$

Taking into account that either gluon one or gluon two can act as a source for the exchanged eikonal gluon, the gluon-gluon heavy quark production cross section is obtained by multiplying Eq. (3.21) by an overall factor.

$$\hat{\sigma}_{gg} = 2\frac{2N^2}{V}\hat{\sigma}_{gq} \quad (3.22)$$

Eqs. (3.21) and (3.22) can be seen to be consistent with the results for $f_{gg}^{(1)}, \overline{f}_{gg}^{(1)}$ plotted in Fig. 2 in the high energy limit.

The scaled impact factors are related in a simple way to the cross-sections for the interaction with a longitudinal gluon. The scaled impact factors are determined by the first moment of the cross-section $\hat{\sigma}(\rho, m^2, a)$ for an off-shell ($q^2 = -\vec{q}_T^2$) eikonal gluon scattering with an on-shell photon.

$$\frac{\pi\alpha_S\alpha e_Q^2}{m^2}j_\gamma(a) = \int_0^1 \frac{d\rho}{\rho}\,\hat{\sigma}_{\gamma g}(\rho, m^2, a), \quad \rho = \frac{4m^2}{s},\quad a = \frac{\vec{q}_T^2}{m^2}. \quad (3.23)$$

The function $\hat{\sigma}(\rho, m^2, 0)$ is the normal short distance cross-section, because the longitudunal projection, Eq. (3.2), can be related to a transverse projector using the Ward identity. The first moment of cross-section $\hat{\sigma}_{gg}$ for a single longitudinal gluon, with off-shellness $q^2 = -\vec{q}_T^2$, scattering with an on-shell gluon to produce a heavy quark pair determines j_g,

$$\frac{\pi\alpha_S^2}{Vm^2}j_g(a) = \int_0^1 \frac{d\rho}{\rho}\,\hat{\sigma}_{gg}(\rho, m^2, a), \quad \rho = \frac{4m^2}{s},\quad a = \frac{\vec{q}_T^2}{m^2}. \quad (3.24)$$

The moments of the scaled impact factors, $\tilde{j}$, are useful for the further elaboration of these results[15]. Using Eq.(3.9) we find,

$$\tilde{j}_\gamma(f) = f \int_0^\infty da\, a^{f-1} j_\gamma(a) = B(1+f, 1-f) B(1-f, 1-f) \left[\frac{(7-5f)}{3(3-2f)}\right] \quad (3.25)$$

where B is the Euler beta function. The corresponding result for gluon-gluon scattering with one line off-shell can be obtained using Eq.(3.18),

$$\tilde{j}_g(f) = f \int_0^\infty da\, a^{f-1} j_g(a) =$$

$$B(1+f, 1-f) B(1-f, 1-f) \left[2N \frac{(2-f)}{(5-2f)(3-2f)} - \frac{1}{6N}\frac{(7-5f)}{(3-2f)} \right] (3.26)$$

The scaled impact factor for direct photon production can be derived in a similar way. If we define the impact factor I_q,

$$I_q^{AB}(\vec{q}_T^2, \vec{p}_T^2) = \delta^{AB} \frac{e_q^2 \alpha \alpha_s}{2N} \frac{\vec{q}_T^2}{\vec{p}_T^2} j_q\left(\frac{\vec{q}_T^2}{\vec{p}_T^2}\right) \quad (3.27)$$

the result for the function $j_q(a)$ is,

$$j_q(a) = \frac{1}{a}\left[\left(\frac{3}{2}+\frac{1}{2a}-\ln a\right)\ln(1-a)^2 + \frac{\pi^2}{3} - 2\,\text{Li}_2(1-a) - 7 + \frac{2}{\sqrt{a}}\ln\left(\frac{1+\sqrt{a}}{1-\sqrt{a}}\right)^2\right] \quad (3.28)$$

The function j_q vanishes in the small a region,

$$j_q(a) = \frac{7}{6} + O(a) \quad (3.29)$$

The necessary weighted integral of this function, subtracted at $\vec{q}_T = 0$ is given by,

$$\int_0^\infty \frac{da}{a}\left[j_q(a) - j_q(0)\,\theta(\mu^2 - a\vec{p}_T^2)\right] = \frac{91}{36} - \frac{7}{6}\ln\frac{\mu^2}{\vec{p}_T^2} \quad (3.30)$$

In contrast to the heavy quark production processes, direct photon production has two sources of collinear divergences. One is the above–mentioned case where the exchanged gluon goes on mass shell and is removed by subtracting the Altarelli–Parisi function for a gluon emitted from a gluon convoluted with the leading order cross section. The other divergence arises when the direct photon is collinear with the emerging quark. This divergence must be absorbed into the fragmentation function of the outgoing quark and is removed by subtracting the Altarelli–Parisi function for the emission of a photon from a quark $P_{\gamma q}$ convoluted with the cross section for gluon–quark scattering.

$$P_{\gamma q}(z) = e_q^2 \left[\frac{1 + (1-z)^2}{z} \right] \tag{3.31}$$

Retaining only the leading power terms we obtain the following relation between perturbative and short distance cross section,

$$\frac{E_3 d\sigma_{qg}^\gamma}{d^{n-1}p_3} = -\frac{1}{\epsilon}\frac{\alpha_S}{2\pi}\left\{ \int \frac{dx_3}{x_3^2} P_{\gamma q}(x_3) \frac{\hat{E} d\hat{\sigma}_{qg}^q(p_3 = \hat{p}_3/x_3)}{d^{n-1}\hat{p}_3} + \int dx_2 \frac{\hat{E} d\hat{\sigma}_{qg}^\gamma(p_2 = x_2 \hat{p}_2)}{d^{n-1}\hat{p}_3} P_{gg}(x_2) \right\} \tag{3.32}$$

After including all the appropriate subtractions the final result for the short distance cross section is,

$$\hat{\sigma}_{qg}(\vec{k}_T > \vec{p}_T) = \frac{e_q^2 \alpha \alpha_S^2}{\vec{p}_T^2} \left[\frac{151}{72} - \frac{7}{6} \ln \frac{\mu^2}{\vec{p}_T^2} \right] \tag{3.33}$$

This formula allows us to extract the coefficients $e_{qg}^{(1)}(\rho)$ and $\bar{e}_{qg}^{(1)}(\rho)$, defined in Eq. (2.9), in the limit $\rho \to 0$. The results are

$$e_{qg}^{(1)}(\rho) \to \frac{151}{288\pi} e_q^2, \quad \bar{e}_{qg}^{(1)}(\rho) \to -\frac{7}{24\pi} e_q^2 \tag{3.34}$$

These results are in agreement with Fig. 5.

A plot of the scaled impact factors is shown in Fig. 6. At low $\vec{q}_T$ the plots exhibit a plateau as determined by Eqs. (3.19, 3.29). This is expected from the discussion of ref. [14]. The fact that the functions j tend to a constant for small a can also be understood physically. A low momentum gluon cannot resolve the various coloured constituents of the upper blob. When the momentum q_T becomes larger than the characteristic momentum in the upper blob, the participating partons are resolved and the cancellation between different diagrams no longer occurs. Consequently at higher values of $\vec{q}_T$ the scaled impact factors j_k make a transition to a $1/a$ behaviour (modulo logarithms). We may estimate the position of this transition by choosing the point at which the scaled impact factor j_k has dropped to half of its plateau value.

$$j_\gamma(6.65) = \tfrac{1}{2} j_\gamma(0), \quad |\vec{q}_T| = 2.6m$$

$$j_g(13.5) = \tfrac{1}{2} j_g(0), \quad |\vec{q}_T| = 3.7m$$

$$j_q(4.75) = \tfrac{1}{2} j_q(0), \quad |\vec{q}_T| = 2.2 |\vec{p}_T| \tag{3.35}$$

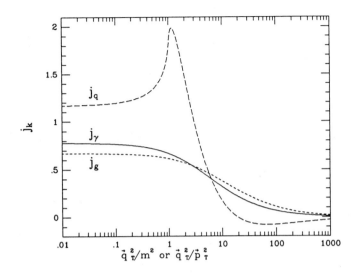

Figure 6: Scaled impact factors j_k

Therefore the scale at which the cancellation between different diagrams ceases to occur because of the injection of an appreciable momentum into the upper part of the graph is a few times the mass or $\vec{p}_T$.

4. Conclusions

We have considered the leading power corrections to three hard scattering processes. The conclusions for the total cross sections are summarised in Table 1, which reports the numerical values of the terms which govern the asymptotic behaviour, as well as the choice of μ which leads to a cancellation of the leading high energy behaviour. A possible conclusion from these results is that the most appropriate choice for μ is given by $\mu = km$ or $\mu = k|\vec{p}_T|$ where k is between 2.5 and 4. With a choice of k in this range the effect of higher order terms in the cross section is reduced. The amount of reduction can be calculated from Table 1. As a by-product of this investigation we have shown that the subtraction scale in the $\overline{MS}$ scheme has a direct physical interpretation, because the values of μ in Table 1 and in Eq. (3.35) are approximately in agreement. The disadvantage of choosing the scale to minimise the high energy correction is apparent from Figs. 1, 2 and 5. Referring back to Eqs. (2.1) and (2.2) we see that such a choice of μ increases the size of the radiative corrections near threshold. In the high energy region the short-dashed curve is positive and the long-dashed curve is negative in all three figures. Choosing μ so that these two terms cancel will lead to an enhancement at threshold, because both curves are positive in this region.

We therefore tentatively propose another possibility. Consider a process with a single incoming line i. The perturbatively calculated cross section σ and the short distance cross section $\hat{\sigma}$ are related by

$$\sigma_i = \hat{\sigma}_j \otimes \Gamma_{ji}(x, \epsilon) \tag{4.1}$$

The symbol $\otimes$ indicates the normal convolution. Some of the leading power correction to the processes considered above can be removed by modifying the factorisation pieces Γ on the incoming lines as follows,

$$\begin{aligned}
\Gamma_{qq}(x,\epsilon) &= \delta(1-x) - \frac{\alpha_S}{2\pi}\left(\frac{1}{\bar{\epsilon}}\right)P_{qq}(x) \\
\Gamma_{qg}(x,\epsilon) &= -\frac{\alpha_S}{2\pi}\left(\frac{1}{\bar{\epsilon}}\right)P_{qg}(x) \\
\Gamma_{gq}(x,\epsilon) &= -\frac{\alpha_S}{2\pi}\left(\frac{1}{\bar{\epsilon}}P_{gq}(x) - k\frac{V}{Nx}\right) \\
\Gamma_{gg}(x,\epsilon) &= \delta(1-x) - \frac{\alpha_S}{2\pi}\left(\frac{1}{\bar{\epsilon}}P_{gg}(x) - k\frac{2N}{x}\right)
\end{aligned} \tag{4.2}$$

where P are the standard one loop Altarelli-Parisi functions and

$$\frac{1}{\bar{\epsilon}} = \frac{1}{\epsilon} + \ln 4\pi - \gamma_E \tag{4.3}$$

The subtractions Γ_{qq} and Γ_{qg} are those of the standard $\overline{MS}$ scheme, but the subtractions Γ_{gq} and Γ_{gg} are new. This is a further modification of the modified minimal subtraction scheme, but it will have little effect on most phenomenology, because few precision experiments are sensitive to the gluon distribution function. A suitable value for the constant k is 2.53. In the initial state factorisation scheme defined by this value of k we obtain

$$c_{\gamma g}^{(1)} \rightarrow -0.108$$

Process	$x_{ij}^{(1)}$	$\overline{x}_{ij}^{(1)}$	$x_{ij}^{(1)}/\overline{x}_{ij}^{(1)}$	μ choice
$c_{\gamma g}^{(1)} : \gamma g \to Q\overline{Q} + X$	0.363	-0.186	-1.95	$\mu = 2.65m$
$f_{gg}^{(1)} : gg \to Q\overline{Q} + X$	0.107	-0.040	-2.69	$\mu = 3.83m$
$e_{qg}^{(1)} : qg \to \gamma q + X$	0.1669	-0.0928	-1.80	$\mu = 2.46 \,|\vec{p_T}|$

Table 1: High energy behaviour of hard scattering cross-sections as defined in Eqs. 2.1, 2.2 and 2.9.

$$f_{gg}^{(1)} \to 0.006$$
$$e_{qg}^{(1)} \to 0.049 \qquad (4.4)$$

This leads to a reduction of the coupling to the QCD pomeron of at least a factor of 3.4. This will be accompanied by a change in the higher order corrections to the gluon structure function as determined in Deep Inelastic Scattering. The effect of our proposed renormalisation scheme is to move the major part of the QCD pomeron into the low x behaviour of the gluon distribution, which is common to all semi-hard processes.

In conclusion, we see that for several cases in which the pomeron couples to hard scattering processes at next to leading order, the large power enhanced correction may be removed by an appropriate choice of scale. This choice makes sense physically at high energy, but it increases the size of the corrections in the region near threshold. A second method is to factorise the correction into the low x behaviour of the gluon distribution function. This modification of the factorisation prescription changes the anomalous dimension at two loops. For gluon distribution functions compared at approximately the same values of μ^2, this modification of the evolution properties will have little significance.

References

1) R. K. Ellis and D. A. Ross, Fermilab preprint, Fermilab-90/19-T, Nuclear Physics B (in press).

2) See, for example, R. K. Ellis, Proceedings of the 1987 Theoretical Advanced Study Institute, (R. Slansky and G. West, Editors).

3) L. V. Gribov, E. M. Levin and M. G. Ryskin, *Phys. Rep.* **100** (1983) 1 .

4) L. N. Lipatov in "Perturbative QCD", World Scientific (1989), A. H. Mueller, (editor).

5) W. A. Bardeen *et al.*, *Phys. Rev.* **D18** (1978) 3998 .

6) G. Altarelli and G. Parisi, *Nucl. Phys.* **B126** (1977) 298 .

7) R. K. Ellis and P. Nason, *Nucl. Phys.* **B312** (1989) 551 .

8) P. Nason, S. Dawson and R. K. Ellis, *Nucl. Phys.* **B303** (1988) 607 .

9) W. Beenakker *et al.*, *Phys. Rev.* **D40** (1989) 54 .

10) L. N. Lipatov and G. V. Frolov, *Sov. J. Nucl. Phys.* **13** (1971) 333 .

11) Ya. Ya. Balitskii and L. N. Lipatov, *Sov. J. Nucl. Phys.* **28** (1978) 822 .

12) H. Cheng and T. T. Wu, *Phys. Rev.* **182** (1969) 1852,1868,1873,1899 .

13) P. Aurenche *et al.*, *Nucl. Phys.* **B297** (1988) 661 .

14) J. C. Collins, D. E. Soper and G. Sterman, *Nucl. Phys.* **B263** (1986) 37 .

15) S. Catani, M. Ciafaloni and F. Hautmann, Cavendish preprint, HEP-90/3, March 1990.

RESUMMATION OF LARGE TERMS IN HARD SCATTERING

Lorenzo Magnea and George Sterman
Institute for Theoretical Physics
State University of New York, Stony Brook, NY 11794-3840

ABSTRACT

A class of the large corrections which are found at one and two loops in the Drell Yan cross section may be resummed to all orders, into a form which exponentiates in the space of moments with respect to the variable $\tau = Q^2/s$. The exponent, which is itself calculable as a power series, includes integrals of the running coupling. The bulk of the large two-loop corrections are reproduced by reexpanding the one-loop exponent, thus lending confidence to the idea that large corrections at higher loops are included in the exponentiation.

1. Large Corrections

In a typical factorization theorem, the cross section for the production of a system of mass Q^2 in the collision of two hadrons $A(p_A)$ and $B(p_B)$ with $(p_A+p_B)^2 = s = Q^2/\tau$ can be expressed as

$$\frac{d\sigma}{dQ^2} = \sigma_0 \sum_{ab} \int_\tau^1 \frac{dz}{z}\, \omega_{ab}^{(f)}(z,\alpha_s(Q^2))\, \mathcal{F}_{ab,AB}^{(f)}(\tau/z)\,, \qquad (1)$$

where the sum goes over all parton types, including quarks, antiquarks and gluons. The function $\mathcal{F}$ is a "parton luminosity",

$$\mathcal{F}_{ab,AB}^{(f)}(\tau/z) = \int dx_a dx_b\, \delta(x_a x_b - \tau/z)\, f_{a/A}(x_a,Q^2) f_{b/B}(x_b,Q^2)\,, \qquad (2)$$

with $f_{a/A}(x_a,Q^2)$ the density for partons of type a and fractional momentum x_a in hadron A. These parton densities must be detemined from experiment. Often this is done in deeply inelastic scattering (DIS), and then applied as above to cross sections with two hadrons in the final state. In this case $f = F$, a DIS structure function, and the corresponding hard part is denoted $\omega_{ab}^{(F)}$. The factor σ_0 may be taken as the Born cross section for the partonic process, in which case $\omega_{ab}^{(f)} \sim \delta(1-z)$, at lowest order in the QCD coupling. The calculation of $\omega^{(f)}(z)$ for any f is implemented entirely in perturbation theory, by calculating the cross section with the external hadrons replaced by quarks or gluons. Such a cross section is only defined by use of an infrared regulator. All regulation-dependence, however, resides in the perturbative expressions for the parton luminosity, Eq. (2), now with $h_1 = q$ and $h_2 = \bar{q}$, say. The hard scattering function, on the other hand, is independent of the nature of the h_i and also of the regulator, and once computed it may be combined with the experimentally measured parton densities to get a real hadronic cross section from Eq. (1).

To lowest order in perturbative QCD, the factorization theorem is equivalent to the parton model. Higher-order corrections in $w_{ab}^{(f)}$ predict deviations from the parton model, and here one is often faced with the problem of large perturbative corrections. In some cases, however, we can identify the origin of large corrections to perturbative cross sections, and use this knowledge to resum the perturbation series while taking them into account to all orders[1,2]. This talk will summarize the results of such a resummation for the Drell-Yan (DY) inclusive cross section $d\sigma/dQ^2$, normalized in terms of DIS structure functions. This is a "classic" case, where large corrections are present at first order in perturbative QCD[3], and are also required by the comparison of the parton model to experiment[4]. Thus, I shall describe an attempt to calculate the "K-factor", which is to be understood as the ratio of the true DY cross section $d\sigma/dQ^2$ to the parton model result.

In DY, large corrections involve only[1] the function $w_{q\bar{q}}^{(f)}(z, \alpha_s(Q^2))$ in Eq. (1), which is of the form

$$w_{q\bar{q}}^{(f)}(z, \alpha_s(Q^2)) = \delta(1-z) + \sum_{n=1}^{\infty} \left(\frac{\alpha_s(Q^2)}{\pi}\right)^n \left(\sum_{m=1}^{2n-1} C_m^{(n)} \mathcal{D}_m(z) \right. \tag{3}$$
$$\left. + C_\delta^{(n)} \delta(1-z) + b_{q\bar{q}}^{(n)}(z)\right).$$

The $b_{q\bar{q}}^{(n)}(z)$ are smooth functions of z, and the $\mathcal{D}_k(z)$ are "plus" distributions, defined by their integral with an arbitrary function $g(z)$,

$$\int_\tau^1 dz \mathcal{D}_k(z) g(z) = \int_\tau^1 dz \left(\frac{\ln^k(1-z)}{1-z}\right)_+ g(z)$$
$$\equiv \int_\tau^1 dz \left(\frac{\ln^k(1-z)}{1-z}\right) (g(z) - g(1)) + g(1)\frac{1}{k+1}\ln^{k+1}(1-\tau). \tag{4}$$

In the $w_{q\bar{q}}^{(f)}$, positive terms in the plus distribution come from real gluon emission and negative terms from virtual gluon corrections.

To one loop the hard part $\omega_{q\bar{q}}^{(F)}$ calculated with DIS distributions is given by

$$\omega_{q\bar{q}}^{(F)}(z,\alpha_s(Q^2)) = \delta(1-z) + \frac{\alpha_s(Q^2)}{2\pi} C_F[4\mathcal{D}_1(z) + 3\mathcal{D}_0(z) \\ + [4\pi^2/3 + 1]\delta(1-z) + \tilde{b}(z)] \ . \tag{5}$$

with $\tilde{b}(z)$ a smooth and relatively small function. In Eq. (1) this one-loop correction to $\omega_{q\bar{q}}^{(F)}$ and the corresponding two-loop correction[5] both give contributions of roughly the same size as the Born cross section for all values of τ. Large corrections at small tau are mostly due to the delta functions in Eq. (3), while τ dependence is mostly due to the $\mathcal{D}_k(z)$'s. The effect of the latter is greatly magnified by the behavior of the parton distributions in Eq. (2), in which the delta function fixes $z = \tau/x_a x_b$. In general the parton distributions decay very rapidly for moderate x, even of the order 0.3 or 0.4. Because z must be less than unity, τ is a lower bound for the product $x_a x_b$, which requires, roughly speaking, the x's to be of the order of the square root of τ. When τ is 0.36, for example, a typical x is of order 0.6, where the structure functions are both small and rapidly decreasing. In this way, the z integral in Eq. (1) gets most of its contribution from z close to unity, even at quite modest values of τ. Thus the plus distributions, which are singular at $z = 1$, can give large corrections even for τ quite far from unity. This accounts for the fact[5,6] that a "leading logarithm" aproximation in τ, in which $\mathcal{F}(\tau/z)$ is approximated by $\mathcal{F}(\tau)$ in Eq. (1) (see Eq. (4)) generally gives a poor approximation to the integral.

Now let us briefly discuss how the plus distributions occur at one loop in the DY cross section. Leading contributions come about when a real gluon becomes soft and nearly parallel to one of the quarks to which it attaches. In DY diagrams, these are the two incoming quarks. In DIS, the structure functions have the same two incoming quarks, but also two outgoing quarks. The divergences associated with real emission from incoming quarks exactly cancel between DY and DIS, but they leave over divergences associated with emission from the outgoing quarks in DIS. These divergences cancel with contributions from virtual corrections on final-state quarks in DIS, which have no analog in the DY process. Large corrections thus arise from regions of momentum space in DIS which are not present in DY[7]. This is the observation that makes it possible to organize these corrections, using the method sketched below.

2. Resummation

To organize and resum the large corrections associated with plus distributions, we use an alternate factorization of the quark-quark DY cross section,

$$\frac{d\sigma_{q\bar{q}}}{dQ^2} = \sigma_0 \int_\tau^1 \frac{dz}{z} \, \omega_{q\bar{q}}^{(\psi,U)}(z, \alpha_s(Q^2)) \, \mathcal{F}_{qq,qq}^{(\psi,U)}(\tau/z) \,, \tag{6}$$

where now

$$\mathcal{F}_{qq,qq}^{(\psi,U)}(\tau/z) = \int dx_a dx_b dw \, \delta(x_a x_b(1-w) - \tau/z) \\ \times \psi_{q/q}(x_a, Q^2) \psi_{q/q}(x_b, Q^2) \, U(w\sqrt{\hat{s}}) \,, \tag{7}$$

which is analogous to Eqs. (1) and (2). Here, the parton density $\psi(x)$ is defined at measured energy,

$$\psi_{q/A}(x) = \frac{1}{6\pi 2^{5/2}} \int_{-\infty}^\infty dy_0 e^{-ixp_0 y_0} < A(p)|\bar{q}(y_0, 0)\gamma^+ q(0)|A(p) > \,, \tag{8}$$

where $|A(p)>$ represents the hadronic state of momentum p^μ. The "observed" quark has a fixed energy xp^0, while its spatial momenta are integrated over. Note that ψ is neither Lorentz nor gauge invariant. We evaluate it in $A_0 = 0$ gauge[1]. Eq. (6) involves a slightly modified version of the parton luminosity, eq. (2). The function $U(x\sqrt{s})$ is a distribution for soft gluons, less their collinear divergences. More specifically, the function $U(x\sqrt{s})$ is defined by the operator relation

$$\sum_n \delta(E_n - x\sqrt{s})| < 0|T(P\exp\left[ig\int_0^\infty d\lambda v_a \cdot A(\lambda v_a)\right] \\ \times P\exp\left[ig\int_0^\infty d\lambda v_b \cdot A(\lambda v_b)\right])|n>|^2 = \\ \int dx_1 dx_2 dx_3 U(x_1\sqrt{s})\delta(x - x_1 - x_2 - x_3) \\ \times \sum_{m_a} \delta(E_{m_a} - x_2\sqrt{s})| < 0|T(P\exp\left[ig\int_0^\infty d\lambda v_a \cdot A(\lambda v_a)\right])|m_a>|^2 \\ \times \sum_{m_b} \delta(E_{m_b} - x_2\sqrt{s})| < 0|T(P\exp\left[ig\int_0^\infty d\lambda v_b \cdot A(\lambda v_b)\right])|m_b>|^2 \,. \tag{9}$$

Here $v_{a,b}$ are unit vectors in the direction of the incoming hadrons A and B. The functions ψ and U are defined to match the phase space of DY, and they exactly absorb[1] all plus distributions in the factorization of $d\sigma/dQ^2$.

This approach may be tested by computing $\omega_{q\bar{q}}^{(\psi,U)}$ to one loop,

$$\omega_{q\bar{q}}^{(\psi,U)}(z',\alpha_s(Q^2)) = \delta(1-z') + \frac{\alpha_s(Q^2)}{\pi}C_F[-2\delta(1-z') + (\frac{1+z'^2}{1-z'})\ln z' + 1] \;, \quad (10)$$

which is to be compared with Eq. (5). As expected, the new hard part is relatively small and includes no plus distributions. Although this result is suggestive, it is not immediately useful, because the function ψ is itself infrared sensitive. To make use of it, we must perform a similar decomposition for DIS.

The factorized form for a DIS structure function analogous to Eqs. (6) and (7) is

$$\begin{aligned}F(x,Q^2) =& |H_{DIS}(Q)|^2 \int_x^1 \frac{dy}{y}\phi(y) \\ & \times \int_0^{y-x}\frac{dw}{1-w}V(wQ)J((y-x-w)Q^2)\;,\end{aligned} \quad (11)$$

which is accurate up to terms which vanish in the high-moment limit. Without going into details, one may note that the function H_{DIS} is ultraviolet dominated, ϕ is analogous to ψ, and V to U, differing only in the phase space which defines them. For the DIS functions, the component of momentum parallel to the incoming hadron takes the place of the center-of-mass energy. The new feature in Eq. (11) is the "jet" subdiagram J, which absorbs collinear singularities associated with the outgoing scattered quark. As suggested above, the bulk of the large corrections are associated with this jet, since it is the feature of DIS which is lacking in DY.

The resummation is most conveniently expressed in terms of moments

$$\tilde{f}(n) = \int_0^1 dx\, x^{n-1}f(x) \quad (12)$$

with respect to $x = -q^2/2p\cdot q$ for the structure function F in DIS, and $\tau = Q^2/s$ for the DY cross section. These moments factorize the convolution form of the DY cross sections in both of Eqs. (1) and (6),

$$\begin{aligned}\int_0^1 d\tau\, \tau^{n-1}\frac{d\sigma}{dQ^2} &= \tilde{\omega}_{q\bar{q}}^{(F)}(n)\tilde{F}^2(n) \\ &= \tilde{\omega}_{q\bar{q}}^{(\psi,U)}(n)\tilde{\psi}^2(n)\tilde{U}(n)\;,\end{aligned} \quad (13)$$

and of the DIS structure function in Eq. (11),

$$\tilde{F}(n) = |H_{DIS}|^2\tilde{J}(n)\tilde{\phi}(n)\tilde{V}(n)\;. \quad (14)$$

Corrections to Eqs. (13) and (14) are of order $1/n$, and do not contribute to the distribution behavior of the inverse transforms[1].

Eqs. (13) and (14) allow us to solve for $\tilde{\omega}^{(F)}(n)$, whose inverse transform is $\omega^{(F)}(z)$,

$$\tilde{\omega}_{q\bar{q}}^{(F)}(n) = \tilde{\omega}_{q\bar{q}}^{(\psi,U)}(n) \big| \frac{\Gamma(Q^2)}{\Gamma(-Q^2)} \big|^2 [\frac{\tilde{\psi}^2(n)\tilde{U}(n)}{\tilde{\phi}^2(n)\tilde{V}(n)}]_{real} \frac{1}{|H_{DIS}|^2 \tilde{J}^2(n)\tilde{V}(n)} , \quad (15)$$

where the subscript *real* implies contributions with at least one gluon in the final state. Each of the functions in Eq. (15) satisfies an evolution equation which can be derived[1] by the methods of Sen[8] and Collins and Soper[9].

The complete expression for the moments of the short-distance function which follows from eq. (22) may now be expressed as[2]

$$\tilde{\omega}_{q\bar{q}}^{(F)}(n,Q^2) = \sigma_0 (1 + \frac{2\alpha_s(Q^2)}{\pi} C_F) \big| \frac{\Gamma(Q^2)}{\Gamma(-Q^2)} \big|^2 E(n) , \quad (16)$$

where n dependence is in the exponential form

$$E(n) = exp\{f(\alpha_s(Q)) - \int_0^1 dx \big(\frac{x^{n-1}-1}{1-x}[\int_0^x dy \frac{g_1(\alpha_s[(1-x)(1-y)Q^2])}{1-y}$$
$$+ g_2(\alpha_s[(1-x)Q^2]) + g_3(\alpha_s[(1-x)^2 Q^2]) \quad (17)$$
$$- \frac{1}{2}\beta \frac{\partial}{\partial g} f(\alpha_s[(1-x)Q^2])])\} .$$

The crucial feature of this result is that the functions f and g_i are power series in the running couplings of the indicated arguments, which organize all logarithms of Q^2 and n. The exponentiation is thus highly nontrivial. To first order, f and the g_i are given by

$$f = -\frac{3\alpha}{2\pi} C_F , g_1(\alpha_s) = g_3(\alpha_s) = \frac{2\alpha_s}{\pi} C_F , g_2(\alpha_s) = -\frac{7\alpha_s}{2\pi} C_F . \quad (18)$$

Essentially, f, g_1 and g_2 are associated with the jet function J described above, and g_3 is associated with the ratio of functions $\tilde{U}$ and $\tilde{V}$ in Eq. (15).

More recently, we have also found that, following the work of Refs. 8 and 9, one also may derive the following expression for the ratio of spacelike and timelike form factors[10],

$$\big| \frac{\Gamma(Q^2)}{\Gamma(-Q^2)} \big| = \big| \frac{i}{2} \int_0^\pi d\theta [G(\bar{\alpha}[e^{i\phi}Q^2]) + \frac{i}{2}\int_0^\theta d\phi \gamma_K(\bar{\alpha}[e^{i\phi}Q^2])]\big| \quad (19)$$

where again G and γ_K are calculable order by order in perturbation theory[11,12],

$$G = \frac{3}{2}C_F \frac{\alpha}{\pi} + \cdots,$$

$$\gamma_K = 2C_F \frac{\alpha}{\pi} + [(\frac{67}{18} - \frac{\pi^2}{6})C_A C_F - \frac{5}{9}n_f C_F](\frac{\alpha}{\pi})^2 + \cdots. \qquad (20)$$

Eqs. (16)-(20) can be inverted[2] to give an expression which includes all plus distributions in $\omega_{q\bar{q}}^{(F)}$.

3. Discussion

If we expand the running couplings in Eq. (17), we get explicit predictions for the n-dependence of $\tilde{\omega}_{q\bar{q}}^{(F)}$ at order $\alpha_s^2(Q^2)$, due to the one-loop calculation alone. Inverting the moments then gives explicit predictions for plus distributions at two loops. These may be compared with the explicit two-loop calculations described in Ref. 5.

This comparision may be made analytically, by comparing the two-loop coefficients $C_i^{(2)}$ of Eq. (3), from Eqs. (17) and (19) using one-loop input in the exponents, with those found at two loops by direct calculation[6]. It may also be done numerically, by generating an effective two-loop "K-factor" as in Fig. 1 from the expanded resummation, and comparing it with the numerical result of the direct calculation[6]. Each of these comparisons is a measure of what might be termed "intrinsic" two-loop corrections, not included in the resummation.

The resummation is exact[1] for the first two leading terms $C_3^{(2)}$ and $C_2^{(2)}$; it generates a surprising 99% of $C_1^{(2)}$, corrections being entirely from the numerically small $(\alpha/\pi)^2$ term in γ_K, Eq. (20). Finally, it generates approximately two-thirds of the third nonleading coefficient $C_0^{(2)}$ and, taking the resummation Eq. (19) of the Sudakov form factor into account, a very satisfying three quarters of the fourth nonleading $C_\delta^{(2)}$ term.

The numerical consequences of this good agreement with the two-loop coefficients are shown in Figs. 1 and 2. The figures give ratios of corrections to the cross section over zeroth order (parton model) results as a function of τ. Following Ref. 5, Fig. 1 is for pion-nucleus scattering at $\sqrt{s} = 19.1$ GeV, and Fig. 2 for proton-antiproton scattering at $\sqrt{s} = 630 GeV$. The details of the computation are described in Ref. 6. In both cases the lower solid line is the ratio computed with the exact one-loop result[3] for $\omega_{q\bar{q}}^{(F)}$, and the dashed line with the explicit one- and two-loop results of Ref. 5, while the upper solid line is computed with the expansions of Eqs. (17) and (19) up to $O(\alpha_s^2)$, using only one-loop results in the exponents. Clearly, the resummation reproduces the bulk of the two-loop K-factor.

(The differences between these curves and those given in Ref. 6 are entirely due to the inclusion here of nonabelian effects in the ratio of Sudakov form factors, via Eq. (17).) The consequences of Eq. (16) for the K-factor at all orders have been discussed in Ref. 2.

Finally, we note that Catani and Trentedue[13] have recently also shown that considerable information on nonleading logarithms may be gleaned from already existing computations of Sudakov effects. In addition, Auranche, Baier and Fontannaz[14] have used optimization techniques to relate first and second order results for the Drell-Yan cross section.

Acknowledgement This work was supported in part by the National Science Foundation under Grant No. Phys-89-08495.

References

1. G. Sterman, *Nucl. Phys.* **B281** (1987) 310.
2. D. Appel, G. Sterman and P. Mackenzie, *Nucl. Phys.* **B309** (1988) 259.
3. G. Altarelli, R. K. Ellis and G. Martinelli, *Nucl. Phys.* **B157** (1979) 461; J. Kubar-Andre and F. E. Paige, *Phys. Rev.* **D19** (1979) 221; B. Humpert and W. L. van Neerven, *Phys. Lett.* **84B** (1979) 327; K. Harada, T. Kaneko and N. Sakai, *Nucl. Phys.* **B155** (1979) 169; (E) **B165** (1980) 545.
4. I. R. Kenyon, *Rep. Prog. Phys.* **45** (1982) 1261; C. Grosso-Pilcher and M. J. Shochet, *Ann. Rev. Nucl. Part. Sci.* **36** (1986) 1.
5. T. Matsuura and W. L. van Neerven, *Z. Phys.* **C38** (1988) 623, T. Matsuura, S. C. van der Marck and W. L. van Neerven, *Phys. Lett.* **B211** (1988) 171; *Nucl. Phys.* **B319**(1989) 570.
6. L. Magnea, Stony Brook preprint ITP-SB-90-28 (1990).
7. G. Parisi, *Phys. Lett.* **90B** (1980) 295; G. Curci and M. Greco, *Phys. Lett.* 92B (1980) 175.
8. A. Sen, *Phys. Rev.* **D24** (1981) 3281.
9. J. C. Collins and D. E. Soper, *Nucl. Phys.* **B193** (1981) 381.
10. L. Magnea and G. Sterman, Stony Brook preprint ITP-SB-90-43 (1990).
11. J. Kodaira and L. Trentedue, *Phys. Lett.* **112B** (1982) 66; **123B** 335 (1983).
12. S. Catani and L. Tentedue, *Nucl. Phys.* **B327** (1989) 323.
13. J. C. Collins, in *Perturbative QCD*, ed. A. H. Mueller (World Scientific, Singapore, 1989).
14. P. Aurenche, R. Baier and M. Fontannaz, Fermilab preprint, FERMILAB-PUB-90/27-T (1990).

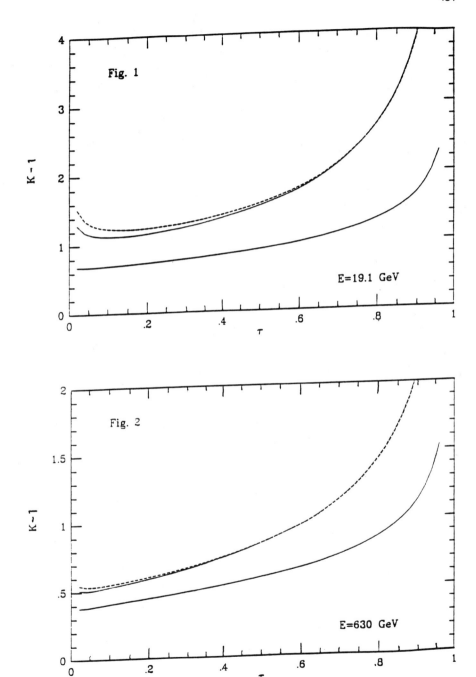

MULTI-PARTON INTERACTIONS AND INELASTIC CROSS SECTION IN HIGH ENERGY HADRONIC COLLISIONS

G. CALUCCI and D. TRELEANI

Dipartimento di Fisica Teorica dell'Università di Trieste, Trieste, I 34014 Italy
INFN, Sezione di Trieste, Italy

ABSTRACT

We discuss the leading contribution of multiple semi-hard partonic interactions to the inelastic cross section in high energy hadronic collisions. Two parton correlations are explicitly taken into account and shown to give a negligible contribution to the integrated cross section.

In high energy hadronic reactions one observes an increasingly large hard component in the interaction[1]. One evidence is the large size of the cross section for production of mini-jets that has been observe at CERN $p\bar{p}$ Collider[2]. Estimates of the rate of production of mini-jets at higher energies show that inclusive cross sections much larger than the total hadronic cross section have to be expected[3]. An inclusive cross section larger than the total one is not inconsistent, since the inclusive cross section counts the multiplicity[4], the indication is therefore that the multiplicity of mini-jets is going to increase much at large energies. One has then to keep into account the possibility of several partonic interactions with exchange of a relatively large momentum in each inelastic event[5]. While all the multiple partonic collisions cancel when evaluating the inclusive cross section[6], the complexity of the interaction shows up when looking to different physical observables[7].

In the present note we want to discuss the simplest quantity that is sensible to the presence of multiple partonic collisions, namely the semi-hard cross section σ_H. One will define as semi-hard cross section the cross section for the events where there is at least one partonic interaction with momentum transfer larger than a given cut-off, the cross section is therefore obtained summing all possible multiple partonic collisions. To discuss σ_H one needs to introduce multiple parton collisions, the problem is that they depend also on multiple parton distributions that are quantities independent of the single parton distributions appearing in the QCD-parton-model. We will show, however, that, although multiple parton distributions are independent of the single parton distributions, since they contain the correlation term, this last one is not seen when looking at the leading contribution to the semi-hard cross section.

Multiple parton collisions are of two qualitatively different kinds. The first kind consists of the disconnected processes: several pairs of partons scatter independently, the collisions being localized in different points in the transverse plane. The second kind is represented by parton rescatterings, namely a parton in the projectile can interact with exchange of relatively large transverse momentum with more than one parton of the target. The two possibilities can then combine. The process is obviously incoherent in the first case, while in the second case the situation is more complex since the two collisions are localized at the same transverse coordinate. The rescattering contribution to the integrated cross section is obtained summing over all possible discontinuities of the rescattering diagrams. As it has been shown in Ref.8, when looking to the integrated cross section, all possible cuts in the rescattering diagram are, at the leading order, proportional each other, the different weights been given by the AGK rules[9]. The sum of all the contributions will, as a consequence, amount to the introduction of an absorptive correction that is given by the iteration of the single scattering term. Introducing the probability for a parton i of the projectile to have a semi-hard interaction with a parton j of the target $\hat{\sigma}_{i,j}$ the absorptive correction to the semi-hard cross section can therefore be estimated consistently. We will then approach the problem of the semi-hard cross section in a probabilistic way.

To deal with multiple parton distributions we will use the technique of the generating functional, often used in the context of inclusive reactions[10].

We start from the exclusive n-parton distribution $W_n(u_1 \ldots u_n)$, where u_i represents the variables (b_i, x_i), being b the transverse partonic coordinate and x the corresponding fractional momentum. The scale for the distributions is given by the cut off p_t^{min} that defines the separation between soft and hard collisions. The distributions are symmetric in the variables u_i and their generating functional is defined as:

$$\mathcal{Z}[J] = \sum_n \frac{1}{n!} \int J(u_1) \ldots J(u_n) W_n(u_1 \ldots u_n) du_1 \ldots du_n. \qquad (1)$$

The conservation of the probability yields the overall normalization condition $\mathcal{Z}[1] = 1$. The logarithm of the same functional is also useful: $\mathcal{Z}[J] = e^{\mathcal{F}[J]}$, with the normalization $\mathcal{F}[1] = 0$. The many body densities, i.e. the inclusive distributions, can then be introduced in the following way:

$$D_1(u) = W_1(u) + \int W_2(u, u')du' + \frac{1}{2}\int W_3(u, u', u'')du'du'' + \ldots$$
$$= \frac{\delta \mathcal{Z}}{\delta J(u)}\bigg|_{J=1} = \frac{\delta \mathcal{F}}{\delta J(u)}\bigg|_{J=1},$$
$$D_2(u_1, u_2) = W_2(u_1, u_2) + \int W_3(u_1, u_2, u')du' + \frac{1}{2}\int W_4(u_1, u_2, u', u'')du'du'' \ldots$$
$$= \frac{\delta^2 \mathcal{Z}}{\delta J(u_1)\delta J(u_2)}\bigg|_{J=1} = \frac{\delta^2 \mathcal{F}}{\delta J(u_1)\delta J(u_2)}\bigg|_{J=1} + \frac{\delta \mathcal{F}}{\delta J(u_1)}\frac{\delta \mathcal{F}}{\delta J(u_2)}\bigg|_{J=1} \quad (2)$$

and so on. The expansion of the functional $\mathcal{F}$ around $J = 1$ gives the correlations $C_n(u_1 \ldots u_n)$, describing how much the distribution deviates from a Poisson distribution, for which in fact $C_n \equiv 0, n \geq 2$:

$$\mathcal{F}[J] = \int D_1(u)[J(u) - 1]du + \sum_{n=2}^{\infty} \frac{1}{n!}\int C_n(u_1 \ldots u_n)[J(u_1) - 1] \ldots$$
$$\ldots [J(u_n) - 1]du_1 \ldots du_n \quad (3)$$

Given the general expressions for the multiparton distributions one can write down an expression for the semi-hard cross section that will take into account all multiple partonic interactions:

$$\sigma_H = \int d\beta \int \sum_n \frac{1}{n!} \frac{\delta}{\delta J(u_1)} \cdots \frac{\delta}{\delta J(u_n)} \mathcal{Z}_A[J]$$
$$\times \sum_m \frac{1}{m!} \frac{\delta}{\delta J'(u'_1 - \beta)} \cdots \frac{\delta}{\delta J'(u'_m - \beta)} \mathcal{Z}_B[J'] \quad (4)$$
$$\times \left\{1 - \prod_{i=1}^{n}\prod_{j=1}^{m}[1 - \hat{\sigma}_{i,j}(u, u')]\right\} \prod dudu'\bigg|_{J=J'=0}$$

where β is the impact parameter between the two interacting hadrons A and B and $\hat{\sigma}_{i,j}$ is the probability for parton i (of A) to have an hard interaction with parton j (of B). The expression for σ_H is obtained summing over all partonic configurations of the two hadrons and for each configuration requiring at least one hard interaction (the factor in curly brackets). In fact here one takes into account all multiparton interactions since any given configuration with n partons of one hadron can interact with any configuration with m partons of the other hadron. Not all multiple collisions are equally likely, however, since the elementary interaction probability $\hat{\sigma}$ is small. We will then work out of the general expression for the cross section the contribution of disconnected partonic collisions. We can therefore require that each parton interacts at most once, but, even with this simplification, the general form of the multiparton correlations makes the treatment exceedingly complicated. We will then consider the case of two body correlations only, so that in Eq.3 $\mathcal{F}[J]$ is given by the first two terms of the expansion. In order to simplify the notation we will avoid in the following writing explicitly the variables u and u' (a part of a few cases where it will be convenient for a better understanding of the argument) and the integration on the impact parameter β. We will also represent the functional derivative $\frac{\delta}{\delta J(u_i)}$ with ∂_i. The semi-hard cross section σ_H is therefore expressed as:

$$\sigma_H = \sum_n \frac{1}{n!} \partial_1 \ldots \partial_n \sum_m \frac{1}{m!} \partial'_1 \ldots \partial'_m \left\{ 1 - \prod_{i,j}^{n,m} [1 - \hat{\sigma}_{ij}] \right\} \mathcal{Z}_A[J] \mathcal{Z}_B[J'] \Big|_{J=J'=0} \quad (5)$$

We now write the term in curly brackets in Eq.5 as:

$$S \equiv 1 - exp \sum_{ij} ln(1 - \hat{\sigma}_{ij}) = 1 - exp\left[-\sum_{ij} \left(\hat{\sigma}_{ij} + \frac{1}{2} \hat{\sigma}_{ij} \hat{\sigma}_{ij} + \ldots \right) \right] \quad (6)$$

and, in order to eliminate all partonic re-interactions we perform the substitutions:

$$S \Rightarrow 1 - exp \sum_{ij} (\hat{\sigma}_{ij}) \Rightarrow \sum_{ij} \hat{\sigma}_{ij} - \frac{1}{2} \sum_{ij} \sum_{k \neq i, l \neq j} \hat{\sigma}_{ij} \hat{\sigma}_{kl} \ldots$$

Keeping into account the symmetry of the derivative operators in Eq.5 one can carry out all the sums. The expression for σ_H then becomes:

$$\sigma_H = exp(\partial) \cdot exp(\partial') \left[1 - exp(-\partial \cdot \hat{\sigma} \cdot \partial') \right] \mathcal{Z}_A[J] \mathcal{Z}_B[J'] \Big|_{J=J'=0}$$
$$= \left[1 - exp(-\partial \cdot \hat{\sigma} \cdot \partial') \right] \mathcal{Z}_A[J+1] \mathcal{Z}_B[J'+1] \Big|_{J=J'=0} \quad (7)$$

With the help of the logarithmic functional $\mathcal{F}$ one can introduce explicitly the partonic correlations. We will limit ourselves, as already anticipated, to the case where all correlation functions C_n with $n > 2$ can be neglected. Eq.7 is then given by:

$$\sigma_H = \left[1 - exp\left\{-\int dudu' \frac{\delta}{\delta J(u)} \hat{\sigma}(u,u') \frac{\delta}{\delta J(u')}\right\}\right] \cdot$$
$$\cdot exp\left\{\int D_A(u)J(u)du + \frac{1}{2}\int C_A(u,v)J(u)J(v)dudv\right\} \qquad (8)$$
$$\cdot exp\left\{\int D_B(u)J(u)du + \frac{1}{2}\int C_B(u,v)J(u)J(v)dudv\right\}\bigg|_{J=J'=0}$$

Eq.8 can be expressed in a compact way by introducing the matrices:

$$\varphi(u) \equiv \begin{pmatrix} J(u) \\ J(u) \end{pmatrix}, \qquad \eta(u) \equiv \begin{pmatrix} D_A(u) \\ D_B(u) \end{pmatrix}$$

$$\mathbf{M}(u,u') \equiv \hat{\sigma}(u,u')\begin{pmatrix} 0 & 1 \\ 1 & 0 \end{pmatrix}, \qquad \mathbf{C}(u,v) \equiv -\begin{pmatrix} C_A(u,v) & 0 \\ 0 & C_B(u,v) \end{pmatrix}$$

so that one can write

$$\sigma_H = \left[1 - exp\left\{-\frac{1}{2}\frac{\delta}{\delta\varphi}^T \mathbf{M}\frac{\delta}{\delta\varphi}\right\}\right] \cdot exp\left\{\eta^T\varphi - \frac{1}{2}\varphi^T\mathbf{C}\varphi\right\}\bigg|_{\varphi=0} \qquad (9)$$

where the upper T is for transposed. One can now eliminate the term linear in φ in the argument of the exponential by means of the shift

$$\varphi \to \chi + \mathbf{C}^{-1}\eta,$$

in such a way that the operator in square parentheses in Eq.9 will act on

$$\mathcal{R} \equiv exp\left\{-\frac{1}{2}\chi^T\mathbf{C}\chi\right\} \cdot exp\left\{\frac{1}{2}\eta^T\mathbf{C}^{-1}\eta\right\}.$$

It is useful to express $\mathcal{R}$ by means of a functional Fourier transform, in this way σ_H is given by:

$$\sigma_H = \left[1 - exp\left\{-\frac{1}{2}\frac{\delta}{\delta\chi}^T \mathbf{M}\frac{\delta}{\delta\chi}\right\}\right] \cdot$$
$$\cdot \frac{(2\pi)^{-\nu}}{\sqrt{det\mathbf{C}}} \int e^{i\lambda^T\chi} \cdot exp\left\{-\frac{1}{2}\lambda^T\mathbf{C}^{-1}\lambda\right\}d\lambda \cdot exp\left\{\frac{1}{2}\eta^T\mathbf{C}^{-1}\eta\right\}\bigg|_{\chi=-\mathbf{C}^{-1}\eta} \qquad (10)$$

where the index ν is symbolic, it remembers that the functional integral is treated as an integral over a space of finite (2ν) dimensions. The det (and the tr symbol in the following expressions) must be understood to operate both on the 2×2 matrices and in the functional u−space. Acting with the operator in square parentheses one introduces, in the integral representation, the factor $1 - exp\left(\frac{1}{2}\lambda^T \mathbf{M}\lambda\right)$. The cross section is therefore represented with a Gaussian functional integral. Performing the functional integral one obtains:

$$\sigma_H = 1 - \frac{exp\left[-\frac{1}{2}\eta^T \mathbf{C}^{-\frac{1}{2}}\left(1 - \mathbf{C}^{\frac{1}{2}}\mathbf{M}\mathbf{C}^{\frac{1}{2}}\right)^{-1}\mathbf{C}^{-\frac{1}{2}}\eta + \frac{1}{2}\eta^T \mathbf{C}^{-1}\eta\right]}{\sqrt{det(1 - \mathbf{C}^{\frac{1}{2}}\mathbf{M}\mathbf{C}^{\frac{1}{2}})}} \quad (11)$$

Eq.11 is a closed expression for the semi-hard cross section, it represents the sum of all the disconnected partonic processes (disconnected from the side of the semi-hard partonic interaction). One can analyze the physical content of Eq.11 explicitating the dependence on the parton averages D and on the correlations C. Let us look first to the exponent:

$$-\frac{1}{2}\eta^T \mathbf{C}^{-\frac{1}{2}}\left[1 - \left(1 - \mathbf{C}^{\frac{1}{2}}\mathbf{M}\mathbf{C}^{\frac{1}{2}}\right)^{-1}\right]\mathbf{C}^{-\frac{1}{2}}\eta =$$

$$= -\frac{1}{2}\eta^T \left[\mathbf{M} + \mathbf{MCM} + \mathbf{MCMCM} + \ldots\right]\eta \quad (12)$$

$$= -\frac{1}{2}\left[2 D_A \hat{\sigma} D_B - D_A \hat{\sigma} C_B \hat{\sigma} D_A - D_B \hat{\sigma} C_A \hat{\sigma} D_B + \ldots\right] \cdot$$

where all the products have to be understood as convolutions. Also the root of the determinant can be written as an exponential:

$$\frac{1}{\sqrt{det(1 - \mathbf{C}^{\frac{1}{2}}\mathbf{M}\mathbf{C}^{\frac{1}{2}})}} = exp\left[-\frac{1}{2}tr\left(\mathbf{MC} + \frac{1}{2}\mathbf{MCMC} + \ldots\right)\right] \quad (13)$$

where all the traces with an odd power of $\mathbf{MC}$ are zero The first term different from zero in the exponent in Eq.13 is:

$$tr(\mathbf{MCMC}) = \int \hat{\sigma}(u_1, u_1') C_A(u_1, u_2) \hat{\sigma}(u_2, u_2') C_B(u_1', u_2') \prod_{i=1}^{2} du_i du_i' + A \leftrightarrow B. \quad (14)$$

One has then two possible structures for the connected parton collisions, the first has the form:

$$\int D_A(u_1)\hat{\sigma}(u_1,u_1')C_B(u_1',u_2')\hat{\sigma}(u_2',u_2)C_A(u_2,u_3)\ldots$$
$$\ldots\hat{\sigma}(u_n,u_n')D_B(u_n')\prod_{i=1}^{n}du_idu_i' \qquad (15)$$

while the second is:

$$\int C_A(u,u_1)\hat{\sigma}(u_1,u_1')C_B(u_1',u_2')\hat{\sigma}(u_2',u_2)C_A(u_2,u_3)\ldots$$
$$\ldots\hat{\sigma}(u_n,u_n')C_B(u_n',u')\hat{\sigma}(u',u)dudu'\prod_{i=1}^{n}du_idu_i' \qquad (16)$$

As shown in the figure (case b) Eq.16 corresponds to a 'closed' graph while Eq.15 is an 'open' graph (case a). One can 'open' a 'closed' graph replacing one C with a DD pair. A 'closed' graph of order $\hat{\sigma}^{2n}$ can be 'opened' in $2n$ ways, so that for each 'closed' graph of order $\hat{\sigma}^{2n}$ there are $2n$ open graphs of the same order, this explains the $\frac{1}{n}$ in the argument on the exponential coming from the expansion of the square root of the determinant in Eq.13. All disconnected scatterings can be obtained expanding the exponential *.

One will notice that, when no correlations are present, everything becomes trivial and one is left only with the term $D_A\hat{\sigma}D_B$ in the exponent. Switching on two body correlations corresponds to take into account all possible structures of the kind represented in Eq.15 and in Eq.16. One then expects that the problem of including more than two body correlations will be a very difficult one to handle. On the other hand, in the limit of a large number of partonic interactions, the hadron becomes 'black' so that keeping terms of order $\hat{\sigma}^2$ in the exponent in Eq.11 is not going to be of much importance in estimating σ_H. One then expects that σ_H can be approximated well by:

$$\sigma_H = \int d\beta\left\{1 - exp\left[-\int D_1^A(u-\beta)D_1^B(u')\hat{\sigma}(u,u')dudu'\right]\right\} \qquad (17)$$

---

* We remark that the whole machinery of the functional integration has been merely used as a device to deal with the combinatorial properties of Eq.9. For this reason the calculations were carried out as if we knew that the eigenvalues of $\mathbf{C}$ are definite positive. If this is not the case we must perform a sort of continuation (in $\mathbf{C}_A$, $\mathbf{C}_B$) in Eq.12 and in Eq.13.

where no correlation is present any more, the reason being not that partonic correlations are a priori small, but rather that they appear with higher powers in the elementary interaction probability $\hat{\sigma}$. In writing Eq.17, one is actually neglecting terms of kind $D_A\hat{\sigma}C_B\hat{\sigma}D_A$ in comparison with terms of kind $D_A\hat{\sigma}D_BD_B\hat{\sigma}D_A$. That is not reasonable when correlations are important, however this attitude can be justified if the first term in the exponent, namely $D_A\hat{\sigma}D_B$, is already large enough to make the exponential close to zero. That will happen when the typical number of partonic collisions is large and the impact parameter is small. For large impact parameters the number of collisions becomes small, and neglecting terms of order $\hat{\sigma}^2$ is then reasonable. As a consequence σ_H is represented well by Eq.17 both for small values of β and for large ones. In the eikonal models for high energy hadronic interaction[11] the contribution from semi-hard partonic interactions is included writing σ_H as in Eq.17.

One nice feature of the expression for σ_H, even in the simplest case as given by Eq.17, we want to point out is that, being obtained keeping unitarity into account, it is infrared safe: when the cut-off p_t^{min} becomes smaller and smaller, while the argument of the exponent becomes very large, σ_H, as given by Eq.17, will rather tend to a finite limiting value. The reason is that, even if the cut off and the c.m. energy are such that the average number of partons D is very large, still the argument of the exponential depends on the hadronic impact parameter β. When β is larger than some typical transverse hadronic scale R, the argument of the exponent becomes zero since there is no overlap between the matter distributions of the two interacting hadrons. σ_H is therefore a regular function of the cut off p_t^{min} for $p_t^{min} \to 0$, its limiting value being πR^2.

References

1. UA5 Coll. G. Alpgard et al., *Phys. Lett.* **121B** (1983) 209; UA5 Coll. G. Alner et al. , Phys. Lett. **138B** (1984) 304; G. Pancheri and Y.N. Srivastava, *Phys. Lett.* **159B** (1985) 69 ; T.K. Gaisser and F. Halzen, *Phys. Rev. Lett.* **54** (1985) 1754.
2. F. Ceradini et al., Study of minimum-bias-trigger events at $\sqrt{s} = 0.2 - 0.9 TeV$ with magnetic and calorimetric analysis CERN-EP/85-196, November 1985, published in *Bari Europhys. High Energy* (1985) p. 363.
3. G. Pancheri and Y.N. Srivastava, *Phys. Lett.* **182B** (1986) 199; S. Lomatçh, F.I. Olness and J.C. Collins, *Nucl. Phys.* **B317** (1989)617.
4. M. Jacob and P.V. Landshoff, *Mod. Phys. Lett.* **A1** (1986) 657.
5. N. Paver and D. Treleani, *Nuovo Cimento* **A70**(1982)215; *Zeit. Phys.***C28** (1985)187.
6. J.L. Cardy and G.A. Winbow, *Phys. Lett.* **52B**(1974)95; C.E. De Tar, S.D. Ellis and P.V. Landshoff, *Nucl. Phys.* **B87** (1975) 176.
7. Ll. Ametller and D. Treleani, *Int. J. Mod. Phys.* **A3** (1988) 521.
8. L.V. Gribov, E.M. Levin and M.G. Ryskin, *Phys. Reports* **100** (1983) 1.
9. V.A. Abramovskii, V.N. Gribov and O.V. Kancheli, *Yad. Fiz.* **18** (1973) 595, (*Sov. J. Nucl. Phys.* **18** (1974) 308).
10. L.S. Brown, *Phys. Rev.* **D5** (1972)748; Y.Akiyama and S. Hori, *Progr. Theor. Phys.* **49**(1973)276.
11. L. Durand and H. Pi, *Phys. Rev. Lett.* **58**(1987)303; A. Capella, J. Tran Thanh Van and J. Kwiecinski, *Phys. Rev. Lett.* **58**(1987)2015; B. Margolis, P. Valin, M.M. Block, F. Halzen and R.S. Fletcher, *Phys. Lett.* **B213** (1988) 221; W.R. Chen and R.C. Hwa, *Phys. Rev.* **D39** (1989)179.

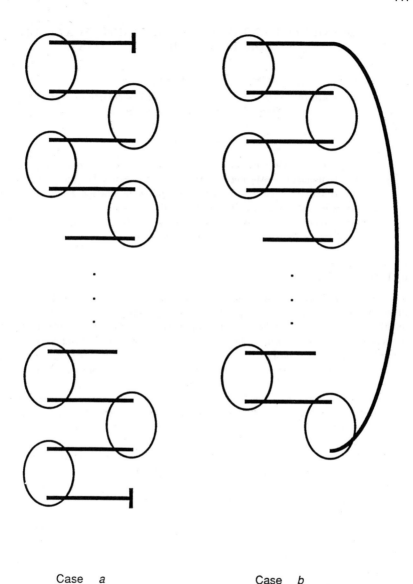

Case a Case b

Figure: Graphical representation of equations 15 and 16 in the text. The vertical lines in 'case a' indicate the average number of partons D, the bubbles the two parton correlation C and the lines connecting the bubbles represent the elementary interaction probability $\hat{\sigma}$.

$1/Q^2$ CORRECTIONS TO DEEPLY INELASTIC SCATTERING AND DRELL-YAN CROSS SECTIONS

Jianwei Qiu and George Sterman

Institute for Theoretical Physics, State University of New York
Stony Brook, New York 11794-3840

ABSTRACT

We calculate the leading $1/Q^2$ corrections to the cross sections of deeply inelastic scattering and Drell-Yan processes. We find that the experimentally measured longitudinal structure function F_L in deeply inelastic scattering can be used to predict the size of another longitudical structure function W_L in Drell-Yan. Possible enhancements of $1/Q^2$ corrections in nuclear scatterings are also discussed.

1. Introduction

For logarithmic leading twist corrections, perturbative QCD has been very successful in interpreting and predicting many different high energy experimental scattering processes. The knowledge of inverse power $1/Q^2$, or higher twist, corrections to these processes will make possible new tests of QCD. Such power-suppressed contributions have already received considerable attention both experimentally and theoretically. It has been reported in this workshop that there is clear experimental evidence for $1/Q^2$ behavior in hadron structure functions[1]. A number of theoretical calculations have also been carried out for leading inverse power corrections to structure functions in deeply inelastic scattering (DIS), by directly using the operator product expansion (OPE)[2], and by analyzing Feynman diagrams[3,4]. The results can be summerized in a form in which the perturbatively calculable short-distance coefficient functions are convoluted with nonperturbative matrix elements of operators involving multi-parton fields. These operators are different from those in the matrix elements of leading contributions. Because of these new nonperturbative matrix elements, perturbative QCD alone can not predict exactly the size of the $1/Q^2$ corrections.

Much of the predictive power of perturbative QCD is contained in factorization theorems and in the universality of the nonperturbative matrix elements[5]. Predictions follow when processes with different hard scatterings but the same nonperturbative matrix elements are compared. For example, in case of the leading contributions, the nonperturbative matrix elements, which are interpreted as quark and gluon distributions, measured in deeply inelastic scattering may be used to normalize the Drell-Yan cross section. Therefore, it is important to study $1/Q^2$ corrections in processes other than deeply inelastic scattering in order to be able to test the theory and to set the normalization. In this talk, we present our leading order calculation of $1/Q^2$ correction to the Drell-Yan cross section, and discuss the relation between nonperturbative matrix elements appearing in both DIS and Drell-Yan.

Before we start to calculate the $1/Q^2$ correction to Drell-Yan cross section, we should ask ourselves whether a factorization theorem applies to the $1/Q^2$ correction at all. If the theorem fails at this power, in principle, we will not be able to calculate such $1/Q^2$ correction consistently. It was suggested by Politzer[6] that factorization might be generally true for $1/Q^2$ corrections to hadonic processes. In particular, we believe that it is possible to give an all order argument in perturbation theory why the program of factorization, to separate the calculable short-distance dependence from nonperturbative matrix elements, can be extended to $1/Q^2$ corrections in a large class of inclusive cross sections[7]. Such an extension of factorization theorems is consistent with the known failure of naive factorization at $O(1/Q^4)$ by noncancelling infrared divergences at two loops[8]. It is also consistent with the heuristic arguments based on classical eletrodynamics[9]. All order arguments will be given elsewhere[7]. Here, we will give the result of our leading order perturbative calculation for $1/Q^2$ corrections to structure functions in DIS, and to the Drell-Yan cross section[10]. We will show explicitly the relationship between $1/Q^2$ corrections in deeply inelastic scattering and in Drell-Yan. We will also discuss the possibility of finding enhanced $1/Q^2$ corrections in hadronic nuclear scatterings.

2. The Leading $1/Q^2$ Corrections in DIS

The method that will be used to separate the *calculable* short-distance hard part from the corresponding long-distance matrix elements in Drell-Yan is a direct extension of techniques which have been applied in DIS. In most previous work, however, calculations of $1/Q^2$ corrections in DIS were carried out in $n \cdot A = 0$ gauge, with n^μ a fixed lightlike vector. To pick out a specific direction in this manner is not very natural for the Drell-Yan cross section, in which there are already two hadron momenta in the initial state. Therefore, we shall outline the calculation in a covariant gauge. Because the Drell-Yan and other hadonic processes lack an OPE description, we shall use a Feynman diagram approach.

The main difficulty encountered with the diagramic approach is that a single diagram may contribute to almost every order corrections in the $1/Q^2$ expansion. For example, the diagram in Fig. 1a can give contributions to *every* order in an $1/Q^2$ expansion of the structure functions in DIS. Therefore, the main task in the Feynman diagram approach is to identify and organize the different Q^2 contributions in every diagram, and then to isolate systematically those contributions in which one is interested. This is the process of factorization.

Fig. 1 gives the quark-gluon diagrams contributing to the leading Q^2 corrections in DIS. In fact, in Feynman gauge, diagrams with any number of gluons contribute to every order in $1/Q^2$ expansion of the hadronic tensor $W^{\mu\nu}$. However, to calculate the hard part at lowest order in α_s, we need only diagrams with at most two gluons. This is because gluons in excess of two contribute at order $1/Q^2$ only through unphysical polarizations, which produce gauge invariant matrix elements, and leave the lowest-order hard parts unchanged. There are also four-quark diagrams contributing to $1/Q^2$ correction[3,4]. It is relatively easy to separate the four-quark partonic parts from their corresponding target matrix elements. We shall therefore concentrate on the quark-gluon diagrams shown in Fig. 1.

In general, higher order diagrams of the type shown in Fig. 1 contribute at leading powers in Q^2 only through unphysical "scalar" polarizations of the gluons participating in the hard process, that is, through the component of the A^μ field proportional to the incoming momentum, while lower order diagrams contribute at inverse power corrections through possible nonvanishing "intrinsic" parton transverse momenta and off-shell spinor structure. To isolate all $1/Q^2$ contributions, our technical steps are (1) to expand both momenta and gauge fields about their components parallel to the momentum of the (initial-state) hadron with which they are associated (We call this the "collinear expansion"); (2) to expand the spinor structures of the partonic part, which will make electromagnetic gauge invariance manifest (here we employ the method of "special propagators" introduced in Ref.[4].); (3) to calculate the factorized short-distance hard part.

Following the above technical steps, we have derived the following form for the leading twist-4 contributions to $W^{\mu\nu}$,

$$W^{\mu\nu}_{twist-4} = \int dx_1 dx_2 dx \mathrm{Tr}[\hat{H}^{\mu\nu}_{\alpha\beta}(x_1 p, xp, x_2 p)\omega^\alpha_{\alpha'}\omega^\beta_{\beta'}\hat{T}^{\alpha'\beta'}(x_1, x, x_2)], \quad (1)$$

where the nonvanishing contributions to $\hat{H}^{\mu\nu}_{\alpha\beta}$ are given by the imaginary part of all tree diagrams of a forward photon-parton scattering process shown in Fig. 2, with all parton propagators not being between the two photon-quark vertices are replaced by corresponding special propagators[4]. $\hat{T}^{\alpha'\beta'}(x_1, x, x_2)$ is a two-quark-two-gluon matrix element. $\omega^\alpha_{\alpha'}$ is a projection operator designed to pick up the noncollinear components of vector indices[10]. After decoupling the remaining spinor and vector sums that link H and T, calculating the hard part, and comparing with the standard decompsition of $W^{\mu\nu}$ in terms of structure functions $F_2(x_B)$ and $F_1(x_B)$, we obtain $1/Q^2$ corrections to the structure functions as[10]

$$\frac{1}{x_B}F_2(x_B) = L.T. + \left(\frac{\Lambda^2}{Q^2}\right)\sum_q e_q^2 \int dx_1 dx dx_2 \Big\{4\delta(x - x_B)T^{(-)}_{qg}(x_1, x, x_2)$$
$$- x_B \frac{\delta(x_2 - x_B) - \delta(x_1 - x_B)}{x_2 - x_1}T^{(+)}_{qg}(x_1, x, x_2)\Big\} \quad (2)$$
$$+ \text{four-quark contribution}$$

and

$$F_L(x_B) = (\alpha_s) L.T. + \left(\frac{\Lambda^2}{Q^2}\right)\sum_q e_q^2 \int dx_1 dx dx_2 4\delta(x - x_B)T^{(-)}_{qg}(x_1, x, x_2), \quad (3)$$

where we define the longitudinal structure function $F_L \equiv (F_2 - 2x_B F_1)/x_B$. x_B is the Bjorken variable, and e_q is the fractional electric charge for quarks of flavor q, the sum running over all flavors. The *two* twist-4 matrix elements are given by

$$T^{(\pm)}_{qg}(x_1, x, x_2) = T_{qg}(x_1, x, x_2) \pm \tilde{T}_{qg}(x_1, x, x_2), \quad (4)$$

with

$$\Lambda^2 T_{qg}(x_1,x,x_2) = \int \frac{p^+ dy_1^-}{2\pi} \frac{p^+ dy_2^-}{2\pi} \frac{p^+ dy^-}{2\pi} e^{ix_1 p^+ y_1^-} e^{i(x-x_1)p^+ y^-} e^{-i(x-x_2)p^+ y_2^-}$$
$$\times \frac{d^{\alpha\beta}}{2p\cdot n} \langle p|T\{\bar{\psi}(0)\gamma\cdot n\, D_\alpha(y_2^-) D_\beta(y^-)\psi(y_1^-)\}|p\rangle,$$

$$\Lambda^2 \tilde{T}_{qg}(x_1,x,x_2) = \int \frac{p^+ dy_1^-}{2\pi} \frac{p^+ dy_2^-}{2\pi} \frac{p^+ dy^-}{2\pi} e^{ix_1 p^+ y_1^-} e^{i(x-x_1)p^+ y^-} e^{-i(x-x_2)p^+ y_2^-}$$
$$\times \frac{i\epsilon^{\alpha\beta}}{2p\cdot n} \langle p|T\{\bar{\psi}(0)\gamma\cdot n\, \gamma_5\, D_\alpha(y_2^-) D_\beta(y^-)\psi(y_1^-)\}|p\rangle.$$
(5)

We have suppressed line integrals of the scaler polarized gluon field. In Eq. (2), L.T. means the standard leading twist contribution, and in Eq. (3), (α_s) L.T. is the leading twist radiative correction.

3. The Leading $1/Q^2$ Corrections to Drell-Yan Cross Section

The procedure for calculating the leading $1/Q^2$ corrections in Drell-Yan is a direct extension of the above. We first identify the sources of all possible leading $1/Q^2$ corrections, and then factorize them into a calculable short-distance hard part times some long-distance matrix elements. The short-distance hard part is calculable, and in general is process dependent. The most important result is the universality of the matrix elements. They have the same operator expressions as those we obtained in DIS.

By ignoring higher order diagrams which contribute to the leading $1/Q^2$ corrections through eikonal lines in the matrix elements, diagrams that may contribute to the hard part of the leading $1/Q^2$ corrections in Drell-Yan are given in Fig. 3 for quark-gluon process, and similar diagrams for four-quark process.

From the standard QCD factorization theorem, the leading power contribution for the Drell-Yan process is given in general by a convolution of two standard twist-2 matrix elements and a calculable short-distance hard part. We find that the order $1/Q^2$ corrections to the leading contribution have a similar form, except one of the two twist-2 matrix elements is repalced by one twist-4 matrix element. From dimensional counting, $1/Q^2$ corrections might also get contributions from a process which has a form of two twist-3 distributions times a calculable hard part. Such process actually vanish because there is no twist-3 distribution for an unpolarized projectile or an unpolarized target. As an example, the diagram shown in Fig. 3f has the right dimension, but it does not contribute at order of $1/Q^2$.

We can group $1/Q^2$ corrections from diagrams shown in Fig. 3 into two sets depending on which side of matrix element is twist-2 or twist-4. These two sets of $1/Q^2$ corrections have the same expression if we exchange p_1 and p_2, k and l. One set of $1/Q^2$ corrections is given by diagrams shown in Fig. 3a, 3b and 3c, when the matrix element in p_2 side is a twist-2 matrix element. To derive and calculate such $1/Q^2$ corrections, we first pick up the leading twist contribution from p_2 side. We expand the parton momentum l in p_2 side around its collinear component, and keep only the leading term. We then decouple the spinor trace. In this way, we can write the contribution from diagrams shown Fig. 3a, 3b and 3c as

$$W^{\mu\nu}_{(a-c)} = \int \frac{d\eta}{\eta}\,\bar{q}(\eta)\left(w^{\mu\nu}_a(\eta p_2) + w^{\mu\nu}_b(\eta p_2) + w^{\mu\nu}_c(\eta p_2)\right), \qquad (6)$$

where $\bar{q}(\eta)$ is a standard twist-2 antiquark distribution of momentum fraction η of a hadron of momentum p_2, $w^{\mu\nu}_i(\eta p_2)$, $i = a, b, c$ are similar to contribution of three diagrams shown in Fig. 1 for DIS, up to a difference in phase space. We can then follow the same procedures used in DIS to separate the hard part from the matrix element of p_1. As a result, the $1/Q^2$ term from $w^{\mu\nu}_i(\eta p_2)$, $i = a, b, c$ is factorized into a form like that in Eq. (1). Similarly, we can carry out the same procedure for the other set of $1/Q^2$ corrections from diagrams in Figs. 3a, 3d and 3e.

Having calculated the factorized hard parts of the hadronic tensor $W^{\mu\nu}$, it is straightforward to derive the Drell-Yan cross sections $d\sigma/d^4q d\Omega$ and/or $d\sigma/d^4q$, including their leading $1/Q^2$ corrections. After integrating over the azimuthal angle of the lepton pair in the photon's rest frame, over the photon's transverse momentum q_T, and its rapidity, we obtain[10]

$$\frac{d\sigma}{dQ^2 d\cos\theta} = \frac{4\pi\alpha^2_{EM}}{9Q^2 S}\left\{W_T(1+\cos^2\theta) + W_L\sin^2\theta\right\}, \qquad (7)$$

where the helicity structure functions W_T and W_L are given by, in Collins-Soper frame[11],

$$W_T = L.T. + \frac{3}{8}\frac{\Lambda^2}{Q^2}\Bigg[\sum\int\frac{d\eta}{\eta}\bar{q}(\eta)\int dx_1 dx dx_2 \Big\{T^{(+)}_{qg}(x_1,x,x_2)$$
$$\times\left(\frac{Q^2}{\eta S}\right)\frac{\delta(x_2-\frac{Q^2}{\eta S})-\delta(x_1-\frac{Q^2}{\eta S})}{x_2-x_1}\Big\} \qquad (8)$$
$$+\eta\leftrightarrow x + \text{four-quark contribution}\Bigg],$$

$$W_L = (\alpha_s)\,L.T. + 3\frac{\Lambda^2}{Q^2}\sum\int\frac{d\eta}{\eta}\bar{q}(\eta)\int dx_1 dx dx_2\,T^{(-)}_{qg}(x_1,x,x_2)\,\delta(x-\frac{Q^2}{\eta S}) \qquad (9)$$
$$+\eta\leftrightarrow x.$$

The sums run over all possible flavors, the fractional electric charge squared for each flavor is suppressed, and two twist-4 distributions $T^{(\pm)}_{qg}$ are the same as those in DIS. The general formulas given in Eqs. (7) to (9) are consistent with the lowest order explicit calculation in pion scatterings done by Berger and Brodsky[12].

Comparing Eqs. (2) and (3) with Eqs. (8) and Eq.(9), it is clear that physical measurable quantities, such as F_T, F_L in DIS and W_T and W_L, are all factorized into a short-distance hard part and some long-distance matrix elements. Once determined, the same matrix element specifies the normalization of both processes.

4. Possible Enhancement of $1/Q^2$ Corrections in Nuclear Scatterings

It is possible that *nuclear* matrix elements of the twist-4 operators above can have A^α dependence with $\alpha > 1$ enhancement. That is, a big nucleus might be a good enviroment to see or test power corrections. The power corrections from one single hard scattering may not be able to resolve the internal nuclear structure, but the double scattering hard process may serve this purpose. By picking up the double scattering hard part[13] and the corresponding twist-4 matrix element involving multiparton fields, we find

$$\frac{1}{x_B}F_2^A(x_B)\bigg|_{DS} = \left(\frac{g^2}{2N}\right)\frac{1}{Q^2}\left[x\frac{d}{dx}C_{qg}(x,A)\right]_{x=x_B} \quad (10)$$
$$+ \text{terms linear in } A,$$

where A is the atomic weight of the target nucleus, and where the new twist-4 nuclear quark-gluon correlation function $C_{qg}(x,A)$ turns out to be

$$C_{qg}(x,A) = \int \frac{dy_1^-}{4\pi}e^{ixp^+y_1^-}\int_0^\infty dy^-\int_{-y^-}^\infty dy_2^-$$
$$\times \Big\{\langle P_A|T\{\bar\psi(0)\gamma\cdot n\, F^{+\alpha}(y^-)F_\alpha^+(y^-+y_2^-)\psi(y_1^-)\}|P_A\rangle \quad (11)$$
$$- i\epsilon^{\alpha\beta}\langle P_A|T\{\bar\psi(0)\gamma\cdot n\,\gamma_5\, F^+{}_\alpha(y^-)F_\beta^+(y^-+y_2^-)\psi(y_1^-)\}|P_A\rangle\Big\}.$$

The sum of quark flavors and the quark fractional charge square are suppressed. Similarly, for Drell-Yan, we obtain

$$\frac{d\sigma^A}{dQ^2}\bigg|_{DS} = -\frac{4\pi\alpha_{EM}^2}{9Q^2\,S}\left(\frac{g^2}{2N}\frac{1}{Q^2}\right)\int\frac{d\eta}{\eta}\bar q(\eta)\left[x\frac{d}{dx}C_{qg}(x,A)\right]_{x=Q^2/(\eta S)} \quad (12)$$
$$+ \text{terms linear in A.}$$

The correlation function $C_{qg}(x,A)$ is the same for both DIS and Drell-Yan. In models where the multiparton correlation functions $C_{qg}(x,A)$ is due to a collection of nucleons, we find that the size of the possible A^α enhanced terms are small compared to the terms linear in A, which is consistent with current experimental data. Somewhat more precise measurements, however, could still detect this effect.

5. Conclusion

We have presented the results of $1/Q^2$ corrections to structure functions F_2 and F_L measured in DIS and helicity distributions W_T and W_L measured in Drell-Yan process. We find that all these $1/Q^2$ corrections can be factorized into a perturbatively calculable short-distance hard part times a nonperturbative matrix element. These matrix elements in different processes are related. That is, once determined, they can be used to make predictions for several processes. We notice that the longitudinal distributions are a good place to see such $1/Q^2$ corrections. This is because in both W_L and F_L, leading twist appears only at order α_s, in contrast to the transverse distributions, which are zeroth order in α_s. Finally, the direct relationship between $1/Q^2$ longitudinal corrections to DIS and to Drell-Yan that we find here, is an interesting prediction of the theory and should be tested experimentally.

Acknowledgement

This work is supported in part by the National Science Foundation, Grant No. PHY-89-08495.

REFERENCES

1. A. Bodek, in this preceeding.
2. R.L. Jaffe and M. Soldate, *Phys. Lett.* **105B**, 467(1981), *Phys. Rev.* **D26**, 49(1982); R.L. Jaffe, *Nucl. Phys.* **B229**, 205(1983).
3. R.K. Ellis, W. Furmanski and R. Petronzio, *Nucl. Phys.* **B207**, 1(1982), **B212**, 29(1983).
4. J. Qiu, *Phys. Rev.* D (in press).
5. For example, A.H. Mueller, *Phys. Report* **73**, 237(1981).
6. H.D. Politzer, *Nucl. Phys.* **B172**, 349(1980).
7. J. Qiu and G. Sterman, in preparation.
8. F.T. Brandt, J. Frenkel, and J.C. Taylor, *Nucl. Phys.* **B312**, 589(1989), and references therein.
9. J.C. Collins, D.E. Soper and G. Sterman, in *"Perturbative QCD"*, edited by A.H. Mueller, World Scientific Co. 1989.
10. J. Qiu and G. Sterman, in preparation.
11. J.C. Collins and D.E. Soper, *Phys. Rev.* **D16**, 2219(1977).
12. E.L. Berger, and S.J. Brodsky, *Phys. Rev. Lett.* **42**, 940(1979).
13. K. Kastella, J. Milana and G. Sterman, *Phys. Rev.* **D39**, 2586(1989).

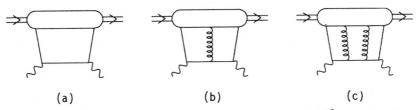

Fig. 1. Feynman diagrams relevant to the hard part of $1/Q^2$ corrections in deeply inelastic scattering.

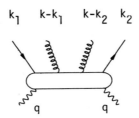

Fig. 2. Photon-parton forward scattering amplitude which gives the hard part of $1/Q^2$ corrections in deeply inelastic scattering. The blob includes all possible two-photon, two-quark and two-gluon tree diagrams.

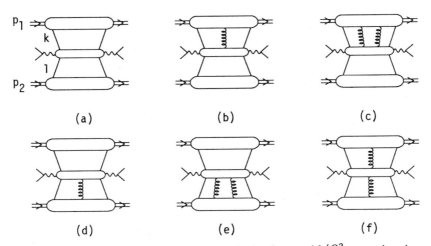

Fig. 3. Feynman diagrams relevant to the hard part of $1/Q^2$ corrections in the Drell-Yan process.

SUMMARY

Experimental Horizons for Structure Function Measurements

Günter Wolf

Deutsches Elektronen Synchrotron, DESY,

Hamburg, Germany

ABSTRACT

The data on structure functions of the nucleon span the Q^2 range up to ~ 200 GeV^2 with a systematic uncertainty of 2-3%. When HERA begins operation the available c.m. energy will increase by a factor of ten and the Q^2 range by a factor of hundred. The expansion of the kinematic region will benefit not only the nucleonic structure functions and the testing of QCD but also the measurement of more elusive structure functions such as that of the gluon, the photon and the pomeron. The high c.m. energy will open the small x region as a new field; x-values as low as 10^{-4} can be reached and saturation of parton densities may become observable.

1. Introduction

The determination of structure functions is a long and arduous task with many years of dedicated experimental and theoretical work. This is illustrated in fig.1.1 which shows the time table for some of the major lepton nucleon scattering experiments contributing today to this field [1.1]. The SLAC experiments are active since 1966. A combined analysis of the SLAC measurements has been presented at this workshop. [1.2] and has resulted in structure functions of excellent precision for Q^2 values up to ~ 30 GeV^2. With the exception of the high Q^2 end, the accuracy of the data in general is limited by systematic uncertainties, which are of the order of 2%. The SLAC data on the electromagnetic structure function F_2 are shown in fig. 1.2 together with the high precision measurements from BCDMS [1.3] which reach Q^2 ~ 200 GeV^2. The systematic error quoted

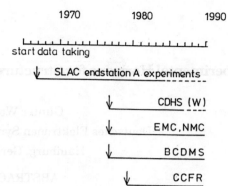

Fig. 1.1 Time chart for some structure function experiments

by BCDMS are typically around 3%. Figure 1.3 indicates the x- Q^2 region which has been explored sofar by lepton-nucleon scattering; the highest Q^2 value is ~200 GeV^2 near x= 0.7, the smallest x value for Q^2 = 10 GeV^2 is about 0.02.

By taking suitable combinations of structure functions and the help of QCD, one can extract the distribution of partons inside the nucleon. As an exapmple fig. 1.4 shows the results of Morfin and Tung [1.4] for the u, d , s and gluon distributions in the proton at Q^2 = 10 GeV^2.

The parton distributions - determined mainly from lepton - nucleon scattering and at Q^2 values between, say 1 and 200 GeV^2 - have been used to calculate [1.5] jet production in pp scattering at high transverse energies E_T (fig.1.5). Good agreement is observed between the theoretical predictions and the data from ISR, UA1, UA2, CDF which reach E_T values as high as 400 GeV. With the Q^2 scale being defined by E_T, $Q^2 = E_T^2/2$, the $\bar{p}p$ data provide an important check on the consistency of QCD with the structure functions and the parton distributions up to Q^2 values of about 80 000 GeV^2.

In the near future, a vast increase of the kinematical domain for structure functions can be expected from experiments at HERA. In the more distant future a new class of fixed target experiments, which are now under discussion for CERN and FNAL, promise high precision measurements. As an example fig. 1.6 shows the setup of the Air Toroid Experiment [1.6] proposed for µp scattering. The major improvements over existing experiments are an accurate measurement (0.1%) of the momentum difference between incoming and scattered µ and high event rates as a result of a very long (30 - 40 m) hydrogen target. The quality of the data expected for F_2 from one year of running is impressive (fig. 1.6). It

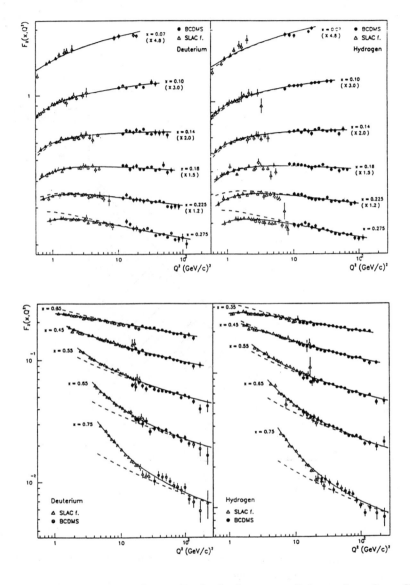

Fig. 1.2 The recent results on F_2 for hydrogen and deuterium from SLAC and BCDMS. The curves show QCD fits in NLO, dashed: perturbative part alone, solid: including higher twists. Figure taken from [1.3].

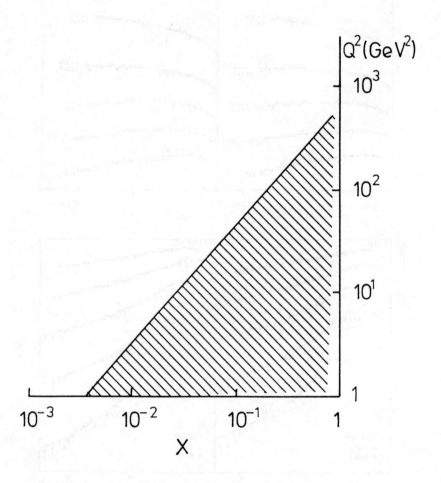

Fig.1.3 The shaded area shows the x - Q^2 for which F_2 has been measured on heavy targets.

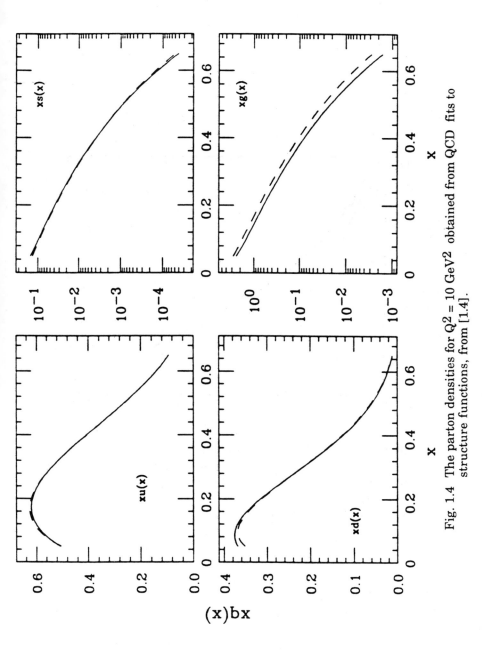

Fig. 1.4 The parton densities for $Q^2 = 10$ GeV^2 obtained from QCD fits to structure functions, from [1.4].

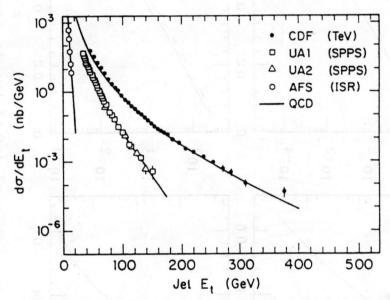

Fig. 1.5 The inclusive jet E_T cross section as measured at the ISR, $S\bar{p}pS$ and Tevatron collider. The curves show the prediction of leading order QCD with the choice $Q^2 = E_T^2/2$. Figure taken from [1.5].

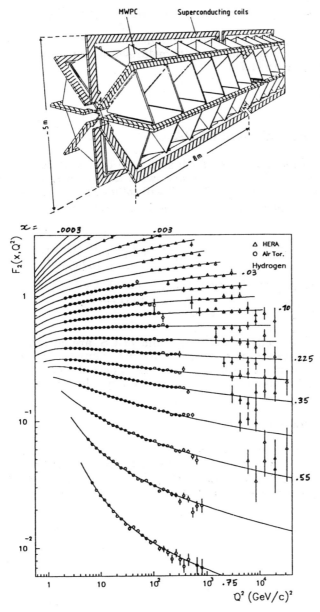

Fig. 1.6 The setup of the proposed Air Toroid experiment together with simulated F_2 data expected from this experiment and at HERA; from [1.6].

is obvious that such a project should be pursued vigorously.

The prospects for measurements of spin dependent structure functions from new experiments with polarised leptons and nucleon have been discussed by Rith [1.7].

2. Structure Functions at HERA

The Q^2 domain over which structure functions can be measured will increase with HERA by roughly two orders of magnitude [2.1]. In HERA electrons of energy $E_e = 30$ GeV will collide with protons of $E_p = 820$ GeV. Some of the characteristic parameters of HERA are:

Center of mass energy squared $s = 4\,E_e\,E_p = 10^5$ GeV^2, $\sqrt{s} = 314$ GeV
Maximum Q^2 $Q^2_{max} = s = 10^5$ GeV^2
Equivalent energy for fixed target $\nu_{max} = 52\,000$ GeV
Design luminosity $L = 1.5\ 10^{31}$ cm^{-2} s^{-1}

Two large collaborations, H1 and ZEUS, are currently installing their detectors. Data taking is expected to start in summer of next year. An integrated luminosity of 100 pb^{-1}/year is expected. For the purpose of forecasting measurements of structure function a luminosity of 500 pb^{-1} will be assumed.

2.1 Cross sections and event rates

The prime processes relevant for the extraction of structure functions are depicted in fig. 2.1: these are neutral current processes (NC) with photon and Z^0 exchange, charged current processes with $W^{\pm}$ exchange (CC) and heavy quark production by current-gluon fusion (CGF).

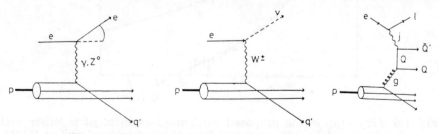

Fig.2.1   Diagrams for NC and CC scattering and current (γ,Z,W)-gluon fusion

The cross sections for the NC and CC processes are related to the structure functions F_1, F_2 and F_3 of the proton:

NC: $$\frac{d^2\sigma(\gamma+Z^0)}{dxdy} = \frac{4\pi\alpha^2}{sx^2y^2}[(1-y)F_2(x,Q^2) + y^2 x F_1(x,Q^2)]$$

CC: $$\frac{d^2\sigma}{dxdy}(e_L^- p \to \upsilon X) = \frac{G_F^2 s}{\pi} \frac{1}{(1+Q^2/M_W^2)^2}$$

$$[(1-y)F_2(x,Q^2) + y^2 x F_1(x,Q^2) + (y - y^2/2) x F_3(x,Q^2)]$$

where G_F is the Fermi coupling constant, $G_F = 1.02 \cdot 10^{-5}/M_p^2$.

The number of events expected from NC and CC scattering are shown in fig. 2.2 for 500 pb^{-1}. The large event rates at low Q^2 stem from photon exchange. At Q^2 values around the mass squared of the Z^0 (m_Z) the contribution from Z - exchange becomes equally important. The requirement of a minimum of 100 events leads to a maximum Q^2 value of 35000 GeV^2 up to which NC measurements are feasible. The event rate for CC scattering at low Q^2 is much smaller than for the NC case; at $Q^2 > m_W$, however, the CC cross section exceeds that for NC scattering. The maximum practical Q^2 value for CC is about 40 000 GeV^2.

2.2 Experimentally Accessible x - Q^2 region

The x - Q^2 region accessible to the experiments depends on the structure of the events and on the detector design. For NC events, the values of x and Q^2 can be determined from the energy and directions of *either* the scattered electron or the current jet. For CC events, where the scattered lepton is a neutrino, x and Q^2 can be measured only with the current jet. Figure 2.3 shows for the nominal beam energies the regions over which x and Q^2 can be measured well from the electron and the jet parameters, respectively. The main limitations stem from the precision with which the electron and jet energies can be measured, and from the size of the beam hole. For NC scattering, structure function measurements should be feasible for basically the full range of x and Q^2 over which there is a sufficient event rate. In the case of CC scattering precise measurements will be difficult for $y > 0.6$ ($y = Q^2/xs$). The well measurable region can be extended by operating HERA at smaller beam energies.

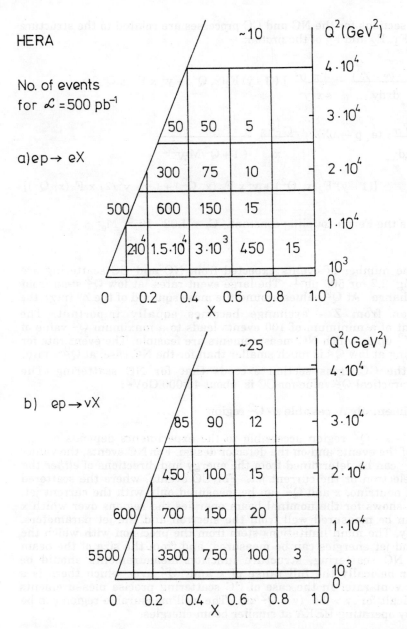

Fig. 2.2 Event rates for NC and CC scattering at HERA with $L = 500$ pb^{-1} calculated with Lund-LEPTO and EHLQ structure functions.

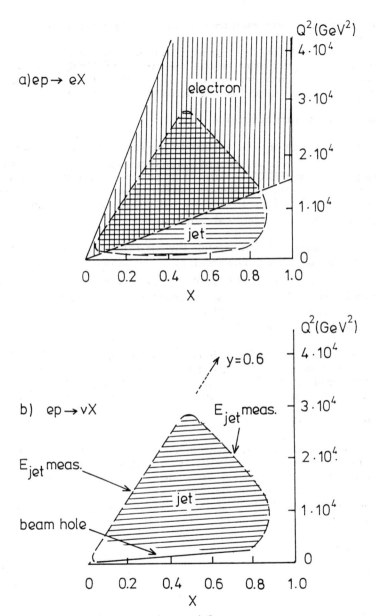

Fig. 2.3 The regions where x and Q^2 can be well measured at HERA:
a) for NC scattering from either the electron or the jets;
b) for CC scattering only from the jets.

2.3 Expected Results on Structure Functions at Standard x Values

By virtue of the large energy range, a multitude of structure functions and parton densities become measurable at HERA. Besides those of the nucleon, these are, for instance, the gluon structure function, the photon structure function and the pomeron structure function.

2.3.1 Nucleon structure functions

The kind of data to be expected from HERA on F_2 is illustrated by fig. 2.4 [2.2]. It shows recent measurements together with a "measurement" from HERA performed at nominal beam energies for 500 pb^{-1}. In order to fill the gap between existing and HERA data a second HERA measurement at E_e = 10 GeV, E_p = 300 GeV and with a luminosity of ~ 100 pb^{-1} would be desirable, as shown in fig 2.4.

As said before, HERA extends the Q^2 range for structure functions by two orders of magnitude. The accuracy which can be expected for α_s is shown in fig. 2.5 [2.3]. The primary goal, however, will not be a measurement of α_s or Λ but a stringent test for the logarithmic Q^2 behavior predicted by QCD at Q^2 values well above the regime of higher twist and target mass effects.

The quality of data for parton densities is indicated by fig. 2.6 which shows also existing results determined for Q^2 values of 11 GeV^2 and 15 GeV^2, respectively [2.4]. The HERA data present about one year of data taking. The curves indicate the predicted evolution of the parton densities.

2.3.2 Gluon structure function

Data from a direct measurement of the gluon structure function $G(x)$ are shown in fig. 2.7 [2.5]; they are very scanty, and for x < 0.05, where $G(x)$ is expected to be large, they are nonexistent. HERA experiments can follow several avenues for a determination of $G(x)$. One involves the longitudinal structure function F_L [2.6] which can be experssed in terms of the structure functions F_1, F_2:

$$F_L(x, Q^2) = F_2(x, Q^2) - 2 x F_1(x, Q^2)$$

The relation with $G(x)$ is as follows:

$$F_L = \alpha_s/2\pi \; x^2 \int_x^1 dy/y^3 \; \{8/3 \; F_2 + 2 \Sigma e_i^2 (y-x) G \}$$

where the e_i are the quark charges.

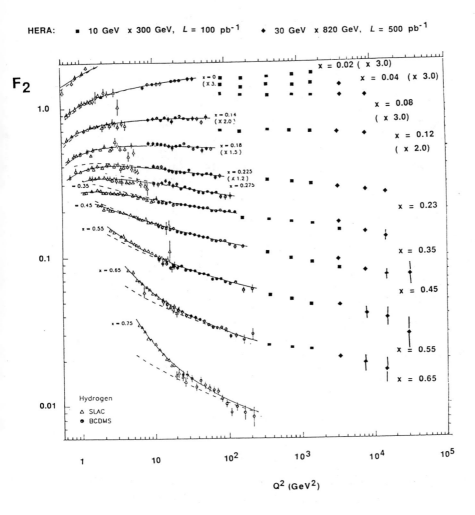

Fig. 2.4 The structure function F_2 for protons as determined by SLAC and BCDMS (from [1.2]) and as expected from HERA experiments at $E_e \times E_p$ = 30 GeV$\times$ 820 GeV (L = 500 pb^{-1}) and at 10GeV x 300 GeV (L=100 pb^{-1}); HERA data from [2.2].

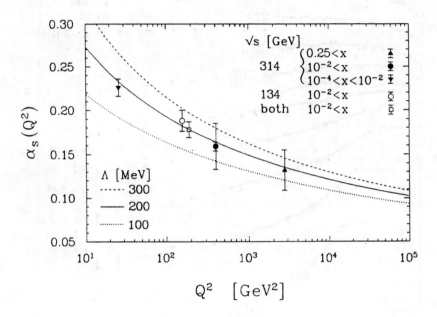

Fig. 2.5 Determination of α_s at HERA for runs at $\sqrt{s}$ = 134 GeV (L=100 pb^{-1}) and 314 GeV (200 pb^{-1}). Figure taken from [2.3].

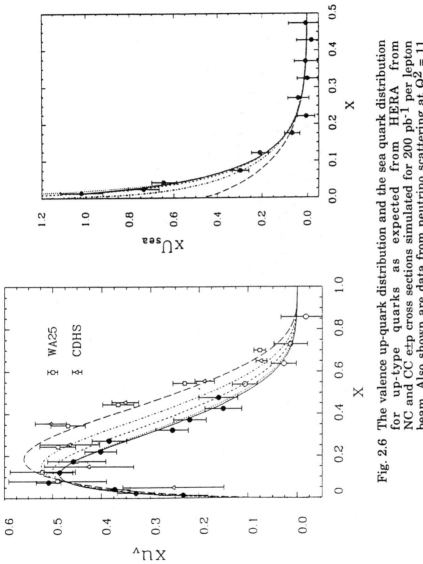

Fig. 2.6 The valence up-quark distribution and the sea quark distribution for up-type quarks as expected from HERA from NC and CC e±p cross sections simulated for 200 pb^{-1} per lepton beam. Also shown are data from neutrino scattering at $Q^2 = 11$ (WA25) and 15 GeV^2 (CDHS). The curves show the expected evolution for $Q^2 = 10, 100, 1000$ and $10\,000$ GeV^2.

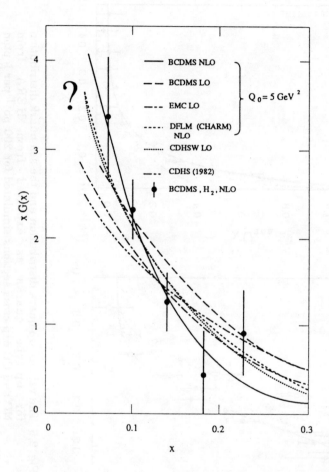

Fig. 2.7 Compilation of the gluon distribution x $G(x,Q^2)$. Figure taken from [2.5].

At small x the contribution from F_2 is small and the integral is expected to be dominated by $G(x)$. This is shown by fig. 2.8a, where the expected contributions from F_2 and from G for two different models (A, B) have been plotted. The $G(x)$ values extracted for the two models are shown in fig. 2.8b. Good sensitivity is expected for $G(x)$ at $x < 0.01$. A similar analysis has been performed in [2.6] leading to the same conclusion.

The gluon structure function determines the size of the cross section for heavy quark production. As an example, we discuss J/Ψ production via [2.7 - 9]

$$e\,p \rightarrow e\,J/\Psi\,X.$$

Two diagrams are expected to provide the dominant contributions, one representing the point-like photon, the other the hadron-like photon (fig. 2.9a). The cross section contributions by the two diagrams has been calculated imposing acceptance cuts for the detector:

$$e\,p \rightarrow e\,J/\Psi\,X,\ J/\Psi \rightarrow \mu^+\mu^-,\ \text{with polar angles}\ \Theta_\mu > 3^0$$

The resulting cross sections are given in fig. 2.9b as a function of the J/Ψ rapidity. The cross sections for both diagrams are large, the sum being of the order of 50 pb; furthermore, negative rapidities are dominated by point-like photons, positive ones by hadronic photons. It should therefore be possible to separate both contributions and to extract $G(x)$ from the data.

Another candidate reaction for a measurement of $G(x)$ is the production of $b\bar{b}$ pairs by photon-gluon fusion [2.10] which has a total cross section of about 4.2 nb or 420 000 events produced per 100 pb^{-1}.

2.3.3 Photon structure function

HERA will permit also a study of the structure function of the photon. In [2.11] production of photons with high transverse momenta (p_T) has been analysed for quasi real photon proton scattering (fig. 2.10a). Perturbative QCD calculations are possible for large p_T. The cross section for a square of the photon-proton c.m. energy, $s_{\gamma p} = 30\,000$ GeV^2 and $p_T = 5$ GeV/c is shown in fig. 2.10b as a function of the rapidity of the photon. For positive rapidities, the contributions from the hadronic structure of the photon dominate. Integration of the virtual photon spectrum yields the following (preliminary) cross sections for HERA:

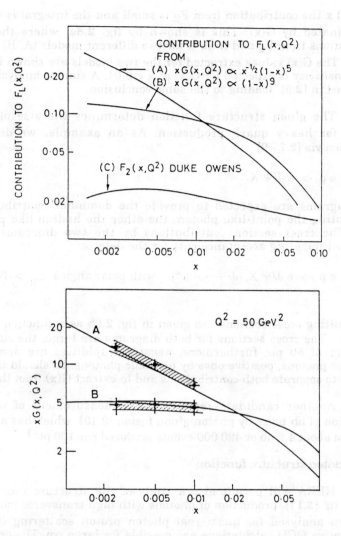

Fig. 2.8 a) The relative contributions to F_L from quark (A) and gluon operators (B,C).
b) The two different gluon distributions shown in a) superimposed on data expected from HERA. The error bars indicate the statistical error corresponding to $L = 100$ pb^{-1}; the shaded band indicates the size of the systematic error.
Figure taken from [2.6].

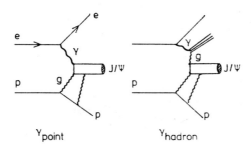

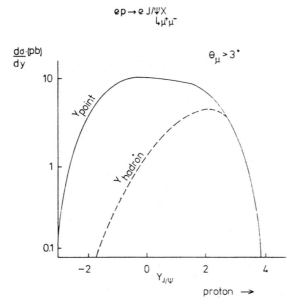

Fig. 2.9 a) Diagrams for the point-like and hadron-like contributions of the photon to the process e p -> e J/ΨX.
b) The cross section at HERA for e p -> e J/ΨX, J/Ψ -> $\mu^+\mu^-$, with the angles of the muons, $\Theta_\mu > 3^0$ relative to the beam axis.

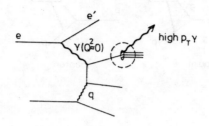

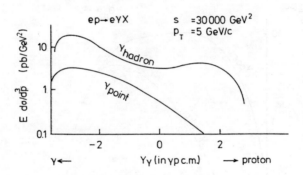

Fig. 2.10 a) On of the many possible diagrams for production of photons with high transverse momenta by e p scattering.
b) The cross section for production of photons with transverse momenta of $p_T = 5$ GeV/c by γ p scattering for a cm energy squared of 30 000 GeV^2. Figure after [2.11].

e p -> e γX , p$_T$(γ) > 5 GeV/c : 430 pb
10 GeV/c : 50 pb
20 GeV/c : 3 pb

3. Small x Physics

The field of physics at small x has been pioneered by Gribov, Levin and Ryskin [3.1]. It deals with scattering processes where Q^2 is large, $Q^2 > Q_0^2$ - so that pertubative QCD can be applied - and where the struck parton carries only a small fraction of the initial energy, x << 1. Since the number of partons grows as x -> 0, and since these partons are confined inside the nucleon, at some x < x_{crit} the partons begin to overlap. Hot spots are formed and the structure functions start to saturate. At this point, the standard Altarelli-Parisi evolution breaks down (fig. 3.1). The value of x_{crit} has not yet been calculated; speculations indicate x_{crit} < 0.001. The authors of ref [3.1] expect that at very high energies such semihard processes are the main source of secondary hadrons from hadron - hadron collisions.

The minimum x value that can be studied depends on the c.m. energy, viz.

$x = Q^2/(ys) = Q^2/s$ for y = 1.

For fixed target experiments at CERN or FNAL s ~ 1000 GeV^2 and therefore only x values as small as 0.01 can be reached for Q^2 = 10 GeV^2. HERA will be the first machine where saturation effects may become observable: the minimum x value is smaller by two orders of magnitude thanks to the much higher c.m. energy: s = 10^5 GeV^2, x_{min} = 10^{-4} at Q^2 = 10 GeV^2.

Can such small values be explored at HERA? Figure 3.2a shows the NC cross sections in pb in bins of x and Q^2. Keeping in mind that the expected yearly luminosity is 100 pb^{-1}, events are abundant at small values of x. For instance, in the bin 10^{-4} > x > 10^{-3} and 10 < Q^2 < 20 GeV^2 there are about 7.10^5 events per year. Another quetion is whether NC events at small x-values can be measured well. This depends on the acceptance of the detector and the trigger. Figure 3.2b shows an educated guess for the well measurable region. It is limited by the requirements:

$\Theta_{current\ jet}$ < 172°, $\Theta_{electron}$ < 172°, $E_{electron}$ > 5 GeV

and by the beam hole. The comparison between figs. 3.2a and b shows that HERA experiments should be well suited to explore the region x > 10^{-4}, Q^2 > 10 GeV^2.

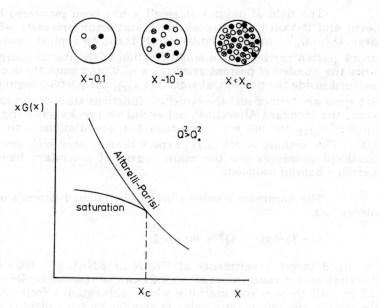

Fig. 3.1 a) The parton densities in the nucleon for different values of x.
b) Structure function for $Q^2 > Q_0^2$ as a function of x exhibiting saturation at $x < x_{crit}$.

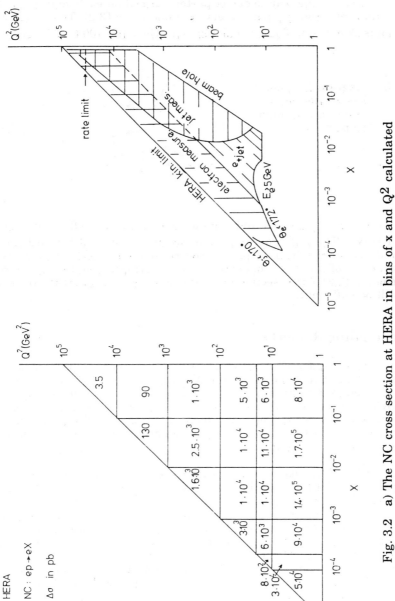

Fig. 3.2 a) The NC cross section at HERA in bins of x and Q^2 calculated with Lund-LEPTO and EHLQ structure functions.
b) The regions where x and Q^2 can be well measured.

A promising way to detect parton saturation early may be photon dissociation into high p_T parton jets as proposed in [3.2]. The diagram in question is sketched in fig. 3.3: a photon with $Q^2 \sim 10 - 100$ GeV^2, $x \ll 1$,

Fig. 3.3   Diagram for photon dissociation on a pomeron into high p_T partons

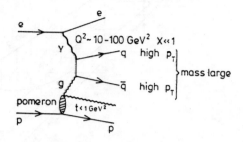

emits two high p_t quarks and a gluon which scatters off a pomeron of $|t| < 1$ GeV^2. The high energy proton which is scattered forward can be detected e.g. in the leading proton spectrometer of ZEUS. Obviously, a measurement of this type should also yield information on the structure of the pomeron [3.3]. Cross section estimates indicate that at HERA the event rate will be sufficient.

4. Concluding Remarks

The impact of a structure function experiment rests on the precision of the data and the x and Q^2 range covered. The existing data have reached systematic errors as small as 2-3%, Q^2 values as high as $\sim$ 200 GeV^2 and x as low as 0.02 in the deep inelastic region. It seems obvious, that a fixed target experiment such as that discussed in [1.6] which aims at substantially smaller systematic uncertainties would have a large impact on the field.

Within a year's time HERA is expected to begin operation. Experiments at HERA have the possibility to push the maximum Q^2 to 30 - 40 000 GeV^2 and the minimum value of x down to 10^{-4} for $Q^2 = 10$ GeV^2. The greatly extended kinematical region will allow for a critical testing of QCD. Not surprisingly for collider experiments, the limiting factor will probably be the event statistics and therefore luminosity will be at a premium at HERA. The possibility to reach very small x values will permit the search for new effects expected from the overlap of partons, a subject which may have a strong impact on understanding formation of hadrons in hard collisions.

While fixed target experiments concentrate on the structure functions F_1, F_2 and F_3, because of the much higher c.m. energy, the variety of other structure functions accessible to experiments at HERA is much larger and includes e.g. those of the gluon, photon and pomeron. It seems clear that HERA will make exciting contributions to the field of structure functions.

5. Acknowledgments

The author would like to thank Profs. W. Bardeen and Wu-Ki Tung for organising this workshop which provided such a concise overview of the field of structure functions. Discussions with J. Bartels, E. Levine and A. Schuler have been very helpful. I want to thank Prof. E. Lohrmann for a critical reading of the manuscript.

6. References

[1.1] For recent reviews of the field see the presentations at this workshop and S.A. Mishra, F. Sciulli, Ann. Rev. Nucl. Part. Sci., Vol. 42 (1989); J. Feltesse, Proc. 1989 Int. Symp. on Electron and Photon Interactions, ed. by M. Riordan, p. 13.
[1.2] A. Bodek, talk presented at this workshop; L.M. Whitlow, thesis, SLAC Report 357 (1990).
[1.3] A. Milsztajn, data presented at this workshop.
[1.4] J.G. Morfin, Wu-Ki Tung, talk presented by J.G. Morfin at this workshop.
[1.5] J. Huth, Proc. 1989 Int. Symp. on Electron and Photon Interactions, ed. by M. Riordan, p. 368
[1.6] C. Guyot et al., see talk by M. Virchaux at this workshop.
[1.7] K. Rith, presented at this workshop.
[2.1] See also talk by A. Schuler at this workshop.
[2.2] J. Blumlein et al., Akad. Wissenschaften, Berlin-Zeuthen, Report 88-01 (1988)
[2.3] J. Blumlein et al, Z. Phys. C 45 (1990) 501
[2.4] G. Ingelman, R. Rückl, Z. Phys. C 44 (1989) 291.
[2.5] J. Feltesse, Proc. 1989 Int. Symp. on Electron and Photon Interactions, ed. by M. Riordan, p. 13.
[2.6] A.M. Cooper - Sarkar et al., Z. Phys. C 39 (1988) 281.
[2.7] E. Gotsman, U. Maor, DESY Report 88-141 (1988).
[2.8] E.L. Berger, D. Jones. Phys. Rev. D23 (1981) 1521; A.D. Martin, C.K. Ng, W.J. Stirling. Phys. Lett. B191 (1987) 200; S.M. Tkaczyk, W.J. Stirling, D.H. Saxon, Proc. HERA workshop Vol 1 (1987) 265.
[2.9] R.S. Fletcher, F. Halzen, R.W. Robinett, Phys. Lett. B225 (1989) 175.
[2.10] A. Ali et al., DESY Report 88-119 (1988), Proc. HERA Workshop, ed. R.D. Peccei, Vol. I, p. 395 (1988).
[2.11] M. Krawczyk, Acta Physica Polonica B21 (1990) and Univ. Warsaw

Report IFT 7/89
[3.1] L.V. Gribov, E.M. Levin, M.G. Ryskin, Phys. Rep. 100 (1983) 1; see also E.M. Levin, talk presented at this workshop.
[3.2] M.G. Ryskin, DESY Report 90-050 (1990).
[3.3] J. Bartels, G. Ingelmann, Phys. Lett. B235 (1990) 175.

PROGRAM

PROGRAM

PROGRAM

THURSDAY, APRIL 26, 1990

OPENING REMARKS	W. Bardeen, Fermilab
	J. Peoples, Fermilab
GENERAL SESSION	W. Bardeen, Fermilab
	(Session Chairman)

Overview of Structure Functions and Experiments — F. Sciulli, Columbia University

Overview of Parton Distributions and QCD Framework — W. K. Tung, Illinois Inst. of Tech.

THEORY SESSION — W. Bardeen, Fermilab (Session Chairman)

Small-X Physics — E. M. Levin, Leningrad Nuc. Phys. Inst.

NLO Parton Distribution Functions — R. K. Ellis, Fermilab

Resummation of Large Terms in Hard Scattering — G. Sterman, SUNY, Stony Brook

DIRECT PHOTON & HEAVY QUARK PRODUCTION (I) — P. Aurenche, LAPP (Session Chairman)

Heavy Quark Production in QCD — P. Nason, CERN

The Gluon Distribution from the Tagged Photon Lab — M. Purohit, Princeton University

New Results on Heavy Quark Production — P. J. Lebrun, Fermilab

Direct Photon Production in QCD — J. Owens, Florida State University

DIRECT PHOTON & HEAVY QUARK PRODUCTION (II)

New Results on Direct Photons from Fixed-Target Experiments — C. M. Bromberg, Michigan State University

New Results on Direct Photons from CDF — R. M. Harris, Fermilab

New Results from UA6 — L. Camilleri, CERN

Discussion — J. Huston, Michigan State University

SPECIAL EVENING THEORY SESSION

How Can the Strangeness Content of the Proton be Both Large and Small? — B. L. Ioffe, ITEP, Moscow

What Can We Learn About Parton Distributions From the Jet Cross Section? — D. Soper, U. of Oregon

QCD Calculations of the Structure Functions of Heavy Mesons	A. Y. Khodjamirian, TPI, U. of Minnesota/Yerevan Phys. Institute
Numerical Results on the Small-X Behavior of the Gluon Density	G. Schuler, DESY
Uncertainties in QCD Analysis of DIS Structure Functions	R. Lednicky, JINR, Dubna

FRIDAY, APRIL 27, 1990

SPIN STRUCTURE OF THE NUCLEON

P. Schuler, Yale
(Session Chairman)

Measurements of Nucleon Spin Structure Functions with Polarized Leptons	K. Rith, Max Planck Institute
Models of Nucleon Spin Structure - What Can We Learn?	R. L. Jaffe, Massachusetts Inst. of Tech.
Probing Spin Structure with Polarized DY Experiments	J. Collins, IIT/Argonne Nat'l Laboratory
Prompt Photon Experiments with Polarized Beams	P. Slattery, University of Rochester
Anomaly, Spin, the Parton Model, and High Energy Reactions	A. Mueller, Columbia University
Discussion	E. Berger, Argonne Nat'l Laboratory

DEEP INELASTIC SCATTERING EXPERIMENTS

M. Shaevitz, Columbia University
(Session Chairman)

Recent Results from DIS Experiments:

NMC	C. Peroni, University of Torino
CCFR	P. Z. Quintas, Columbia University
E665	S. Aid, University of Maryland
New Analysis of SLAC Experiments	A. Bodek, Rochester University
Combined Fits to BCDMS & SLAC Structure Functions	A. Milsztajn, CEN Saclay
Probing Hadron Structure with Neutrino DIS Experiments	S. Mishra, Columbia University
Probing Hadron Structure with Muon DIS Experiments	M. Virchaux, CEN Saclay
QCD Tests at HERA	J. Bluemlein, Inst. HEP Adw., DDR
Discussion	J. Morfin, Fermilab

SATURDAY, APRIL 28, 1990

W/Z AND LEPTON-PAIR PRODUCTION AND
PARTON DISTRIBUTIONS

L. Pondrom, U. of Wisconsin
(Session Chairman)

- Parton Distributions from W/Z Hadroproduction — F. Halzen, University of Wisconsin
- W/Z and Lepton-Pair Production at the Tevatron — J. Hauser, Fermilab
- Measurement of $W \to \mu + \nu$ /$Z \to \mu + \nu$ Ratio at UA1 — M. W. van de Guchte, CERN
- QCD Studies with the UA2 Detector — G. F. Egan, University of Melbourne
- Drell-Yan Pair Production in Fixed-Target Experiments — J. P. Rutherfoord, University of Arizona
- Global Analysis of Parton Distributions — J. Morfin, Fermilab
- Global Analysis of Parton Distributions — R. Roberts, Rutherford Laboratory
- Discussion — J. Huth, Fermilab

THEORY

D. Soper, U. of Oregon
(Session Chairman)

- Phenomenological Consequences of MRS Distributions — A. Martin, Durham University
- Multi-Parton Interactions and Inelastic Cross Section — D. Treleani, IFT/INFN, Trieste
- $1/Q^2$ Corrections to DIS and DY Cross Sections — J. Qiu, SUNY, Stony Brook

CONCLUSION

- Experimental Horizon — G. Wolf, DESY

LIST OF PARTICIPANTS

PARTICIPANTS LIST

NAME	INSTITUTION
Adams, Mark R.	University of Illinois, Chicago
Aid, Silhacene	University of Maryland
Aivazis, Michael	Illinois Institute of Technology
Appel, Jeffrey A.	Fermi National Accelerator Laboratory
Arnold, Raymond G.	American University
Artru, Xavier	IPN, Lyon
Baer, Helmut W.	Los Alamos National Laboratory
Baker, Mark	Massachusetts Institute of Technology
Baker, Winslow F.	Fermi National Accelerator Laboratory
Bamberger, Andreas	University of Freiburg
Barbaro-Galtieri, A.	Lawrence Berkeley Laboratory
Bardeen, William A.	Fermi National Accelerator Laboratory
Baum, Guenter	University of Bielefeld
Baur, Ulrich	University of Wisconsin
Bazizi, Kamel	U. of California, Riverside
Berger, Edmond L.	Argonne National Laboratory
Bernardi, Gregorio	University of Paris
Bernstein, Robert	Fermi National Accelerator Laboratory
Bhat, Pushpalatha	Fermi National Accelerator Laboratory
Bhatti, Anwar Ahmad	University of Washington
Bianco, Stefano	INFN, Frascati
Blusk, Steven	University of Pittsburgh
Bodek, Arie	University of Rochester
Bodwin, Geoffrey	Argonne National Laboratory
Borcherding, Fred	Fermi National Accelerator Laboratory
Borner, Harald P.	Oxford University

NAME	INSTITUTION
Brasse, F. W.	DESY
Bromberg, Carl M.	Michigan State University
Brown, Chuck	Fermi National Accelerator Laboratory
Brown, David S.	Michigan State University
Buckley, Elizabeth	Rutgers University
Budd, Howard	University of Rochester
Burleson, George R.	New Mexico State University
Byrum, Karen	University of Wisconsin
Camilleri, Leslie	CERN
Carlson, Carl E.	College of William and Mary
Chang, C. C.	University of Maryland
Charchula, Krzysztof	DESY
Cheng, Ta-Pei	University of Missouri
Close, Frank E.	Rutherford Appleton Laboratory
Cobau, William G.	Michigan State University
Coester, Fritz	Argonne National Laboratory
Collins, John	IIT/Argonne National Laboratory
Conrad, Janet	Harvard University
Crittenden, James A.	Bonn University
Cummings, Mary Anne	University of Michigan
D'Agostini, Giulio	U. of Rome/IFT/INFN, Trieste
Dauwe, Loretta J.	University of Michigan
de Mello, Joao	Fermi National Accelerator Laboratory
Dimitroyannis, Dimitrios A.	University of Maryland
Dreyer, Thomas	University of Freiburg
Duff, Adam	University of Wisconsin
Dukes, Edmond Craig	University of Virginia
Egan, Gary F.	University of Melbourne

NAME	INSTITUTION
Ellis, R. Keith	Fermi National Accelerator Laboratory
Esaibegyan, Sergei	Yerevan Physics Institute
Fang, Guang Yin	Harvard University
Ferrero, Maria Itala	University of Torino
Ficenec, John R.	Virginia Polytechnic Inst. & State Univ.
Finley, David	Fermi National Accelerator Laboratory
Flaugher, Brenna	Fermi National Accelerator Laboratory
Fletcher, Robert S.	University of Wisconsin
Foxbush, Michael	Texas A & M University
Frisch, Henry	University of Chicago
Garino, Gerald	Northwestern University
Geesaman, Donald F.	Argonne National Laboratory
Gialas, Ioannis	Columbia University
Giele, W.	Fermi National Accelerator Laboratory
Ginther, George	University of Rochester
Glover, Nigel	Fermi National Accelerator Laboratory
Greeniaus, L. Gordon	University of Alberta/TRIUMF
Hagopian, Sharon	Florida State University
Halliwell, Clive	University of Illinois, Chicago
Halzen, Francis	University of Wisconsin
Harris, Robert M.	Fermi National Accelerator Laboratory
Hasell, Doug	York University
Hatcher, Robert	Michigan State University
Hauser, Jay	Fermi National Accelerator Laboratory
Heinrich, Joel	Princeton University
Holt, Roy J.	Argonne National Laboratory
Huston, Joey	Michigan State University
Huth, John	Fermi National Accelerator Laboratory

NAME	INSTITUTION
Igo, George	U. of California, Los Angeles
Introzzi, Gianluca	University of Pavia
Ioffe, Boris	ITEP, Moscow
Jackson, Harold	Argonne National Laboratory
Jaffe, R .L.	Massachusetts Institute of Technology
Jarmer, John J.	Los Alamos National Laboratory
Kearns, Edward	Harvard University
Keller, Stephane	University of Wisconsin
Khodjamirian, Alexander	U. of Minnesota/Yerevan Physics Institute
Kinney, Edward R.	Argonne National Laboratory
Kovacs, Eve	Fermi National Accelerator Laboratory
Kuhlmann, Steve	Argonne National Laboratory
Kumano, Shunzo	Indiana University
Ladinsky, Glenn	Illinois Institute of Technology
Lamoureux, Jodi	University of Wisconsin
Lassila, Ken	Iowa State University
Lebrun, Paul J.	Fermi National Accelerator Laboratory
LeCompte, Thomas J.	Northwestern University
Lednicky, Richard	JINR, Dubna
Levin, Evgeny M.	Leningrad Nuclear Physics Institute
Limentani, Silvia	University of Padova
Lomatch, Susanne	Illinois Institute of Technology
Londergan, Tim	Indiana University
Long, Kenneth	Rutherford Appleton Laboratory
Maas, Peter	University of Wisconsin
Madden, Patrick B.	U. of California, San Diego
Magill, Stephen R.	University of Illinois, Chicago
Mallik, Usha	University of Iowa

NAME	INSTITUTION
Mangano, Michelangelo	INFN, Pisa
Mani, Sudhindra	University of Pittsburgh
Mannel, Eric	University of Notre Dame
Maor, Uri	University of Illinois, Champaign
Margolis, Bernard	McGill University
Markosky, Leigh	Fermi National Accelerator Laboratory
Martin, Alan D.	University of Durham
Martin, John	University of Toronto
Mathiot, Jean-Francois	IPN, Orsay
Maul, Andre	Michigan State University
Maung, Khin Maung	Hampton University
Melanson, Harry	Fermi National Accelerator Laboratory
Migneron, Roger	University of Western Ontario
Mikamo, S.	KEK, Japan
Miller, Donald H.	Northwestern University
Milsztajn, Alain	DPhPE, CEN Saclay
Mishra, Sanjib R.	Columbia University
Moore, Craig	Fermi National Accelerator Laboratory
Moreno, Matias	Fermilab/UNAM, Mexico
Morfin, Jorge	Fermi National Accelerator Laboratory
Mortimer, Thomas	Imperial College/U. of London
Mueller, Alfred H.	Columbia University
Mulders, P. J.	NIKHEF
Naples, Donna	University of Maryland
Nason, Paolo	CERN
Nelson, Ken	University of Virginia
Ogawa, Satoru	Fermi National Accelerator Laboratory
Ohashi, Yuji	Argonne National Laboratory

NAME	INSTITUTION
Mangano, Michelangelo	INFN, Pisa
Mani, Sudhindra	University of Pittsburgh
Mannel, Eric	University of Notre Dame
Maor, Uri	University of Illinois, Champaign
Margolis, Bernard	McGill University
Markosky, Leigh	Fermi National Accelerator Laboratory
Martin, Alan D.	University of Durham
Martin, John	University of Toronto
Mathiot, Jean-Francois	IPN, Orsay
Maul, Andre	Michigan State University
Maung, Khin Maung	Hampton University
Melanson, Harry	Fermi National Accelerator Laboratory
Migneron, Roger	University of Western Ontario
Mikamo, S.	KEK, Japan
Miller, Donald H.	Northwestern University
Milsztajn, Alain	DPhPE, CEN Saclay
Mishra, Sanjib R.	Columbia University
Moore, Craig	Fermi National Accelerator Laboratory
Moreno, Matias	Fermilab/UNAM, Mexico
Morfin, Jorge	Fermi National Accelerator Laboratory
Mortimer, Thomas	Imperial College/U. of London
Mueller, Alfred H.	Columbia University
Mulders, P. J.	NIKHEF
Naples, Donna	University of Maryland
Nason, Paolo	CERN
Nelson, Ken	University of Virginia
Ogawa, Satoru	Fermi National Accelerator Laboratory
Ohashi, Yuji	Argonne National Laboratory

NAME	INSTITUTION
Olness, Fredrick I.	University of Oregon
Orr, Robert S.	University of Toronto
Owens, Jeff	Florida State University
Papavassiliou, Vassili	Yale University
Park, Sung	Ohio State University
Pauletta, Giovanni	University of Udine
Peoples, John	Fermi National Accelerator Laboratory
Peroni, Cristiana	University of Torino
Pi, Bo	Michigan State University
Pipkin, Francis M.	Harvard University
Pitzl, Daniel D.	U. of California, Santa Cruz
Pondrom, Lee	University of Wisconsin
Predazzi, Enrico	University of Torino
Purohit, Milind	Princeton University
Qiu, Jianwei	SUNY, Stony Brook
Quintas, Paul Z.	Columbia University
Radeztsky, Scott	University of Wisconsin
Ratti, Sergio	University of Pavia
Read, A. Lincoln	Fermi National Accelerator Laboratory
Reeder, Don D.	University of Wisconsin
Repond, Jose	Argonne National Laboratory
Revel, Daniel	Weizmann Institute
Riesselmann, Kurt	University of Wisconsin
Rith, Klaus	Max-Planck Institute
Ritz, Steven	Columbia University
Roberts, R. G.	Rutherford Appleton Laboratory
Rock, Stephen	American University
Rodning, Nathan L.	University of Alberta

NAME	INSTITUTION
Rohaly, Tim	University of Pennsylvania
Romanowski, Thomas A.	Ohio State University
Rosati, Marzia	McGill University
Rosner, Jonathan L.	University of Chicago
Rutherfoord, John P.	University of Arizona
Saito, Naohito	Fermilab
Salvarani, Alexandro	U. of California, San Diego
Scarpine, Victor	University of Illinois, Champaign
Schellman, Heidi	Fermi National Accelerator Laboratory
Schmitt, Michael	Harvard University
Schuler, Gerhard A.	University of Hamburg
Schuler, Peter	Yale University
Sciulli, Frank	Columbia University
Segel, Ralph E.	Northwestern University
Seligman, Bill	Columbia University
Shaevitz, Michael	Columbia University
Shrauner, J. Ely	Washington University
Slattery, Paul	University of Rochester
Smith, J.	SUNY, Stony Brook
Smith, Richard P.	Fermi National Accelerator Laboratory
Smith, Wesley	University of Wisconsin
Snow, Gregory	University of Michigan
Soper, Dave	University of Oregon
Stairs, Douglas G.	McGill University
Stanco, Luca	INFN, Padova
Stanfield, Kenneth C.	Fermi National Accelerator Laboratory
Stange, Alan	University of Wisconsin
Staude, Arnold	University of Munich

NAME	INSTITUTION
Stelzer, Tim	University of Wisconsin
Sterman, George	SUNY, Stony Brook
Strayer, Michael R.	Oak Ridge National Laboratory
Sugano, Katsuhito	Argonne National Laboratory
Sutherland, Mark	University of Toronto
Talaga, Richard	Argonne National Laboratory
Tanaka, Nobuyaki	Los Alamos National Laboratory
Tollestrup, Alvin	Fermi National Accelerator Laboratory
Treleani, Daniele	University of Trieste
Trudel, Anne	TRIUMF
Tung, T. Y.	Northwestern University
Tung, Wu-Ki	Illinois Institute of Technology
Ukegawa, Fumihiko	University of Tsukuba
Underwood, David G.	Argonne National Laboratory
van de Guchte, Maarten Wisse	CERN
Varelas, Nikos	University of Rochester
Virchaux, Marc	DPhPE, CEN Saclay
Wagoner, David E.	Prairie View A & M Univ.
Wang, Ming-Jer	Case Western/LANL
Weerasundara, Dhammika	University of Pittsburgh
Weerts, Harry	Michigan State University
Wei, Chao	Shandong University
Werlen, Monique	Lausanne University
Wilhelm, Mathias O.	Freiburg University
Williams, Anthony G.	Florida State University
Wilson, Richard	Harvard University
Wimpenny, Stephen J.	U. of California, Riverside
Wolf, Guenter	DESY

NAME	INSTITUTION
Woodside, Jeffrey	Oklahoma State University
Wu, Jianshi	Oak Ridge National Laboratory
Wu, Zhongxin	Yale University
Yeh, G. P.	Fermi National Accelerator Laboratory
Yokosawa, Akihiko	Argonne National Laboratory
Yoshida, R.	Northwestern University
Yu, Zhou	Northwestern University
Yuan, Chien-Peng	Argonne National Laboratory
Zanetti, Anna M.	INFN, Trieste
Zeidman, Ben	Argonne National Laboratory
Zhang, Yangling	Carnegie-Mellon University
Zielinski, Marek	University of Rochester
Zioulas, George	McGill University

AUTHOR INDEX

AUTHOR INDEX

AUTHOR INDEX

Aid, S.	58
Artru, X.	395
Bachmann, K. T.	50
Ballocchi, G.	291
Bernasconi, A.	291
Bernstein, R. H.	50
Blair, R. E.	50
Bluemlein, J.	139
Bodek, A.	50, 67
Borcherding, F.	50
Borner, H. P.	180
Breedon, R.	291
Bromberg, C.	272
Budd, H.S.	50
Calucci, G.	432
Camilleri, L.	291
Collins, J. C.	378
Cool, R. L.	291
Cox, P. T.	291
Cushman, P.	291
Dasu, S.	67
de Barbaro, P.	50
Dick, L.	291
Dukes, E. C.	291
Ebonque, A.	291
Egan, G. F.	224, 297
Ellis, R. K.	405
Fisk, H.E.	50
Foudas, C.	50
Gabioud, B.	291
Gaille, F.	291
Giacomelli, P.	291
Grosso-Pilcher, C.	291
Halzen, F.	187
Harriman, P. N.	338, 346
Harris, R. M.	278
Hauser, J.	205
Hubbard, D.	291
Jeanneret, J.-B.	291
Joseph, C.	291
Keller, S.	187
King, B. J.	50
Kubischta, W.	291
Lamm, M. J.	50
Lebrun, P.	254
Lednicky, R.	169
Lefmann, W. C.	50
Leung, W. C.	50
Loude, J.-F.	291
Magnea, L.	423
Malamud, E.	291
Marsh, W.	50
Martin, A. D.	338, 346
Mekhfi, M.	395
Melese, P.	291
Merritt, F. S.	50
Merritt, K. W. B.	50
Milsztajn, A.	76
Mishra, S. R.	50
Mishra, S. R.	84
Morel, C.	291
Morfin, J.	305
Oberson, P.	291
Oltman, E.	50
Oreglia, M. J.	50
Overseth, O.	291
Owens, J. F.	264
Pages, J. L.	291
Peroni, C.	41
Perroud, J.-P.	291
Purohit, M. V.	247
Qiu, J.	442
Quintas, P. Z.	50
Rabinowitz, S. A.	50
Riodan, E. M.	67
Rith, K.	361
Roberts, R. G.	338, 346
Rock, S.	67
Ruegger, D.	291
Rusack, R.	291
Rutherfoord, J. P.	234
Sakumoto, W. K.	50
Schellman, H.	50
Schuler, G. A.	139
Schumm, B. A.	50
Sciulli, F.	3

Seligman, W. G. 50
Shaevitz, M. H. 50
Singh, V. 291
Slattery, P. F. 384
Smith, W. H. 50
Snow, G. R. 291
Sozzi, G. 291
Sterman, G. 423, 442
Stirling, W. J. 338, 346
Studer, L. 291
Tran, M.-T. 291
Treleani, D. 432
Tung, W. K. 18
Vacchi, A. 291
Valenti, G. 291
van de Guchte, M. W. 215
Virchaux, M. 124
von Dardel, G. 291
Werlen, M. 291
Whitlow, L. S. 67
Wilson, R. 165
Wolf, G. 453
Yovanovitch, D. D. 50